D0860282

OPERATIONAL AMPLIFIERS
Applications, Troubleshooting, and Design

OPERATIONAL AMPLIFIERS
Applications, Troubleshooting, and Design

DAVID A. BELL
Lambton College of Applied Arts and Technology
Sarnia, Ontario, Canada

PRENTICE HALL, Englewood Cliffs, New Jersey 07632

Library of Congress Cataloging-in-Publication Data

Bell, David A.
Operational amplifiers : applications, troubleshooting, and design
David A. Bell.
p. cm.
Includes index.

1. Operational amplifiers. I. Title.
TK7871.58.06B45 1990
621.39'5—dc20 89-8594
CIP

Editorial/production supervision: Karen Winget
Cover design: Diane Saxe
Manufacturing buyer: David Dickey

Printed in the United States of America
10 9 8 7 6 5 4 3 2

ISBN 0-13-635178-6

Prentice-Hall International (UK) Limited, *London*
Prentice-Hall of Australia Pty. Limited, *Sydney*
Prentice-Hall Canada Inc., *Toronto*
Prentice-Hall Hispanoamericana, S.A., *Mexico*
Prentice-Hall of India Private Limited, *New Delhi*
Prentice-Hall of Japan, Inc., *Tokyo*
Simon & Schuster Asia Pte. Ltd., *Singapore*
Editora Prentice-Hall do Brasil, Ltda., *Rio de Janeiro*

To the memory of my wife Evelyn

Contents

PREFACE **xiii**

1 INTRODUCTION TO OPERATIONAL AMPLIFIERS **1**

Objectives, 1
Introduction, 2
1-1 Operational Amplifier Description, 2
1-2 Basic Operational Amplifier Circuit, 5
1-3 The 741 1C Operational Amplifier, 7
1-4 The Voltage Follower Circuit, 9
1-5 The Noninverting Amplifier, 13
Review Questions, 17
Problems, 17
Laboratory Exercises, 18

2 OPERATIONAL AMPLIFIER PARAMETERS **20**

Objectives, 20
Introduction, 21
2-1 Input and Output Voltage, 21
2-2 Common Mode and Supply Rejection, 23
2-3 Offset Voltages and Currents, 26
2-4 Input and Output Impedances, 29
2-5 Slew Rate and Frequency Limitations, 31
Review Questions, 33
Problems, 33
Laboratory Exercises, 34

3 OP-AMPS AS DC AMPLIFIERS **37**

Objectives, 37
Introduction, 38
3-1 Biasing Operational Amplifiers, 38
3-2 Direct-Coupled Voltage Followers, 41
3-3 Direct-Coupled Noninverting Amplifiers, 46

3-4 Direct-Coupled Inverting Amplifiers, 50
3-5 Summing Amplifiers, 54
3-6 Difference Amplifier, 57
Review Questions, 62
Problems, 63
Computer Problems, 64
Laboratory Exercises, 65

4 OP-AMPS AS AC AMPLIFIERS 67

Objectives, 67
Introduction, 68
4-1 Capacitor-Coupled Voltage Follower, 68
4-2 High Z_{in} Capacitor-Coupled Voltage Follower, 71
4-3 Capacitor-Coupled Noninverting Amplifier, 74
4-4 High Z_{in} Capacitor-Coupled Noninverting Amplifier, 75
4-5 Capacitor-Coupled Inverting Amplifier, 78
4-6 Setting the Upper Cutoff Frequency, 78
4-7 Capacitor-Coupled Difference Amplifier, 80
4-8 Use of a Single-Polarity Supply, 81
Review Questions, 85
Problems, 86
Computer Problems, 88
Laboratory Exercises, 88

5 OP-AMP FREQUENCY RESPONSE AND COMPENSATION 90

Objectives, 90
Introduction, 91
5-1 Op-amp Circuit Stability, 91
5-2 Frequency and Phase Response, 93
5-3 Frequency Compensating Methods, 98
5-4 Manufacturer's Recommended Compensation, 101
5-5 Op-amp Circuit Bandwidth, 107
5-6 Slew Rate Effects, 112
5-7 Stray Capacitance Effects, 116
5-8 Load Capacitance Effects, 120
5-9 Z_{in} Mod Compensation, 123
5-10 Circuit Stability Precautions, 126
Review Questions, 127

Problems, 128
Computer Problems, 130
Laboratory Exercises, 130

6 MISCELLANEOUS OP-AMP LINEAR APPLICATIONS 132

Objectives, 132
Introduction, 133
6-1 Voltage Sources, 133
6-2 Current Sources and Current Sinks, 137
6-3 Precision Current Sink and Source Circuits, 140
6-4 Current Amplifiers, 142
6-5 DC Voltmeter, 144
6-6 AC Voltmeter, 145
6-7 Linear Ohmmeter Circuit, 148
6-8 Instrumentation Amplifier, 151
Review Questions, 157
Problems, 158
Computer Problems, 160
Laboratory Exercises, 160

7 SIGNAL PROCESSING CIRCUITS 162

Objectives, 162
Introduction, 163
7-1 Precision Half-Wave Rectifiers, 163
7-2 Precision Full-Wave Rectifiers, 167
7-3 Limiting Circuits, 172
7-4 Clamping Circuits, 178
7-5 Peak Detectors, 182
7-6 Sample-and-Hold Circuits, 185
Review Questions, 189
Problems, 190
Computer Problems, 191
Laboratory Exercises, 191

8 DIFFERENTIATING AND INTEGRATING CIRCUITS 193

Objectives, 193
Introduction, 194

8-1 Differentiating Circuit, 194
8-2 Differentiator Design, 195
8-3 Differentiating Circuit Performance, 197
8-4 Integrating Circuit, 200
8-5 Integrator Design, 202
8-6 Integrating Circuit Performance, 204
Review Questions, 207
Problems, 207
Computer Problems, 208
Laboratory Exercises, 208

9 OP-AMP NONLINEAR CIRCUITS 210

Objectives, 210
Intdroduction, 211
9-1 Op-amps in Switching Circuits, 211
9-2 Crossing Detectors, 213
9-3 Inverting Schmitt Trigger Circuit, 218
9-4 Noninverting Schmitt Trigger Circuits, 222
9-5 Astable Multivibrator, 227
9-6 Monostable Multivibrator, 229
Review Questions, 233
Problems, 234
Computer Problems, 235
Laboratory Exercises, 235

10 SIGNAL GENERATORS 237

Objectives, 237
Introduction, 238
10-1 Triangular/Rectangular Wave Generator, 238
10-2 Waveform Generator Design, 241
10-3 Phase Shift Oscillator, 244
10-4 Oscillator Amplitude Stabilization, 247
10-5 Wien Bridge Oscillator, 249
10-6 Signal Generator Output Controls, 253
Review Questions, 255
Problems, 256
Computer Problems, 257
Laboratory Exercises, 257

11 ACTIVE FILTERS 259

Objectives, 259
Introduction, 260
11-1 All-Pass Phase Shifting Circuits, 260
11-2 First-Order Low-Pass Active Filter, 264
11-3 Second-Order Low-Pass Filter, 266
11-4 First-Order High-Pass Filter, 269
11-5 Second-Order High-Pass Filter, 270
11-6 Third-Order Low-Pass Filter, 272
11-7 Third-Order High-Pass Filter, 274
11-8 Bandpass Filter, 276
11-9 State-Variable Bandpass Filter, 283
11-10 Bandstop Filter, 286
Review Questions, 288
Problems, 289
Computer Problems, 290
Laboratory Exercises, 290

12 DC VOLTAGE REGULATORS 292

Objectives, 292
Introduction, 293
12-1 Voltage Regulator Basics, 293
12-2 Voltage Follower Regulator, 295
12-3 Adjustable Output Regulator, 298
12-4 Precision Voltage Regulator, 302
12-5 Output Current Limiting, 306
12-6 Plus-Minus Supplies, 312
12-7 Integrated Circuit Voltage Regulators, 313
Review Questions, 319
Problems, 320
Computer Problems, 321
Laboratory Exercises, 321

GLOSSARY 323

ANSWERS TO ODD-NUMBERED PROBLEMS 327

APPENDIX 1-1 LM741 Frequency-Compensated Operational Amplifier 331

APPENDIX 1-2 LM 108 and LM 308 Operational Amplifier **335**

APPENDIX 1-3 LF353 BIFET Operational Amplifier **336**

APPENDIX 1-4 LM715 Operational Amplifier **338**

APPENDIX 1-5 LM709 Operational Amplifier **339**

APPENDIX 1-6 723 Precision Voltage Regulator **340**

APPENDIX 2-1 Typical Standard Resistor Values **341**

APPENDIX 2-2 Typical Standard Capacitor Values **343**

INDEX **345**

Preface

Graduates of electronic technology courses should understand the use of operational amplifiers, be able to select op-amps for particular applications, and know how to calculate the values of components that must be connected externally.

The basic operational amplifier is a simple device, and its circuit applications are much easier to understand than similar discrete component transistor circuits. Many op-amp text books offer few practical design examples, and yet op-amp circuit design is astonishingly simple; most design calculations involve little more than Ohm's law and the capacitive impedance equation.

Beginning with a description of the basic operational amplifier circuit, the text moves immediately to the industry standard 741 op-amp. Use of the 741 as a voltage follower, noninverting amplifier, and inverting amplifier is explained. Operational amplifier characteristics and parameters are next investigated, and various op-amp types are compared. Chapters 3 and 4 show how easy it is to design (Bipolar and BIFET) op-amps into practical direct-coupled and capacitor-coupled amplifier circuits. Topics covered are biasing requirements, calculation of resistor and capacitor values for specified gain and frequency response, selection of standard value components, and input and output impedances.

In most textbooks the very important topic of op-amp frequency compensation tends to be either glossed over or presented in an almost incomprehensible manner. In Chapter 5 the use of the manufacturer's recommended compensating components is demonstrated. Practical design examples are given, precautions for avoiding circuit instability are listed, and additional compensating techniques are discussed.

Succeeding chapters treat a wide variety of op-amp applications and op-amp related integrated circuits. Topics covered include: linear applications, nonlinear circuits, signal processing, signal generators, filters, and voltage regulators. In addition to explaining the operation of each circuit, practical design examples are offered throughout.

David Bell

OPERATIONAL AMPLIFIERS
Applications, Troubleshooting, and Design

CHAPTER 1

Introduction to Operational Amplifiers

Objectives

- Sketch the circuit symbol for an operational amplifier (op-amp) and identify all terminals.
- Sketch the input and output (internal) circuit components of an op-amp. Explain *input bias current, output impedance,* and *voltage gain.*
- Sketch and compare typical IC op-amp packages showing the terminal numbering systems.
- Draw a basic (three-transistor) circuit diagram of an op-amp. Identify all terminals and explain the circuit operation.
- Sketch the following operational amplifier internal circuit components and explain the operation of each: *constant current tail, complementary emitter follower, current mirror.*
- State typical values of *voltage gain, input resistance, output resistance, and input bias current* for a 741 op-amp.
- Sketch the following op-amp circuits; in each case show the connection of the basic op-amp circuit and explain its operation: *voltage follower, noninverting amplifier, inverting amplifier*
- Derive the voltage gain equations for noninverting and inverting amplifiers. Given the circuit components, calculate the voltage gains.

INTRODUCTION

Operational amplifiers are very high gain integrated circuit amplifiers with two high-impedance input terminals and one low-impedance output. The inputs are identified as *noninverting input* and *inverting input* because of the way in which they affect the output voltage. The basic circuit of an operational amplifier consists of a differential amplifier input stage and an emitter-follower output stage. By the use of externally connected resistors, an operational amplifier can be employed for a great many applications. The simplest of these are the voltage follower, the noninverting amplifier, and the inverting amplifier.

1-1 OPERATIONAL AMPLIFIER DESCRIPTION

Circuit Symbol and Terminals

The circuit symbol for an operational amplifier (op-amp) illustrated in Fig. 1-1 shows there are two input terminals, one output, and two supply terminals. Because plus-minus supply voltages are normally used, the supply terminals are identified as $+V_{CC}$ (the positive supply terminal) and $-V_{EE}$ (the negative supply terminal). Typical supply voltages for operational amplifiers range from ± 9 V to ± 22 V.

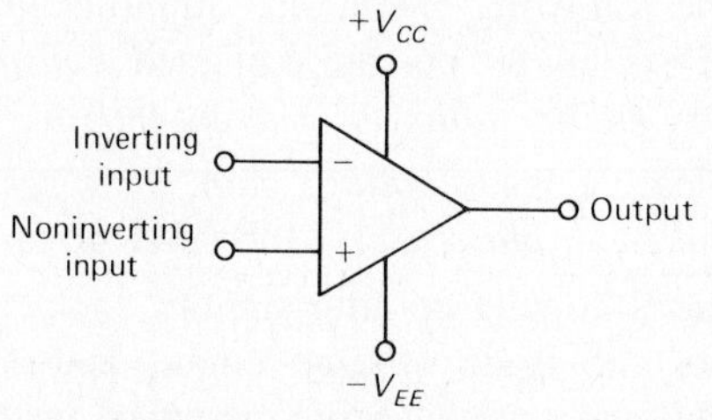

$+V_{CC}$

Inverting input

Noninverting input

Output

$-V_{EE}$

Figure 1-1 Circuit symbol for an operational amplifier (op-amp). There are two supply terminals ($+V_{CC}$ and $-V_{EE}$), two high-impedance inputs (*inverting input* and *noninverting input*), and one low-impedance output.

The two input terminals of the operational amplifier are designated *inverting input* (identified with a minus sign) and *noninverting input* (plus sign). When a positive-going voltage is applied to the inverting input, the output voltage tends to move in a negative direction. Conversely, a negative-going voltage signal at the inverting input causes the output voltage to move in a positive direction. A positive-going input at the noninverting terminal produces a positive-going output and a negative going input to the noninverting terminal produces a negative-going output. Thus, any input signal at the inverting input terminal produces an inverted output, and any input to the noninverting terminal generates a noninverted output.

Currents, Impedances, and Voltage Levels

As illustrated in Fig. 1-2, the input terminals of an operational amplifier are usually connected to the bases of two transistors. (Sometimes the gate terminals of Field Effect transistors are involved.) Because each transistor requires some base current to maintain it in an operating state, a low-level direct current must be provided to each

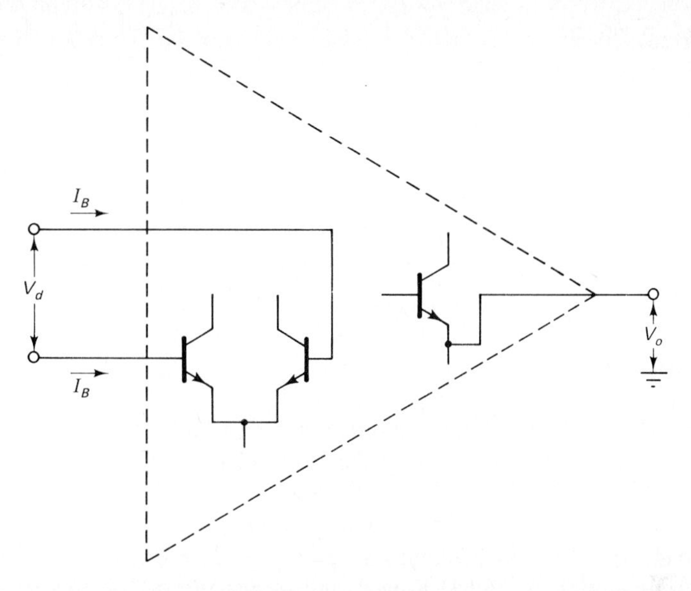

Figure 1-2 The input terminals of an operational amplifier are normally the base terminals of two transistors (or the gates of two FETs). The output is an emitter follower stage for low impedance.

op-amp input terminal. Typically the dc input current is 500 nA or less. The circuit arrangement also affords high input impedances, usually 1 MΩ or greater.

The output stage of an operational amplifier is normally an emitter-follower for low output impedance. (As explained in Section 1-3, the actual circuit is more complicated than the simple emitter-follower illustrated in Fig. 1-2.) The maximum output current that can be supplied is typically 25 mA and the output impedance is usually around 75 Ω.

In some applications both input terminals of an operational amplifier are biased close to ground level and the output remains at ground level until an input signal is applied. With most operational amplifiers the output voltage can be driven to within 1 V of V_{CC} and $-V_{EE}$ before output *saturation* occurs. An op-amp with a ±15 V supply could typically produce a ±14 V maximum output swing.

Voltage Gain

Like other amplifiers, the voltage gain of an operational amplifier is defined as (output voltage)/(input voltage). The output voltage (V_o) is usually measured between the output terminal and ground. However, the op-amp input voltage (V_d) is the *differential input voltage,* or the difference between voltage levels at the input terminals (see Fig. 1-2). Thus, if the inverting input terminal voltage is zero (or ground level) and the noninverting input voltage is 100 μV above ground, the differential input voltage is the difference between the two; that is, 100 μV. Similarly, if the inverting input level is −100 μV and the noninverting input is + 100 μV, then the difference input to the amplifier is 200 μV.

Voltage gains of 200 000 are common with IC operational amplifiers. This means, for example, that to produce a 5 V output, the required voltage difference at the input terminals is

$$V_d = \frac{5\ \text{V}}{200\ 000} = 25\ \mu\text{V}$$

Such voltage gains are usually much too high for most applications. However, they are very suitable as the *open-loop* gain for circuits using negative feedback. Externally connected feedback resistors are normally used to stabilize the circuit voltage gain to a desired value.

Packaging

Some typical operational amplifier packages are illustrated in Fig. 1-3. The plastic *dual-in-line (DIP)* package in Fig. 1-3(a) has eight terminals, only five of which are normally used for a basic operational amplifier. The TO-5 metal can-type package shown in Fig. 1-3(b) also uses only five of its eight terminals. As with other semiconductor devices, a metal can package can normally dissipate more heat than a plastic container. But DIP packages are usually the least expensive and they can be more compact than metal can containers. This last point is further illustrated by the (four amplifier) quad DIP package shown in Fig. 1-3(c).

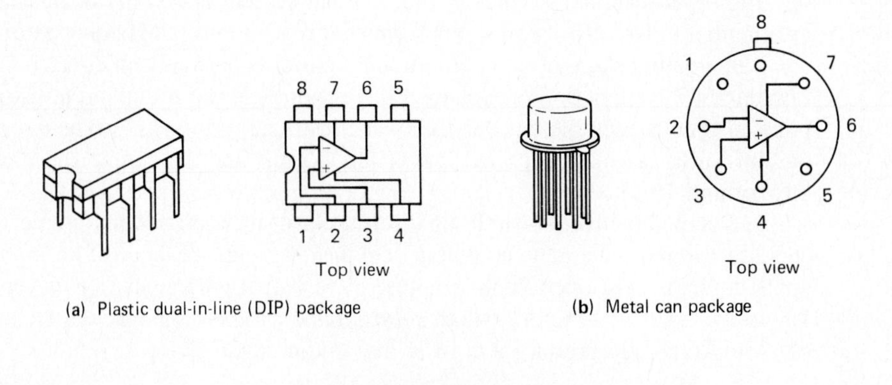

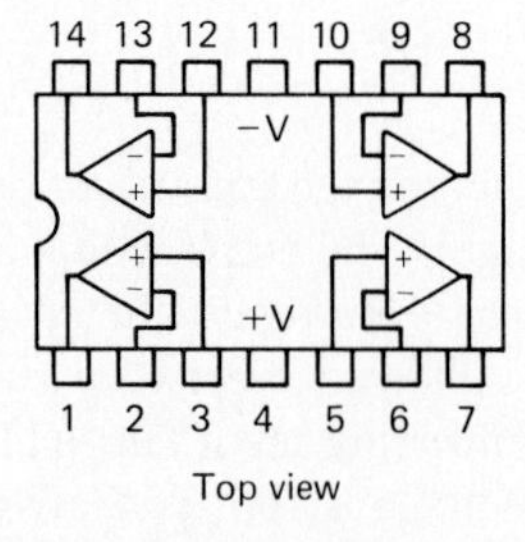

Figure 1-3 Operational amplifier packages. Single op-amps are available in plastic *dual-in-line* (*DIP*) packages and in metal can enclosures. Dual and quad DIP packages are also available. The terminal numbers shown are for a 741 op-amp.

1-2 BASIC OPERATIONAL AMPLIFIER CIRCUIT

The basic circuit of an operational amplifier is illustrated in Fig. 1-4. *Plus* and *minus* supply voltages ($+V_{CC}$ and $-V_{EE}$) are provided, and the two input terminals are grounded. Transistors Q_1 and Q_2 constitute a *differential amplifier,* which produces a voltage change at the collector of Q_2 when a difference input voltage is applied to the bases of Q_1 and Q_2. Transistor Q_3 operates as an emitter-follower to provide a low output impedance. As illustrated, the dc output voltage level at the emitter of Q_6 is

$$V_o = V_{CC} - V_{RC} - V_{BE3}$$

or

$$V_o = V_{CC} - (I_{C2}R_C) - V_{BE} \qquad (1\text{-}1)$$

Assume that Q_1 and Q_2 are *matched transistors,* that is they have equal V_{BE} levels and equal current gains. Then, with both transistor bases at ground level, the emitter currents are equal, and both I_{E1} and I_{E2} flow through the common emitter resistor, R_E. The total emitter current could be calculated as

$$I_{E1} + I_{E2} = \frac{V_{RE}}{R_E}$$

With Q_1 and Q_2 bases grounded,

$$V_{RE} = V_{EE} - V_{BE}$$

so,

$$I_{E1} + I_{E2} = \frac{V_{EE} - V_{BE}}{R_E} \qquad (1\text{-}2)$$

To investigate the circuit operation, assume that $V_{CC} = +10$ V, $V_{EE} = -10$ V, $R_E = 4.7$ kΩ, $R_C = 6.8$ kΩ, and all transistors have $V_{BE} = 0.7$ V.

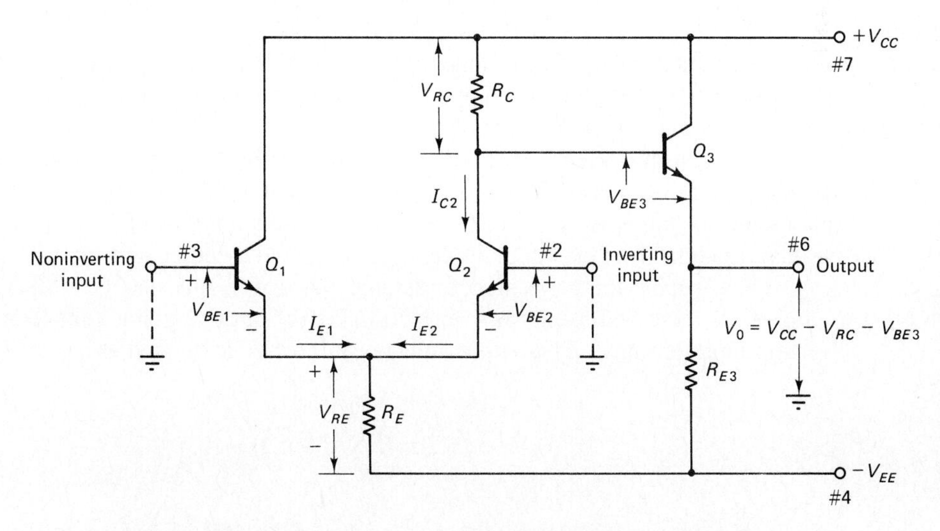

Figure 1-4 The basic circuit of an operational amplifier has a differential amplifier input stage and an emitter follower output.

With both input terminals at ground level,

Eq. 1-2, $$I_{E1} + I_{E2} = \frac{V_{EE} - V_{BE}}{R_E} = \frac{10\text{ V} - 0.7\text{ V}}{4.7\text{ k}\Omega}$$

$$\approx 2\text{ mA}$$

therefore, $$I_{E1} = I_{E2} = 1\text{ mA}$$

and $$I_{C2} \approx I_{E2} \approx 1\text{ mA}$$

Eq.1-1, $$V_o = V_{CC} - (I_{C2}R_C) - V_{BE}$$

$$\approx 10\text{ V} - (1\text{ mA} \times 6.8\text{ k}\Omega) - 0.7\text{ V}$$

$$\approx 2.5\text{ V}$$

If a positive-going voltage is applied to the noninverting input terminal, Q_1 base is pulled up by the input voltage, and its emitter terminal tends to follow the input signal. Since Q_1 and Q_2 emitters are connected together, the emitter of Q_2 is also pulled up by the positive-going signal at the noninverting input terminal. The base voltage of Q_2 is fixed at ground level so the positive-going movement at its emitter causes a reduction in its base-emitter voltage (V_{BE2}). The result of the reduction in V_{BE2} is that its emitter current is reduced and consequently its collector current is reduced.

For convenience, assume that the positive-going input at the base of Q_1 reduces I_{C2} by 0.2 mA, (i.e., from 1 mA to 8.0 mA). This gives a new level of output voltage

Eq. 1-1, $$V_o = V_{CC} - (I_{C2}R_C) - V_{BE}$$

$$= 10\text{ V} - (0.8\text{ mA} \times 6.8\text{ k}\Omega) - 0.7\text{ V}$$

$$\approx 3.9\text{ V}$$

The output voltage has changed from +2.5 V to +3.9 V, a change of +1.4 V. It is seen that a positive-going signal at the noninverting input terminal has produced a positive-going output voltage.

Now consider what occurs when the noninverting terminal is grounded and a positive-going input is applied to the inverting input terminal. In this case, Q_2 base is pulled up, the base-emitter voltage of Q_2 is increased, and that of Q_1 is reduced by a similar amount. This results in an increase in I_{E2} and a consequent increase in I_{C2}.

Once again, for convenience, assume that a 0.2 mA change occurs in I_{C2}. Thus, I_{C2} is increased from 1 mA to 1.2 mA by the positive-going voltage at the inverting input terminal. The output voltage can now be calculated as

Eq. 1-1, $$V_o = V_{CC} - (I_{C2}R_C) - V_{BE}$$

$$= 10\text{ V} - (1.2\text{ mA} \times 6.8\text{ k}\Omega) - 0.7\text{ V}$$

$$\approx 1.1\text{ V}$$

The output voltage has now changed from its original level of +2.5 V to approximately +1.1 V, a change of −1.4 V. Therefore, the positive-going signal at the inverting input terminal produced a negative-going voltage change at the output.

It has been shown that a basic operational amplifier circuit consists of a differential amplifier stage with two (inverting and noninverting) input terminals and a voltage follower output stage. The differential amplifier offers a high impedance at both input terminals and it produces voltage gain. The output stage gives the op-amp a low output impedance. A practical operational amplifier circuit is much more complex than the basic circuit. This is discussed in the next section.

1-3 THE 741 IC OPERATIONAL AMPLIFIER

Practical Op-Amp Circuitry

As well as the input and output stages discussed in Section 1-2, the actual circuit of a practical IC operational amplifier must have an intermediate stage to produce the required very high voltage gain. The circuitry also has to be considerably more complex than the simple arrangements illustrated in Fig. 1-4. For example, with the differential amplifier stage in Fig. 1-4, any change in $-V_{EE}$ would produce a change in the voltage drop across emitter resistor R_E. This would result in an alteration in I_{E1}, I_{E2}, and I_{C2}. The change in I_{C2} would alter V_{RC} and thus affect the level of the dc output voltage. In fact, the variation in $-V_{EE}$ would have an effect similar to an input voltage. This can be minimized by replacing the emitter resistor with the *constant current circuit* (or *constant current tail*) illustrated in Fig. 1-5(a). A constant voltage

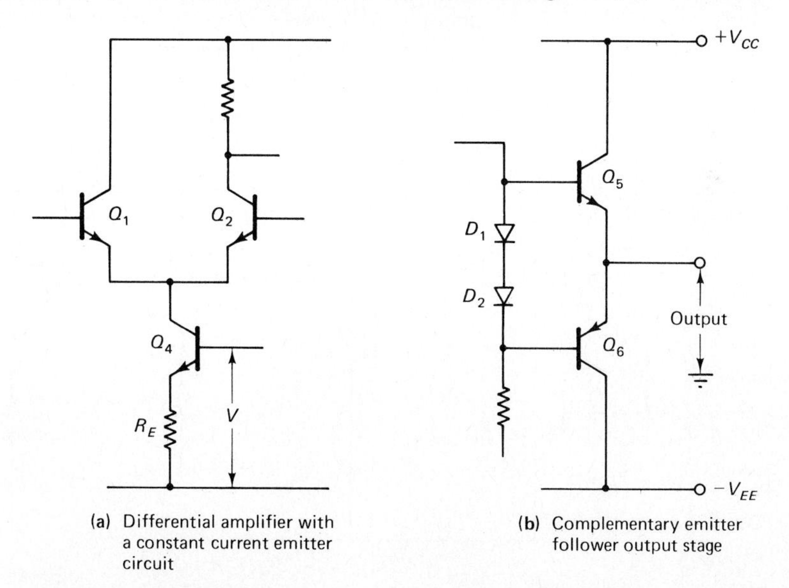

(a) Differential amplifier with a constant current emitter circuit

(b) Complementary emitter follower output stage

Figure 1-5 Inclusion of a *constant current tail* in the differential amplifier, and use of a *complementary emitter follower* improves the performance of the basic op-amp circuit in Fig. 1-4.

drop is maintained across resistor R_E by providing a constant voltage (V) at Q_4 base. Now, any change in the supply voltage is developed across the collector-emitter terminals of Q_4, and the emitter currents of Q_1 and Q_2 are not affected.

The simple emitter follower output stage in Fig. 1.4 is not suitable for a practical operational amplifier circuit. Refer to Fig. 1-4 once again and note that when the base voltage of Q_3 is increased, the emitter voltage (the output) must follow it because the emitter is always V_{BE} below the base level. However, when the base voltage is changed in a negative direction, especially if the change occurs very quickly, the base-emitter junction of Q_3 could become reverse-biased. This would result in a loss of the required low output impedance, and in distortion of the output voltage. The problem is eliminated by the use of a *complementary emitter follower*.

Figure 1-5(b) shows the circuit of a complementary emitter follower. Note that transistor Q_5 is an *npn* device while Q_6 is a *pnp* transistor. Both transistors are biased *on* by the voltage drop across diodes D_1 and D_2. When a signal causes the base voltages to rise or fall, both bases rise or fall together and consequently both (output) emitters rise or fall at the same time. A positive-going base voltage would once again produce a positive-going output which is V_{BE} below the level of Q_5 base. Similarly, a negative-going base voltage pulls down the base of Q_6, and the emitter of Q_6 follows its base keeping the transistor *on* regardless of the fall time of the base voltage. Thus, the output voltage always follows changes in the base voltage of Q_5 and Q_6, and a low output impedance is maintained.

Integrated Circuit Techniques

One reason why most IC operational amplifier circuits look quite complex is that several special techniques, unsuited to discrete component circuits, are employed in integrated circuits. Also, transistors are frequently substituted in place of diodes and resistors.

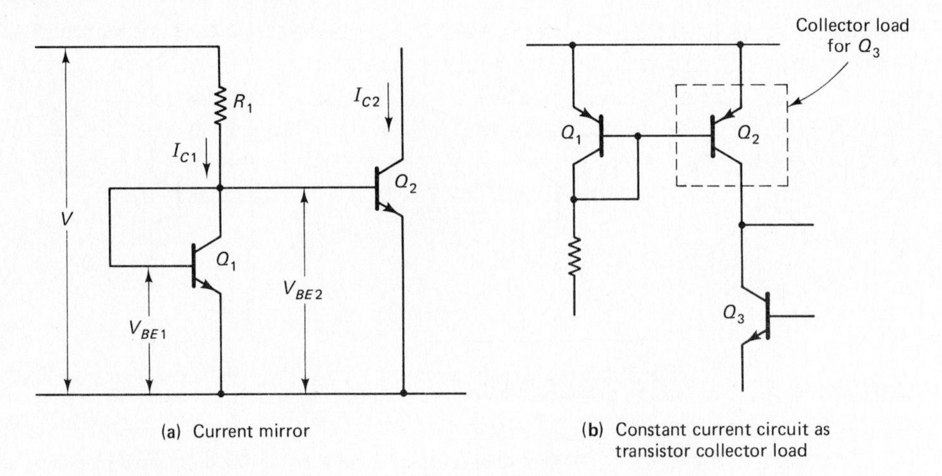

Figure 1-6 The *current mirror* is one IC technique that takes advantage of closely matched transistors. Using the current mirror as a constant current circuit to replace a collector resistor gives a very high voltage gain.

Figure 1-6(a) shows a circuit known as a *current mirror,* which is widely used in linear IC's. Because transistor Q_1 has its base and collector terminals directly connected, it behaves as a diode with the collector terminal functioning as the anode and the emitter as the cathode. The device current I_{C1} can be calculated as, $I_{C1} = (V - V_{BE1})/R_1$. IC fabrication methods result in closely matched components. Therefore, the I_C/V_{BE} characteristics of transistors Q_1 and Q_2 are very similar. Consequently, with V_{BE2} equal to V_{BE1}, I_{C2} is likely to equal I_{C1}. In this case, transistor Q_2 functions as a constant-current source without any need for an emitter resistor and without the voltage drop that would exist across the emitter resistor.

A current-mirror type of constant current source is substituted in place of a collector resistor for Q_3 in Fig. 1-6(b). This arrangement passes the necessary dc collector current for Q_3, but gives an effective collector ac load of approximately $1/h_{oe}$. Because this is a very large quantity, it results in a very high voltage gain for this stage.

741 IC Op-Amp Circuit

The complete circuit of the commonly used 741 IC operational amplifier is illustrated in Fig. 1-7. Clearly, it cannot be readily understood in terms of discrete component circuitry. However, it can be seen that Q_1 and Q_2 are the input transistors of the differential amplifier stage, and Q_{14} and Q_{20} are the complementary emitter follower output transistors.

A thorough understanding of the circuit details of the 741 (or any other integrated circuit) is not required in order to use the device. The operation of the IC can be simply thought of in terms of the basic circuit in Fig. 1-4. What is required is a knowledge of op-amp biasing methods and of how to select the required external components for a given application. Suitable supply voltages and the correct terminal connections must be used. A knowledge of the limitations of the IC is also important.

The very high voltage gain of many operational amplifiers causes high frequency oscillations to occur when feedback is employed. Externally connected compensating components are often required to prevent unwanted oscillations (see Chapter 5). The 741 op-amp is internally compensated to minimize the risk of oscillations. However, the compensation also has the effect of limiting the high frequency performance of the amplifier.

Some of the parameters listed on the 741 op-amp date sheet in Appendix 1-1 are shown in the data sheet portion illustrated in Fig. 1-8. Other parameters are discussed in Chapter 2.

1-4 THE VOLTAGE FOLLOWER CIRCUIT

The IC operational amplifier lends itself to a wide variety of applications. The very simplest of these is the *voltage follower* circuit illustrated in Fig. 1-9(a). The inverting input terminal is connected directly to the output terminal and the signal is applied to the noninverting input terminal. The output voltage now faithfully follows

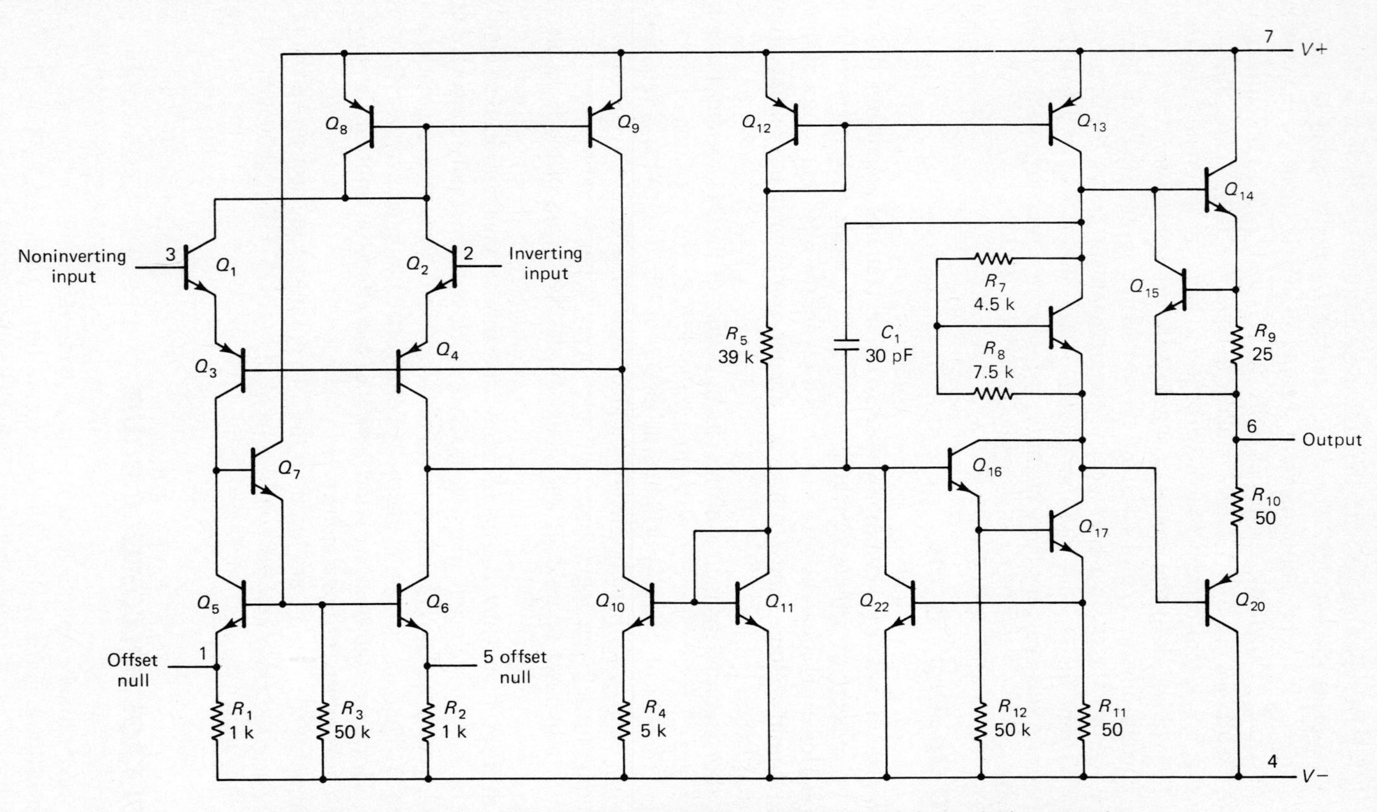

Figure 1-7 Complete circuit of a 741 operational amplifier. Q_1 and Q_2 are the input transistors for the differential amplifier. Q_{14} and Q_{20} constitute the complementary emitter-follower output stage.

Electrical characteristics (V_S = ±15 V, T_A = 25°C unless otherwise specified)

Parameters	Conditions	Min.	Typ.	Max.	Units
Input resistance		0.3	2.0		MΩ
Large signal voltage gain	$R_L \geq 2\ \text{k}\Omega$, $V_{OUT} = \pm 10$ V	50 000	200 000		
Output resistance			75		Ω

Figure 1-8 Portion of 741 op-amp data sheet specifying input resistance, output resistance, and voltage gain.

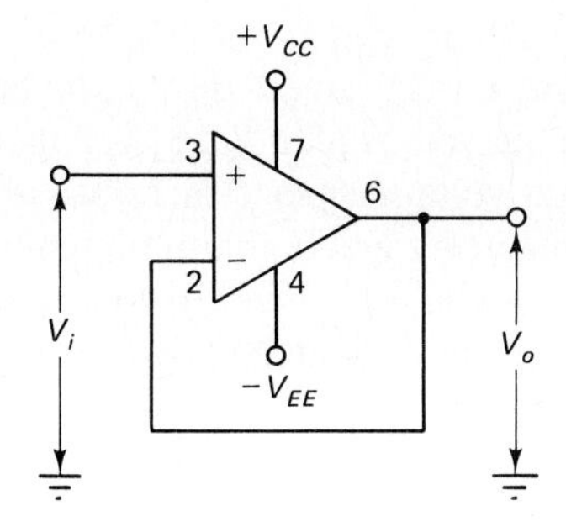

(a) Voltage follower circuit

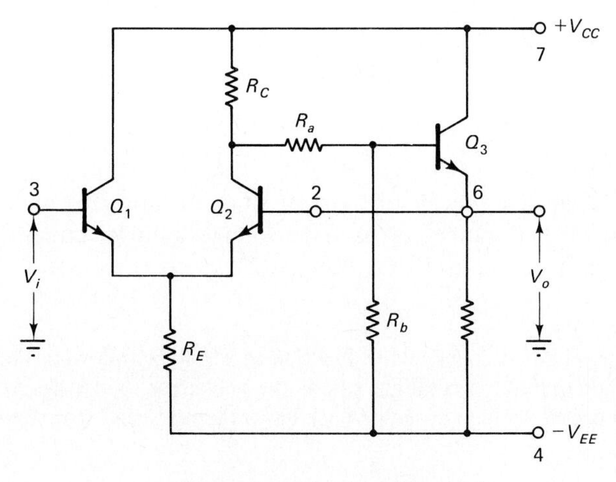

(b) Basic operational amplifier circuit connected as a voltage follower

Figure 1-9 In a *voltage follower* circuit, the op-amp output terminal is directly connected to the inverting input terminal. As the input voltage changes, the output changes to keep the inverting input voltage equal to the noninverting input voltage.

the input, giving the circuit a gain of 1. The voltage follower has a very high input impedance and a very low output impedance.

To understand how the voltage follower operates consider the basic op-amp circuit reproduced in Fig. 1-9(b). As in Fig. 1-9(a), the output (terminal 6) is connected to the inverting input terminal (terminal 2). Note that one difference between the basic op-amp circuit already studied and that in Fig. 1-9(b) is that bias resistors (R_a and R_b) are now included at the base of Q_3. This arrangement allows the dc output voltage to be lower than that at the collector of Q_2.

With terminal 3 (noninverting input) grounded, terminal 2 and the output must also be at ground level. If the output voltage level were to move slightly above ground level, the base voltage of Q_2 would be higher than that at the base of Q_1 and consequently I_{C2} would increase. The increased level of I_{C2} would increase the voltage drop across R_C and thus push the output voltage down until V_{B2} and V_{B1} were once again equal. Similarly, if the output were to fall below ground level V_{B2} would become lower than V_{B1}. This would lower I_{C2}, reduce the voltage drop across R_C, and move the output voltage back up to ground level.

The above reasoning can also be applied to show that if the input voltage (at terminal 3) was +1 V, or −1 V, or almost any other level, the output voltage would be virtually the same as the input. Thus, the output *follows* the input.

In the voltage follower there is a very small difference between the input voltage and the output voltage due to the very high internal voltage gain of the operational amplifier. To produce an output voltage change there has to be a voltage difference (or *differential input*) at the inverting and noninverting input terminals. This means that, because the inverting terminal is connected to the output, there is a small difference between the input and output voltage levels. To produce an output closely equal to V_i, the differential input is

$$V_d = \frac{V_i}{M} \tag{1-3}$$

where M is the internal voltage gain (or *open-loop gain*) of the operational amplifier. The precise output voltage is

$$V_o = V_i - V_d$$

or

$$V_o = V_i - \frac{V_i}{M} \tag{1-4}$$

where V_i/M is an error voltage. Obviously, the largest value of internal gain gives the smallest possible error. Other sources of error in voltage follower circuits are discussed in Chapters 2 and 3.

Example 1-1

A 741 IC operational amplifier is connected to function as a voltage follower. The input signal is 1 V. Assuming that the amplifier gain is the only error source, determine the output voltage that occurs with an op-amp which has (a) typical gain, (b) minimum gain.

Solution

From Fig. 1-8, the 741 has a large signal voltage gain of M = 50 000 (minimum), 200 000 (typical)

(a) Eq. 1-4,
$$V_o = V_i - \frac{V_i}{M} = 1\text{ V} - \frac{1\text{ V}}{200\,000}$$
$$= 0.999\,995\text{ V}$$

(b)
$$V_o = 1\text{ V} - \frac{1\text{ V}}{50\,000}$$
$$= 0.999\,98\text{ V}$$

1-5 THE NONINVERTING AMPLIFIER

The *noninverting amplifier* in Fig. 1-10(a) and (b) behaves in a similar way to the voltage follower circuit. The difference is, instead of all of the output being fed directly back to the input, only a portion is fed back. The output voltage is potentially divided across resistors R_2 and R_3 before it is applied to the inverting input.

Once again, consider the conditions that exist when the noninverting input terminal is grounded. As for the voltage follower circuit, the inverting input terminal must also be at ground level; otherwise any voltage difference would be amplified to move the inverting input terminal back to ground level. Because the junction of R_2 and R_3 is connected to the inverting input terminal, the voltage at this point must also be ground level, and consequently there is zero voltage drop across R_3. With V_{R3} equal to zero, the current I_2 which flows through R_2 and R_3 must also be zero, (neglecting the very small bias current into terminal 2). Thus, there is zero voltage drop across R_2, and this means that the output voltage is equal to the input voltage, which is at ground level.

Now suppose that a +100 mV input is applied to terminal 3. As already explained, the output voltage will move to the level that makes the feedback voltage to terminal 2 equal to the input voltage. Because the feedback voltage is developed across resistor R_3,

$$V_{R3} = V_i = I_2 R_3$$

$$V_o = I_2(R_2 + R_3)$$

voltage gain
$$A_v = \frac{V_o}{V_i} = \frac{I_2(R_2 + R_3)}{I_2 R_3}$$

or
$$A_v = \frac{R_2 + R_3}{R_3} \tag{1-5}$$

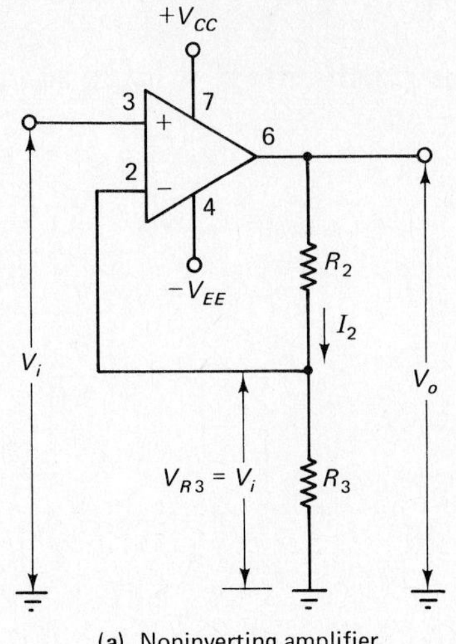

(a) Noninverting amplifier

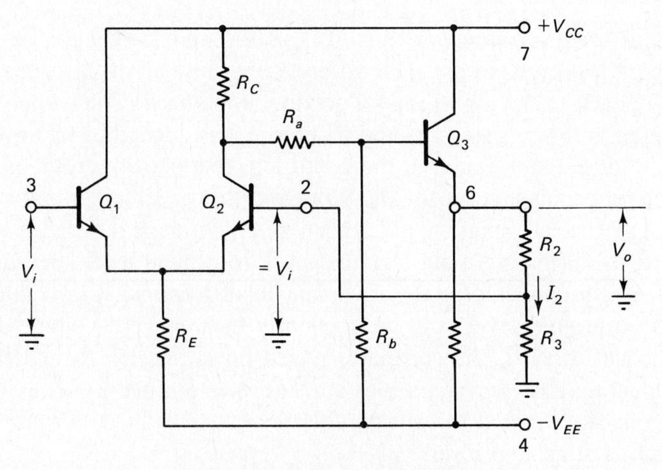

(b) Basic op-amp circuit connected as a noninverting amplifier

Figure 1-10 A *noninverting amplifier* operates in the same way as a voltage follower, except that the output voltage is potentially divided before it is fed back to the inverting input terminal. The circuit voltage gain is $A_v = (R_2 + R_3)/R_3$.

Example 1-2

An op-amp noninverting amplifier, as illustrated in Fig. 1-10, has $R_2 = 8.2\ \text{k}\Omega$ and $R_3 = 150\ \Omega$. Calculate the amplifier voltage gain, and determine a new resistance value for R_3 to give a voltage gain of 75.

Solution

Eq. 1-5,

$$A_v = \frac{R_2 + R_3}{R_3} = \frac{8.2\ \text{k}\Omega + 150\ \Omega}{150\ \Omega}$$

$$= 55.7$$

$$A_v = \frac{R_2 + R_3}{R_3} = \frac{R_2}{R_3} + 1$$

giving,
$$R_3 = \frac{R_2}{A_v - 1} = \frac{8.2\text{ k}\Omega}{75 - 1}$$
$$\approx 111\ \Omega$$

1-6 THE INVERTING AMPLIFIER

The inverting amplifier illustrated in Fig. 1-11(a) is so named, because an input applied via R_1 to the inverting input terminal causes the output to go in a negative direction when the input is positive-going, and vice versa.

In Fig. 1-11(b) the basic op-amp circuit is shown connected as an inverting amplifier. Note that, as in Fig. 1-11(a), R_2 is connected between the output and the inverting input terminal; also that R_1 is connected between the signal V_i and the inverting input terminal. This arrangement looks very similar to the noninverting amplifier in Fig. 1-10(b). In fact, if the bottom terminal of R_1 is grounded in Fig. 1-11(b) and terminal 3 is grounded in Fig. 1-10(b), the circuits are identical. As already discussed in Section 1-5, when the input is grounded, the output of the op-amp circuit is also at ground level.

If an input voltage V_i of +1 V is applied to R_1 in Fig. 1-11(b), the base of Q_2 is driven above the level of Q_1 base. I_{C2} is increased, thus increasing the voltage drop across R_C and driving the output voltage down. The output voltage falls until V_{B2} is again equal to V_{B1}. Because the base of Q_1 is grounded, the base voltage of Q_2 (terminal 2) will always be maintained at ground level regardless of the level of V_i. For this reason the noninverting terminal in this type of circuit is termed a *virtual ground* or *virtual earth*.

Note from the above explanation that V_o is moved in a negative direction when V_i goes positive. Similarly, when V_i goes negative, V_o has to move in a positive direction in order to return the inverting input terminal to ground level.

Now return to Fig. 1-11(a), and recall that the voltage at the inverting input terminal always remains close to ground level, and because the junction of R_1 and R_2 is connected to the inverting input terminal that junction always remains at ground level. With V_i at one end of R_1 and ground potential at the other end, V_i appears across R_1. Also with V_o at one end of R_2 and ground at the other end, V_o is developed across R_2. Ignoring the very small bias current flowing into the op-amp inverting input terminal, I_1 effectively flows through R_1 and R_2. The input and output voltages can now be expressed as

$$V_i = I_1 R_1$$

and
$$V_o = -I_1 R_2$$

and voltage gain
$$A_v = \frac{V_o}{V_i} = \frac{-I_1 R_2}{I_1 R_1}$$

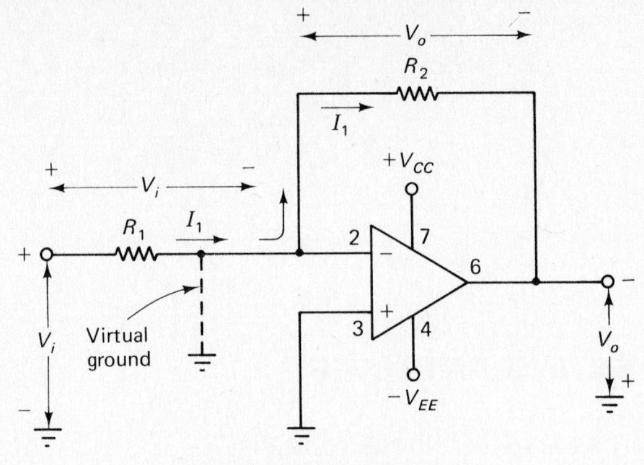

(a) Inverting amplifier

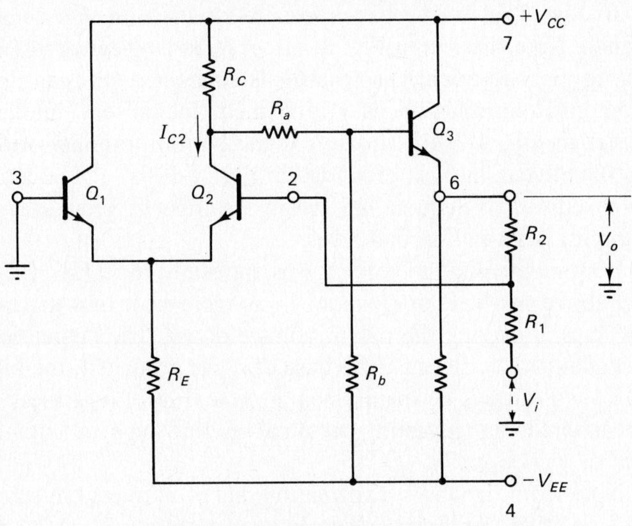

(b) Basic op-amp circuit connected as an inverting amplifier

Figure 1-11 With the *inverting amplifier,* the signal voltage is applied via R_1 to the inverting input terminal. This circuit is essentially the same as a noninverting amplifier with the noninverting input grounded and the signal applied to the bottom of the potential divider.

or

$$A_v = \frac{-R_2}{R_1} \tag{1-6}$$

The minus sign indicates that this is an inverting amplifier. The sign can be ignored when it is convenient to do so.

Example 1-3

An op-amp inverting amplifier, as in Fig. 1-11, has $R_1 = 270\ \Omega$ and $R_2 = 8.2\ \text{k}\Omega$. Determine the voltage gain and calculate a new resistance value for R_1 to give a voltage gain of 60.

Solution

Eq. 1-6,

$$A_v = \frac{R_2}{R_1} = \frac{8.2\ \text{k}\Omega}{270\ \Omega}$$

$$= 30.4$$

$$R_1 = \frac{R_2}{A_v} = \frac{8.2\ \text{k}\Omega}{60}$$

$$= 137\ \Omega$$

REVIEW QUESTIONS

1-1. Sketch the circuit symbol for an operational amplifier and identify all terminals by name and number.

1-2. Draw a sketch to show the basic input and output components of an operational amplifier. Briefly discuss the input bias current, input impedance, output impedance, and voltage gain.

1-3. Sketch typical packaging arrangements for an operational amplifier, identify the terminals, and briefly discuss the merits of each package type.

1-4. Draw the basic circuit diagram of an operational amplifier; identify all terminals and briefly explain how the circuit operates.

1-5. Sketch the circuit of a differential amplifier stage which uses a constant current tail. Explain.

1-6. Draw a circuit diagram for a complementary emitter follower output stage. Explain.

1-7. Draw the circuit diagram of a current mirror and explain its operation. Show how a current mirror can be used to improve the stage gain of an amplifier.

1-8. Briefly discuss the 741 IC operational amplifier and state typical parameters for the 741.

1-9. Sketch an op-amp voltage follower circuit. Also, sketch a basic operational amplifier circuit connected as a voltage follower. Explain the operation of the voltage follower.

1-10. Sketch an op-amp noninverting amplifier circuit. Also, sketch a basic operational amplifier circuit connected to function as a noninverting amplifier. Explain the operation of the noninverting amplifier and derive an equation for its voltage gain.

1-11. Sketch an op-amp inverting amplifier circuit. Also, sketch a basic operational amplifier circuit connected to function as an inverting amplifier. Explain the operation of the inverting amplifier and derive an equation for its voltage gain.

PROBLEMS

1-1. A 741 operational amplifier is connected to function as a voltage follower. If the input voltage is 750 mV and the amplifier gain is the only error source, calculate the output voltage for an amplifier; (a) with the specified minimum voltage gain, (b) with the specified typical voltage gain.

1-2. An LM 308 operational amplifier (data sheet in Appendix 1-2) is substituted in place of the 741 op-amp in Problem 1-1. Calculate the output voltage for cases (a) and (b) once again.

1-3. An operational amplifier voltage follower is to operate with a minimum input signal of 200 mV. If the error in the output voltage due to amplifier gain is not to exceed 0.005%, determine the minimum voltage gain required for the operational amplifier.

1-4. A voltage follower circuit using an LM308 operational amplifier is to reproduce the signal input with a maximum error of 10 μV due to the amplifier gain. Calculate the minimum acceptable signal amplitude.

1-5. An op-amp noninverting amplifier (as in Fig. 1-10) has resistors of R_2 = 22 kΩ, and R_3 = 120 Ω. Calculate the output voltage produced by a 75 mV input.

1-6. An op-amp noninverting amplifier is to have a voltage gain of 101. If R_3 (in Fig. 1-10) is 180 Ω, determine a suitable resistance value for R_2.

1-7. A 120 mV signal is to produce a 12 V output from an op-amp noninverting amplifier. A 15 kΩ resistor is available for use as R_2 (in Fig. 1-10). Determine a suitable resistance value for R_3.

1-8. An op-amp noninverting amplifier (as in Fig. 1-10) has R_2 = 27 kΩ, and R_3 = 390 Ω. Calculate the amplifier voltage gain, and determine the voltage gain that results if the resistor positions are reversed.

1-9. An op-amp inverting amplifier (as in Fig. 1-11) has resistors of R_1 = 120 Ω, and R_2 = 22 kΩ. Calculate the output voltage produced by a 50 mV input.

1-10. An op-amp inverting amplifier is to have a voltage gain of 150. If R_2 (in Fig. 1-11) is 33 kΩ, determine a suitable resistance value of R_1.

1-11. Calculate the voltage gain of an op-amp inverting amplifier (as in Fig. 1-11) which has R_1 = 680 Ω, and R_2 = 39 kΩ. Also determine the new voltage gain if the resistor positions are reversed.

1-12. An op-amp inverting amplifier (as in Fig. 1-11) has a 0.5 V input signal, and its output is to be 9 V. A 12 kΩ resistor is available for use as R_2. Calculate a suitable resistance value for R_1.

LABORATORY EXERCISES

1-1 Voltage Follower

1. Connect a 741 op-amp to function as a voltage follower, as illustrated in Fig. 1-9(a). Use a supply of ±9 V to ±15 V.
2. Connect a dual-trace oscilloscope to monitor the input and output dc voltage levels.
3. Using an adjustable dc voltage source, set the input voltage to +1 V, +2 V, and +3 V, in turn. Observe the output voltage level in each case.
4. Repeat Procedure 3 using input levels of −1 V, −2 V, and −3 V.
5. Remove the dc source from the input, and apply a ±5 V, 1 kHz sinusoidal signal.
6. Measure the circuit output voltage, and note the phase relationship between input and output.
7. Adjust the signal amplitude to 2 V, and 3 V, in turn, and measure the output in each case.

1-2 Noninverting Amplifier

1. Connect a 741 op-amp to function as a noninverting amplifier, as illustrated in Fig. 1-10(a). Use a supply of ±9 V to ±15 V, and resistors of R_2 = 8.2 kΩ and R_3 = 150 Ω, as in the first part of Example 1-2.
2. Connect a dual-trace oscilloscope to monitor the input and output dc voltage levels.
3. Using an adjustable dc voltage source, set the input voltage to +50 mV and +100 mV, in turn. Measure the output voltage level and calculate the voltage gain in each case.
4. Repeat Procedure 3 using input levels of −50 mV and −100 mV.
5. Remove the dc source from the input and apply a ±25 mV, 1 kHz sinusoidal signal.
6. Measure the output and note the phase relationship between input and output.
7. Adjust the signal amplitude to ±50 mV. Measure the output and calculate the voltage gain.
8. Change R_3 to approximately 111 Ω, as in the second part of Example 1-2. (Use two series-connected 56 Ω resistors.)
9. Repeat Procedure 7.

1-3 Inverting Amplifier

1. Connect a 741 op-amp to function as an inverting amplifier, as illustrated in Fig. 1-11(a). Use a supply of ±9 V to ±15 V and resistors of R_1 = 270 Ω and R_2 = 8.2 kΩ, as in the first part of Example 1-3.

2–7. Same as Procedures 2 through 7 for Laboratory Exercise 1-2

8. Change R_1 to approximately 137 Ω, as in the second part of Example 1-3. (Use two series-connected 68 Ω resistors.)
9. Repeat Procedure 7.

CHAPTER 2

Operational Amplifier Parameters

Objectives

- Explain the limitations on the input and output voltage of an op-amp. State typical op-amp input and output voltage ranges.
- Calculate the effects of input and output voltage ranges upon op-amp circuits.
- Define *common mode voltage, common mode voltage gain, common mode rejection ratio,* and *supply voltage rejection* for op-amps.
- Define *input offset voltage* and *input offset current* for op-amps, and calculate the effects of input offset voltages and currents on op-amp circuits.
- Show how offset voltages can be produced by bias resistors. Explain *offset nulling* and show how it can be accomplished.
- Write equations for the input impedance and output impedance of an op-amp circuit using negative feedback and calculate the input and output impedances of various op-amp circuits.
- Explain *output short-circuit current* for an op-amp and state a typical level of op-amp output short-circuit current.
- Define *slew rate* for an op-amp, and show how op-amp circuit output waveforms can be affected by slew rate.
- Sketch and explain a typical open-loop gain/frequency response graph for an op-amp.

INTRODUCTION

Operational amplifier circuits have performance limits dependent upon the particular type of operational amplifier used. Each op-amp has a maximum input voltage range and a maximum output voltage range. Unwanted outputs can occur as a result of bias voltage changes, supply voltage variations, and ac ripple on the supply voltage. Input bias currents and mismatch of op-amp input transistors can also produce unwanted output voltages. The input impedance of an operational amplifier is normally very high and the output impedance is very low. When negative feedback is used, Z_{in} becomes much higher and Z_{out} becomes much lower than without feedback. There is a limit on how fast an operational amplifier output can be made to change and a limit on the highest signal frequency that may be employed.

2-1 INPUT AND OUTPUT VOLTAGE

Input Voltage Range

The basic operational amplifier circuit connected to function as a voltage follower (see Section 1-4) is reproduced in Fig. 2-1. As illustrated, the circuit might be designed to have a V_{CE} of 5 V across transistors Q_2 and Q_4. With a ± 10 V supply and the bases of Q_1 and Q_2 at ground level, the voltage drops across R_2 and R_4 would then be 5.7 V and 4.3 V respectively. Now consider what occurs when the input voltage increases or decreases and the output follows the input.

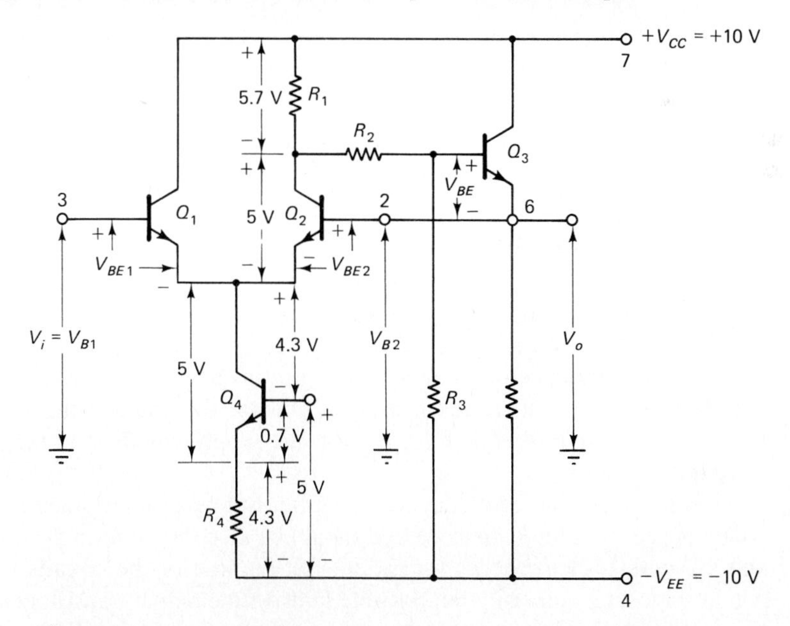

Figure 2-1 There are limits to the input and output voltage ranges of an operational amplifier. In the basic op-amp circuit these limits occur when any of the transistors approach saturation or cutoff.

If the input voltage at Q_1 base goes down to -4 V, the output terminal and Q_2 base also goes down to -4 V as the output follows the input. This means that the emitter terminals of Q_1 and Q_2 are pushed down from -0.7 V to -4.7 V. Consequently, the collector of Q_4 is pushed down by 4 V, reducing V_{CE4} from 5 V to 1 V. Although Q_4 might still be operational with a V_{CE} of 1 V, it is close to saturation. It is seen that there is a limit to the negative-going input voltage that can be applied to the operational amplifier if the circuit is to continue to function correctly.

There is also a limit to positive-going input voltages. When V_{B1} goes to $+4$ V in the circuit of Fig. 2-1, the voltage drop across resistor R_1 must be reduced to something less than 1 V, in order to move V_{B2} and V_{E3} up by 4 V to follow the input. This requires a reduction in I_{C2} to a level that makes Q_2 approach cutoff. The input voltage cannot be allowed to become large enough to drive Q_2 into cutoff.

The maximum positive-going and negative-going input voltage that may be applied to an operational amplifier is termed its *input voltage range*. The 741 data sheet in Appendix 1-1 lists the typical input voltage range as ± 13 V when using a ± 15 V supply (see Fig. 2-2).

Electrical characteristics ($V_S = \pm 15$ V, $T_A = 25°$C unless otherwise specified)

Parameters	Conditions	Min.	Typ.	Max.	Units
Input voltage range		±12	±13		V
Common mode rejection ratio	$R_S \leqslant 10$ kΩ	70	90		dB
Supply voltage rejection ratio	$R_S \leqslant 10$ kΩ		30	150	μV/V
Output voltage swing	$R_L \geqslant 10$ kΩ	±12	±14		V
	$R_L \geqslant 2$ kΩ	±10	±13		V

Figure 2-2 Portion of 741 op-amp data sheet listing input voltage range, common mode rejection ratio, supply voltage rejection ratio, and output voltage swing.

Output Voltage Range

With the voltage follower circuit discussed above, the maximum output voltage swing is limited by the input voltage range. However, where the op-amp is connected to function as either a noninverting or inverting amplifier, the output voltage may be much larger than the input. Just how far the output voltage can swing in a positive or negative direction depends on the supply voltage and the op-amp output circuitry.

Referring to the complementary emitter follower output stage in Fig. 1-5(b), it would appear that the output voltage should be able to rise until Q_5 is near saturation and fall until Q_6 approaches saturation. But because of the circuits that control the output stage, it is normally not possible to drive the output transistors close to saturation levels. A rough approximation for most operational amplifiers is that the maximum output voltage swing is approximately equal to 1 V less than the supply voltage.

For the 741 op-amp with a supply of ±15 V, the data sheet lists the *output voltage swing* as typically ±14 V when $R_L \geq 10$ kΩ (see Fig. 2-2). This means the specification applies only where the equivalent resistance of all resistors connected to the output is not less than 10 kΩ. With lower resistance loads, the output voltage range is reduced.

2-2 COMMON MODE AND SUPPLY REJECTION

Common Mode Rejection

Refer again to the basic operational amplifier circuit in Fig. 1-9(b) reproduced in Fig. 2-3 with the feedback connection from output to input deleted. The two input terminals are connected together and both are raised to 1 V above ground level. This is known as a *common mode input*. Note that there is no differential input; both input terminals are at the same potential. So ideally the output should be zero.

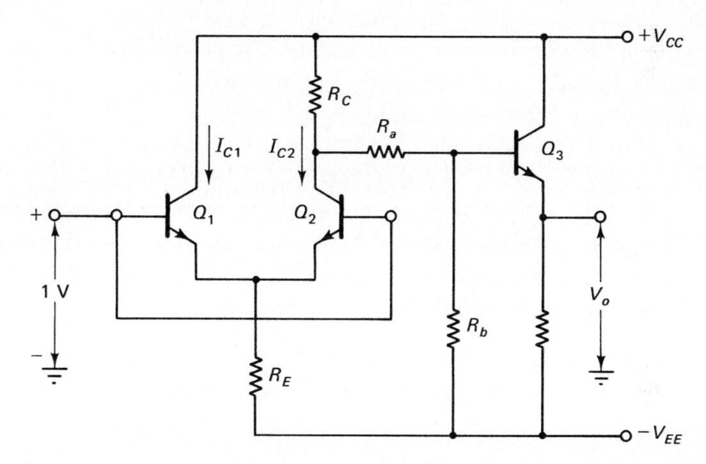

Figure 2-3 Basic op-amp circuit with the two input terminals connected together, and a common mode input voltage applied. The common mode rejection ratio (CMRR) is the open-loop gain divided by the common mode gain. CMRR = M/A_{cm}.

Because the base voltages of Q_1 and Q_2 are raised to 1 V above ground, the voltage drop across emitter resistor R_E is increased by 1 V, and, consequently, I_{C1} and I_{C2} are increased. The increased level of I_{C2} produces an increased voltage drop across resistor R_C, which results in a change in the output voltage at the emitter of Q_3. Similarly, if a −1 V common mode input is applied, I_{C2} falls, and again a change is produced at the circuit output. So, as well as the open-loop (differential input) gain M, each op-amp has a *common mode voltage gain* A_{cm}. The common mode gain is the output voltage change due to the common mode input divided by the common mode input voltage.

$$A_{cm} = \frac{V_{o(cm)}}{V_{i(cm)}}$$

The above discussion refers to the basic operational amplifier circuit. A practical operational amplifier has additional circuitry, such as the constant current tail in Fig. 1-5(a), to minimize the effects of common mode inputs. However, even with such circuitry, common mode signals still have some effect on the output. The success of the op-amp in rejecting common mode inputs is defined in the *common mode rejection ratio (CMRR)*. This is the ratio of the open-loop gain M to the common mode gain A_{cm}.

$$CMRR = \frac{M}{A_{cm}} \tag{2-1}$$

The $CMRR$ is usually expressed as a decibel quantity on the op-amp data sheet.

$$CMRR = 20 \log \frac{M}{A_{cm}} \text{ dB} \tag{2-2}$$

The effect of op-amp common mode gain is modified by feedback, just as the open-loop differential gain is modified by feedback to give a closed-loop gain. Consider the noninverting amplifier circuit in Fig. 2-4. With the input terminal grounded, the circuit output should also be at ground level. Now suppose a sine wave signal is picked up at both inputs, as illustrated. This is a common mode input. The output voltage should tend to be

$$V_{o(cm)} = A_{cm} \times V_{i(com)}$$

However, any output voltage will produce a feedback voltage across resistor R_2, which results in a differential voltage at the op-amp input terminals. The differential input produces an output which tends to cancel the output voltage that caused the feedback. The differential input voltage required to cancel $V_{o(cm)}$ is,

$$V_d = \frac{V_{o(cm)}}{M} = \frac{A_{cm} \times V_{i(cm)}}{M}$$

V_d is also the feedback voltage developed across R_2. So,

$$V_d = \frac{V_{o(cm)} \times R_2}{R_1 + R_2}$$

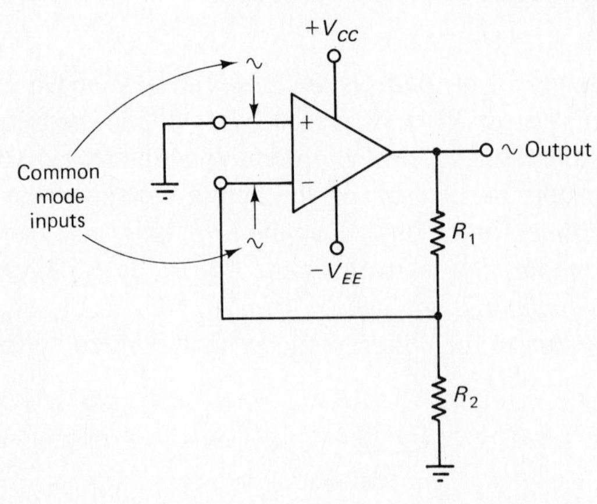

Figure 2-4 A common mode input voltage appears at both input terminals. It is amplified by the common mode gain, but the gain is affected by negative feedback.

or,
$$\frac{V_{o(cm)} \times R_2}{R_1 + R_2} = \frac{A_{cm} \times V_{i(cm)}}{M}$$

giving,
$$V_{o(cm)} = \frac{A_{cm} V_{i(cm)}}{M} \times \frac{R_1 + R_2}{R_2}$$

or,
$$V_{o(cm)} = \frac{V_{i(cm)}}{CMRR} \times A_v \qquad (2\text{-}3)$$

Example 2-1

A 741 op-amp is used in a noninverting amplifier with a voltage gain of 50. Calculate the typical output voltage that would result from a common mode input with a peak level of 100 mV.

Solution

From the 741 data sheet (see Fig. 2-2),

typical common mode rejection ratio = 90 dB

From Eq. 2-2,
$$CMRR = \text{antilog}\,\frac{90\text{ dB}}{20}$$
$$= 31\,623$$

From Eq. 2-3,
$$V_{o(cm)} = \frac{V_{i(cm)}}{CMRR} \times A_v = \frac{100\text{ mV}}{31\,623} \times 50$$
$$= 158\ \mu\text{V}$$

Power Supply Voltage Rejection

At the beginning of Section 1-3 it is explained that, with the basic operational amplifier circuit in Fig. 1-4, a variation in $-V_{EE}$ could have essentially the same effect as an input voltage change. The constant current tail in Fig. 1-5(a) is offered as a means of countering such supply voltage changes. However, even with such circuitry, variations in V_{CC} and V_{EE} do produce some changes at the output. The *power supply rejection ratio (PSRR)* is a measure of how effective the operational amplifier is in dealing with variations in supply voltage.

If a variation of 1 V in V_{CC} or V_{EE} causes the output to change by 1 V, then the supply voltage rejection ratio is *1 V per volt* (1 V/V). If the output changes by 10 mV when one of the supply lines changes by 1 V, then the supply rejection ratio is 10 mV/V.

For the 741 operational amplifier, the supply voltage rejection ratio is specified as typically 30 μV/V (see Fig. 2-2). For the LM108 (Appendix 1-2), the supply voltage rejection is expressed in decibels.

Example 2-2

A 741 operational ampifier uses a ±15 V supply with a 2 mV, 120 Hz ripple voltage superimposed. Calculate the amplitude of the output voltage produced by the power supply ripple.

Solution

$$V_{o(\text{rip})} = V_{s(\text{rip})} \times PSRR$$
$$= 2 \text{ mV} \times (30 \ \mu\text{V/V})$$
$$= 60 \text{ nV}$$

2-3 OFFSET VOLTAGES AND CURRENTS

Input Offset Voltage

Refer once again to the basic operational amplifier circuit connected to function as a voltage follower as illustrated in Fig. 2-1. As already discussed, the output terminal and the inverting input terminal follow the voltage at the noninverting input. For the output voltage to be exactly equal to the input, transistors Q_1 and Q_2 must be perfectly matched. The output voltage can be calculated as,

$$V_o = V_i - V_{BE1} + V_{BE2} \tag{2-4}$$

With $V_{BE1} = V_{BE2}$, and $V_i = 0$,

$$V_o = V_i = 0$$

Now suppose that the transistors are not perfectly matched and that $V_{BE1} = 0.7$ V while $V_{BE2} = 0.6$ V. With the input at ground level,

$$V_o = 0 - 0.7 \text{ V} + 0.6 \text{ V}$$
$$= -0.1 \text{ V}$$

This unwanted output is known as an *output offset voltage*. To set V_o to ground level, the input would have to be raised to + 0.1 V. This is termed an *input offset voltage* (V_{os}). Although transistors in integrated circuits are very well matched, there is always some input offset voltage. The typical input offset voltage is listed as 1 mV on the 741 data sheet (see Fig. 2-5).

Electrical characteristics (V_S = ±15 V, T_A = 25°C unless otherwise specified)

Parameters	Conditions	Min.	Typ.	Max.	Units
Input offset voltage	$R_S \leqslant 10 \text{ k}\Omega$		1.0	5.0	mV
Input offset current			20	200	nA
Input bias current			80	500	nA

Figure 2-5 Portion of 741 op-amp data sheet listing input offset voltage, input offset current, and input bias current.

Input Offset Current

Another result of the input transistors of an operational amplifier not being perfectly matched is that, as well as the transistor base-emitter voltages being unequal, the current gain (h_{FE}) of one transistor may not be exactly equal to that of the other. Thus, when both transistors have equal levels of collector current, the base current in one might be 1 μA while the other has a base current of 1.2 μA. The difference in these two input current levels is known as the *input offset current* (I_{os}).

When an operational amplifier is connected as a simple voltage follower circuit, the input offset current has no effect. But some circuits have two equal-value resistors in series with the input terminals and in this case, the input offset current produces unequal voltage drops across these resistors (see Fig. 2-6). The difference in the resistor voltage drops behaves as a differential input voltage which produces an output offset voltage.

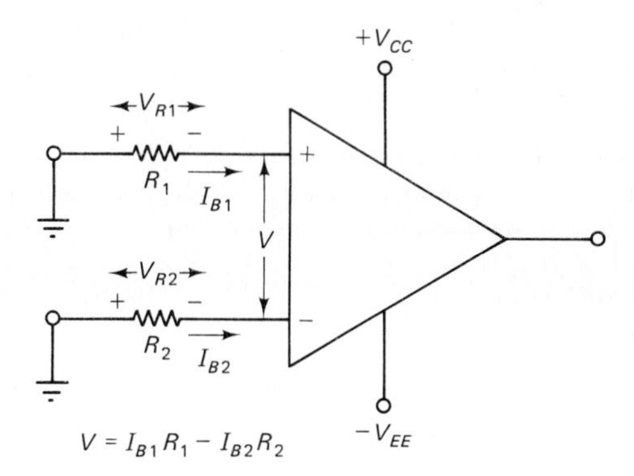

Figure 2-6 Input bias currents to an op-amp produce voltage drops across resistors connected at the input. A difference in the bias currents, known as the *input offset current*, produces unequal resistor voltage drops which result in an unwanted input voltage.

The typical input offset current for the 741 operational amplifier is specified as 20 nA, (see Fig. 2-5).

Offset Nulling

One method of dealing with input offset voltage and current is illustrated in Fig. 2-7(a), which shows a low-resistance potentiometer (R_p) connected at the emitters of Q_1 and Q_2. Adjustment of R_p alters the total voltage drop from each base to the common point at the potentiometer moving contact. Because an offset voltage is produced by the input offset current, this adjustment can null the effects of both input offset current and input offset voltage.

Figure 2-7(b) shows the manufacturer's recommended method of offset nulling for a 741. A 10 kΩ potentiometer is connected to *offset nulling terminals* 1 and 5 and its moving contact is connected to the negative supply line. (Note the offset nulling terminals in Fig. 1-7.) The potentiometer is adjusted to null the output offset voltage to zero, thus nulling both input offset current and input offset voltage.

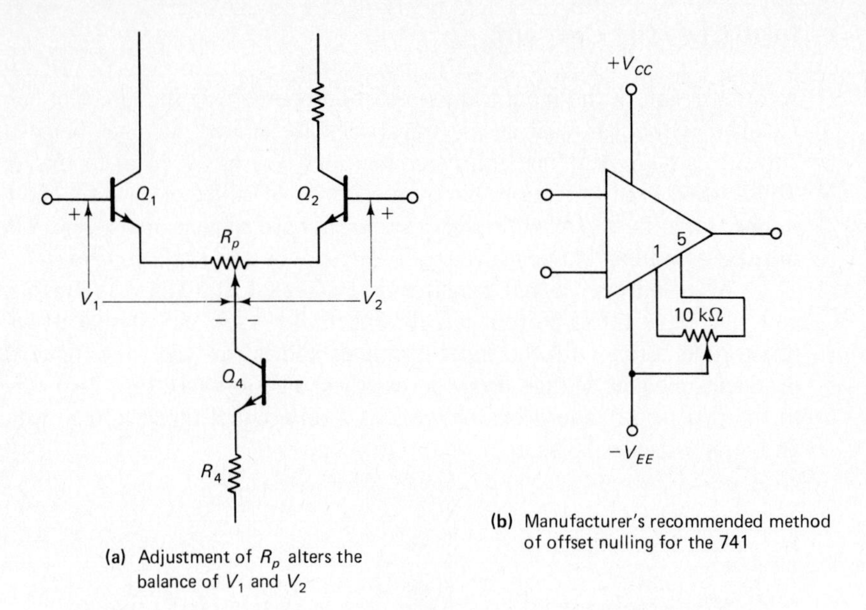

Figure 2-7 When the op-amp input transistors are not perfectly matched, the differences in base-emitter voltages produce an *input offset voltage*. An appropriately connected potentiometer provides adjustment for offset nulling.

Resistor Tolerance Effect

The discussions of offset voltages and currents assumed that either there were no resistors at the op-amp input terminals, or else that exactly equal resistors were connected to the input terminal. Most operational amplifier circuits have resistors at their input terminals and sometimes those resistors may not have equal resistance values. But, even when care is taken to use equal value resistors, the resistor tolerance might be almost as effective in producing an output offset voltage as the operational amplifier input offset voltage and input offset current. This is demonstrated in Example 2-3.

Example 2-3

The circuit in Fig. 2-6 uses a 741 operational amplifier and has $R_1 = R_2 = 22\ \text{k}\Omega$ with a resistor tolerance of $\pm 20\%$. Determine the maximum input offset voltage due to, (a) the 741 specified input offset voltage, (b) the 741 input offset current, (c) the resistor tolerance.

Solution

(a) From the 741 data sheet, $V_{i(\text{offset})} = 5\ \text{mV maximum}$

(b) From the 741 data sheet, $I_{i(\text{offset})} = 200\ \text{nA maximum}$

$$V_{i(\text{offset})} = I_{i(\text{offset})} \times (R_1 \text{ or } R_2)$$

$$= 200\ \text{nA} \times 22\ \text{k}\Omega$$

$$= 4.4\ \text{mV}$$

(c) From the 741 data sheet, $I_B = 500$ nA maximum

Because of the ±20% resistor tolerance, the worst case error could be

$$R_1 = 22\ \text{k}\Omega + 20\% = 26.4\ \text{k}\Omega$$

and

$$R_2 = 22\ \text{k}\Omega - 20\% = 17.6\ \text{k}\Omega$$

For this situation,

$$V_{i(\text{offset})} = I_B R_1 - I_B R_2 = I_B(R_1 - R_2)$$

$$= 500\ \text{nA}\ (26.4\ \text{k}\Omega - 17.6\ \text{k}\Omega)$$

$$= 4.4\ \text{mV}$$

2-4 INPUT AND OUTPUT IMPEDANCES

Input Impedance

The input impedance offered by any operational amplifier is substantially modified by its application. With all linear applications, some form of negative feedback is provided by externally connected components. From negative feedback* theory, the impedance at the op-amp input terminal becomes

$$Z_{in} = (1 + M\beta)\, Z_i \qquad (2\text{-}5)$$

where, Z_i = the op-amp input impedance without negative feedback
M = op-amp open-loop gain
β = feedback factor = 1 for a voltage follower

Note that Eq. 2-5 applies to a noninverting amplifier. As explained in section 3-4, it does not apply to an inverting amplifier.

Example 2-4

Calculate the minimum input impedance of a 741 operational amplifier employed as a voltage follower.

Solution

From the 741 data sheet in Appendix 1-1,

$$R_{i(\text{min})} = 0.3\ \text{M}\Omega, \qquad M_{(\text{min})} = 50\,000$$

Eq. 2-5,

$$Z_{in} = (1 + M\beta)\, Z_i$$

$$= [1 + (50\,000 \times 1)]\ 0.3\ \text{M}\Omega$$

$$= 15\,000\ \text{M}\Omega$$

The impedance of signal sources connected at the input of an operational amplifier circuit (see Fig. 2-8) should be very much smaller than the amplifier input

* David A. Bell, *Electronic Devices and Circuits*, 3rd Ed. (Englewood Cliffs, NJ: Prentice-Hall, Inc., 1986), p. 356.

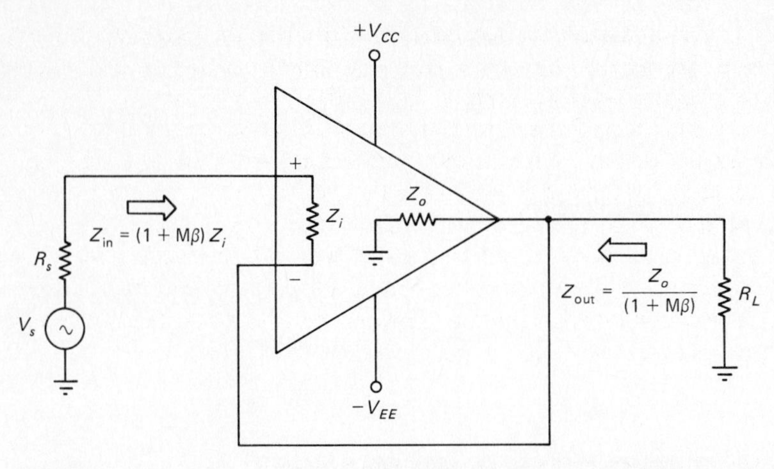

Figure 2-8 Signal source resistances should be very much smaller than the op-amp circuit input impedance. Load resistances should be much larger than the circuit output impedance.

impedance (to avoid a loss of signal across R_s). This is not difficult in the case of a circuit like the one investigated in Example 2-4. But there are other circuits in which the input impedance is reduced by the presence of externally connected components.

Output Impedance

The typical output resistance specified for the 741 op-amp is 75 Ω. Any stray capacitance in parallel with this is certain to have a much larger reactance than 75 Ω. So 75 Ω is also effectively the amplifier output impedance.

Like the input impedance, the output impedance of the op-amp is affected by negative feedback.

$$Z_{out} = \frac{Z_o}{1 + M\beta} \tag{2-6}$$

where Z_o = op-amp output impedance without negative feedback
M = op-amp open-loop gain
β = feedback factor

Example 2-5

Calculate the typical output impedance of a 741 op-amp connected to function as a voltage follower.

Solution

Eq. 2-6, $$Z_{out} = \frac{Z_o}{1 + M\beta} = \frac{75\ \Omega}{1 + (200\ 000 \times 1)}$$

$$\approx 0.004\ \Omega$$

Load impedances connected at the output of an operational amplifier should be much larger than the circuit output impedance (see Fig. 2-8). This is to avoid any significant loss of output as a voltage drop across Z_{out}.

There is another limit to the load that may be connected at the output of an operational amplifier; this is the *output short circuit current*. Internal current-limiting circuitry is included in most operational amplifier to protect the op-amp from damage that might result from a short-circuit at the output. The 741 output short circuitry current is specified as 25 mA. The maximum output current should always be less than the short-circuit current for satisfactory operation. Where a larger output current is required, a power transistor can be connected at the output. This is discussed in Chapter 6.

2-5 SLEW RATE AND FREQUENCY LIMITATIONS

Slew Rate

The *slew rate (S)* of an operational amplifier is the maximum rate at which the output voltage can change. When the slew rate is too slow for the input, distortion results. This is illustrated in Fig. 2-9, which shows a sine wave input to a voltage follower producing a triangular output waveform. The triangular wave results because the op-amp output simply cannot move fast enough to follow the sine wave input.

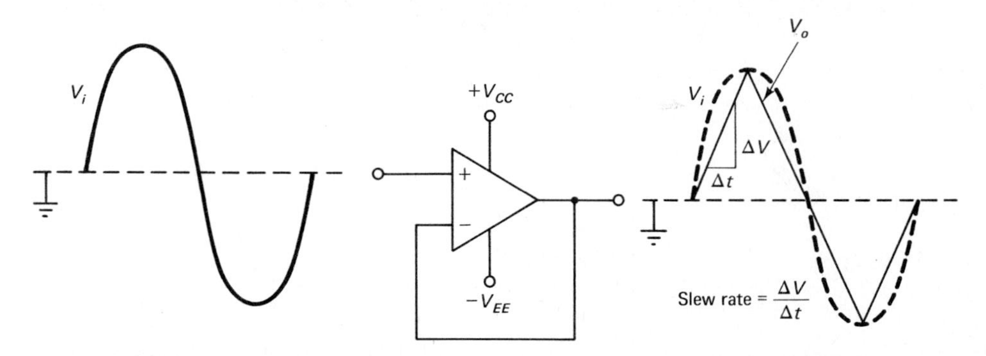

Figure 2-9 The *slew rate* defines the maximum rate of change of operational amplifier output voltage. When the input voltage changes too quickly, output waveform distortion results.

The typical slew rate of the 741 op-amp is specified as *0.5 V per microsecond* (see Fig. 2-10). This means that 1 μs is required for the output to change by 0.5 V. The equation relating time, voltage change, and the slew rate is,

$$t = \frac{\Delta V_o}{S} \tag{2-7}$$

Electrical characteristics (V_S = ±15 V, T_A = 25°C unless otherwise specified)

Parameters	Conditions	Min.	Typ.	Max.	Units
Slew rate	$R_L \geqslant 2\ k\Omega$		0.5		V/μs

Figure 2-10 Portion of 741 op-amp data sheet specifying the slew rate.

A 10 V output change from a 741 typically requires a minimum time of,

$$t = \frac{10\ \text{V}}{0.5\ \text{V}/\mu\text{s}} = 20\ \mu\text{s}$$

Slew rate effects and the relationship between slew rate and circuit cutoff frequency are treated in detail in Chapter 5.

Frequency Limitations

Figure 2-11 shows the graph of the open-loop gain (M) plotted versus frequency (f) for a 741 operational amplifier. It is seen that M is 100 dB when the signal frequency is 1 Hz. At 10 Hz the gain has fallen below 100 dB, and M continues to fall as the signal frequency increases. Note that frequency is plotted to a logarithmic base and that M falls linearly as f increases logarithmically.

at $\quad f = 100\ \text{Hz}, \quad M \approx 80\ \text{dB}$

at $\quad f = 1\ \text{kHz}, \quad M \approx 60\ \text{dB}$

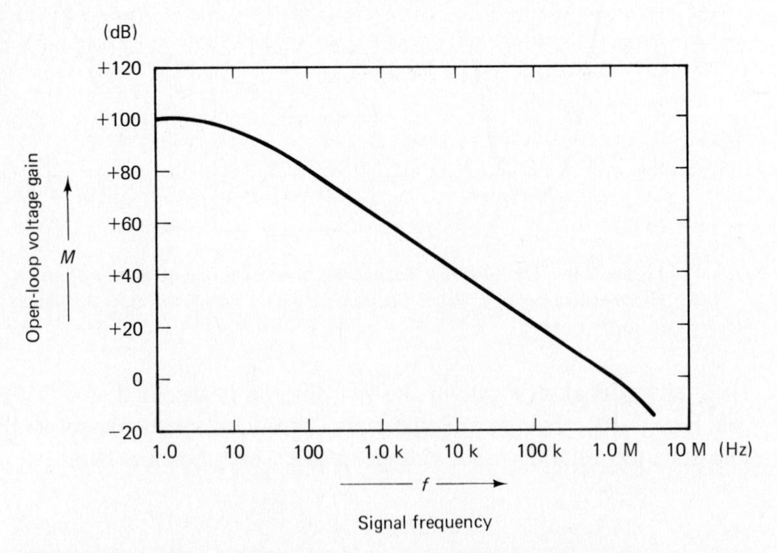

Figure 2-11 Plot of open-loop gain M versus signal frequency for a 741 op-amp. The available open-loop gain falls as the signal frequency increases.

M falls by 20 dB when f increases from 100 Hz to 1 kHz. The ten times increase in frequency is termed a *decade*. So, the rate of fall of the gain is said to be *20 dB per decade*.

Figure 2-11 shows that that M falls to zero at approximately 800 kHz. Where an internal gain equal to or greater than 80 dB is required for a particular application, it is available with a 741 only for signal frequencies up to approximately 100 Hz. An internal gain greater than 20 dB is possible for signal frequencies up to approximately 90 kHz. Other operational amplifiers maintain substantial internal gain to much higher frequencies than the 741.

REVIEW QUESTIONS

2-1. Discuss the limitations on the input voltage range of an operational amplifier. Draw the necessary circuit diagrams to explain the limitations. State a typical op-amp input voltage range.

2-2. Explain the limitations on the output voltage range of an operational amplifier. Draw the necessary circuit diagrams to support your explanation. State a typical op-amp output voltage range.

2-3. Explain common mode voltage, common mode voltage gain, and common mode rejection ratio for operational amplifiers. State a typical common mode rejection ratio.

2-4. Discuss supply voltage rejection in operational amplifiers and state a typical supply voltage rejection ratio.

2-5. Explain input offset voltage and state a typical input offset voltage level for an operational amplifier.

2-6. Explain input offset current and state a typical input offset current level for an operational amplifier. Discuss offset nulling.

2-7. Explain how input offset voltages can be produced by errors in bias resistors.

2-8. State a typical input resistance for an operational amplifier. Write the equation for the input impedance of an op-amp circuit using negative feedback. Identify each quantity in the equation.

2-9. State a typical output resistance for an operational amplifier. Write the equation for the output impedance of an op-amp circuit using negative feedback. Identify each quantity in the equation. Explain output short circuit current.

2-10. Sketch an illustration to show the effect of operational amplifier slew rate. Explain. State a typical op-amp slew rate.

2-11. Sketch a typical gain versus frequency graph for an operational amplifier. Explain.

PROBLEMS

2-1. An operational amplifier has a specified input voltage range of ± 8 V and an output voltage range of ± 14 V when the supply voltage is ± 15 V. Calculate the maximum output voltage that can be produced, (a) when the op-amp is used as a voltage follower and (b) when it is used as an amplifier with a voltage gain of 2.

2-2. An op-amp voltage follower has a dc input voltage of $+6$ V with an ac signal of

±3 V superimposed. Determine the minimum input voltage range for a suitable operational amplifier.

2-3. An LM108 operational amplifier (data sheet in Appendix 1-2) with a supply of ±15 V is employed as an amplifier with a voltage gain of 25. Calculate the maximum ac input signal that can be applied if the output is to remain undistorted. The load resistance is 10 kΩ.

2-4. An operational amplifier circuit with a closed-loop gain of 100 has a common mode output of 5 μV when the common mode input is 5 mV. Determine the common mode rejection ratio.

2-5. An LM308 operational amplifier circuit with a closed-loop gain of 33 has a common mode input of 1.5 V. Calculate the maximum output voltage this might produce.

2-6. A 1 μV supply ripple is present at the output of an operational amplifier circuit when the supply voltage has a 25 mV ripple. Determine the power supply rejection ratio.

2-7. An LM108 operational amplifier (see Appendix 1-2) has a ±15 V supply with a 3 mV ac ripple. Calculate the maximum level of the output ripple voltage.

2-8. A 741 operational amplifier circuit has a minimum output signal level of 100 mV. Output ripple voltage produced by the ripple on the supply voltages is not to exceed 0.1% of the minimum output signal level. Calculate the maximum permissible supply voltage ripple.

2-9. The circuit in Fig. 2-6 uses an LM108 operational amplifier and the resistor values are $R_1 = R_2 = 33$ kΩ. Determine the maximum input offset voltage due to, (a) the op-amp specified input offset voltage and (b) the input offset current.

2-10. The resistors in the circuit in Problem 2-9 have tolerances of ±10%. Calculate the maximum input offset voltage due to the resistor tolerance (at 25°C), (a) when an LM108 is used and (b) when the op-amp is an LM308.

2-11. Determine the minimum input impedance of an LM108 operational amplifier, (a) when employed as a voltage follower, and (b) when used as a noninverting amplifier with a voltage gain of 30.

2-12. Recalculate the input impedance for Problem 2-11 when an LM308 is substituted in place of LM108.

2-13. Calculate the typical output resistance for a 715 op-amp, (a) when employed as a voltage follower, and (b) when used as a noninverting amplifier with a gain of 50.

2-14. Calculate the typical input and output impedances of a noninverting amplifier with a voltage gain of 25, (a) using a 741 op-amp, and (b) using a 715.

2-15. A 741 operational amplifier has a ±15 V supply. Calculate the typical time required for the op-amp output to move from its negative extreme to its positive extreme.

2-16. Repeat Problem 2-15 for an LF 353 op-amp.

2-17. Repeat Problem 2-15 for a 715 op-amp, (a) with $A_v = 100$, and (b) with $A_v = 1$ noninverting.

LABORATORY EXERCISES

2-1 Offset Voltage

1. Connect a 741 op-amp to function as an inverting amplifier as illustrated in Fig. 1-11(a). Use $R_1 = 10\ \Omega$, $R_2 = 100\ \Omega$, and a supply voltage of ±9 V to ±15 V.

2. Ground the input terminal and connect a dc voltmeter at the output. (The voltmeter is to measure millivolts, so a digital instrument is most suitable.)
3. Record the measured output voltage, and calculate the input offset voltage, ($V_{offset} = V_o \times R_1/R_2$). Compare this to the specified input offset voltage for a 741.
4. Calculate the input offset voltage due to the specified maximum input offset current, ($I_{offset} \times 10\ \Omega$). Compare this to the measured input offset voltage to check that it has not introduced a significant error.

2-2 Input Bias and Offset Currents

1. Connect a 741 voltage follower circuit as in Fig. 1-9(a) with a nulling potentiometer as in Fig. 2-7(b). Use a supply of ±9 V to ±15 V.
2. Ground the voltage follower input and connect a digital dc voltmeter at the output.
3. Adjust the nulling potentiometer to give zero output voltage. If the output offset cannot be completely nulled, note the level of V_o.
4. Switch *off* the supply, and insert a 1 MΩ resistor in series with the (grounded) noninverting input terminal.
5. Switch the supply *on* again and note the change in output voltage (from the V_o nulled level). Calculate the input bias current at the op-amp noninverting input. ($I_+ = \Delta V_o/1\ \text{M}\Omega$).
6. Switch *off* the supply, and reconnect the *same* 1 MΩ resistor in series with the inverting input terminal (between input and output terminals). Directly ground the noninverting input once again.
7. Switch the supply *on* again, and note the new change in V_o (from the nulled level). Calculate the input bias current at the op-amp inverting input. ($I_- = \Delta V_o/1\ \text{M}\Omega$).
8. Calculate the input offset current, ($I_{offset} = I_+ - I_-$). Compare the input bias and offset currents to the specified quanties for the 741.

2-3 Input and Output Voltage Ranges

1. Connect a 741 voltage follower circuit as in Fig. 1-9(a) using a 100 kΩ resistor in series with each input terminal. Use a supply of ±15 V.
2. Connect a sine wave signal generator to the voltage follower input and a dual-trace oscilloscope to monitor the op-amp input (right at the op-amp terminal) and output waveforms.
3. Apply a 100 Hz sine wave input, increasing the signal amplitude until the peaks of the input waveform just begin to flatten.
4. Measure the positive and negative input voltage peaks on the oscilloscope to determine the op-amp input voltage range. Compare the measured range to the specified range for the 741.
5. Reconnect the 741 op-amp as an inverting amplifier [Fig. 1-11(a)], using $R_1 = 10\ \text{k}\Omega$, $R_2 = 100\ \text{k}\Omega$, and a supply of ±15 V.
6. Connect a sine wave signal generator to the amplifier input, and a dual-trace oscilloscope to monitor the signal and output waveforms.
7. Apply a 100 Hz sine wave input, increasing the signal amplitude until the peaks of the output waveform just begin to flatten.

8. Measure the positive and negative output peaks on the oscilloscope to determine the output voltage range. Compare the measured range to the specified range for the 741.

2-4 Open-Loop Voltage Gain

1. Connect the 741 op-amp circuit illustrated in Fig. 2-12 using the components and supply voltages shown. Also, connect voltmeters to measure V_o and V_3.

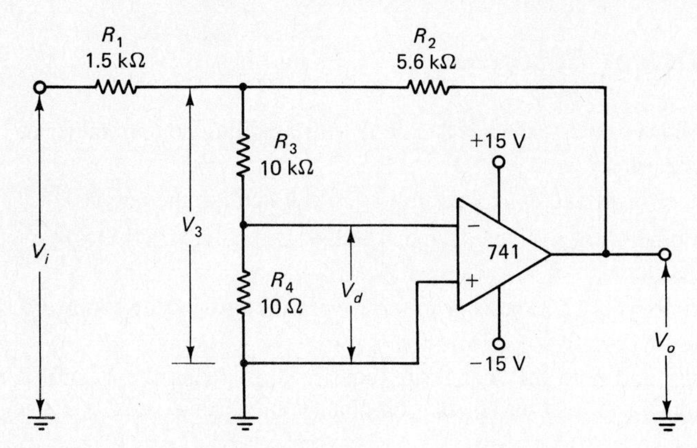

Figure 2-12 Circuit for measuring op-amp open-loop gain.

2. Connect a dc voltage source at the circuit input, adjusting the voltage to give $V_o = 10$ V.
3. Record the level of V_3, and calculate the op-amp differential input voltage, ($V_d = V_3/1000$).
4. Calculate the op-amp open-loop voltage gain, ($M = V_o/V_d$).

CHAPTER 3

OP-AMPS as DC Amplifiers

Objectives

- Explain the requirement for uninterrupted current paths at op-amp input terminals.
- Calculate the maximum bias resistor values that should be used with a bipolar op-amp and explain the design approach for potential divider bias at the input of a bipolar op-amp.
- Sketch the input stage of a BIFET op-amp and discuss how the input parameters differ from those of a bipolar op-amp.
- Discuss the design approach for selection of resistor values for biasing a BIFET op-amp.
- Sketch the following direct-coupled op-amp circuits and explain the operation of each circuit:
  voltage follower, noninverting amplifier, inverting amplifier, noninverting summing circuit, inverting summing circuit, difference amplifier.
- *Easily* design each of the above circuits using bipolar and BIFET op-amps.
- Analyze each of the above circuits to determine input impedances, output impedances, voltage gains, and other performance criteria.
- Show how op-amp circuits may be connected to use a single-polarity supply.

INTRODUCTION

IC operational amplifiers make excellent direct-coupled amplifiers. The design of such circuits involves little more than calculation of resistor values for a potential divider. All op-amp circuits must provide a current path for the input bias current to each terminal of the operational amplifier, and all potential divider circuits should use current levels which are much higher than the op-amp input current. The design approach for circuits using BIFET op-amps differs slightly from that normally employed for bipolar op-amps.

3-1 BIASING OPERATIONAL AMPLIFIERS

Bias Current Paths

Like other electronic devices and integrated circuits, operational amplifiers must be correctly biased if they are to function properly. As already discussed, the inputs of most operational amplifiers are the base terminals of the transistors in a differential amplifier. Base currents must flow into these terminals for the transistors to be operational. Consquently, the input terminals must be directly connected to suitable dc bias voltage sources.

For many applications, the most appropriate dc bias voltage level for the op-amp input terminals is approximately halfway between the positive and negative supply voltages. One of the two input terminals is usually connected in some way to the op-amp output to facilitate negative feedback. The other input might be biased directly to ground via a signal source (see Fig. 3-1(a)). Base current I_{B1} flows into the op-amp via the signal source while I_{B2} flows from the output terminal as illustrated.

Figure 3-1(b) shows a situation in which resistor R_1 is included in series with the inverting terminal to match signal source resistance R_S in series with the noninverting terminal. The op-amp input currents produce voltage drops $I_{B1}R_S$ and $I_{B2}R_1$

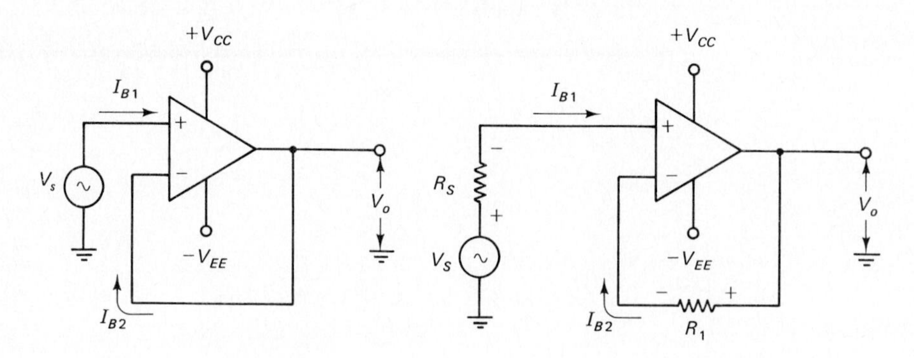

Figure 3-1 Op-amp biasing must be so arranged that base currents I_{B1} and I_{B2} flow into the input terminals.

across the resistors. R_S and R_1 should be selected as equal resistors so that the resistor voltage drops are approximately equal. Any difference in these voltage drops will have the same effect as an input offset voltage (see Section 2-3).

Maximum Bias Resistor Values

If very small resistance values are selected for R_S and R_1 in the circuit in Fig. 3-1(b) the voltage drops across them will be small. On the other hand, if R_S and R_1 are very large, the voltage drops $I_{B1}R_S$ and $I_{B2}R_1$ might be several volts. For good bias stability, the maximum voltage drop across these resistors should be much smaller than the typical forward-biased V_{BE} level for the op-amp input transistors. Usually, the resistor voltage drop is made at least ten times smaller than V_{BE}.

Let
$$I_{B(\max)}R_{(\max)} \approx \frac{V_{BE}}{10} \approx 0.07 \text{ V}$$

From the 741 data sheet in Appendix 1-1, $I_{B(\max)} = 500$ nA.

Therefore
$$R_{(\max)} \approx 0.07 \text{ V}/500 \text{ nA} \approx 140 \text{ k}\Omega.$$

This is a maximum value for the bias resistors for a 741 operational amplifier. Where other op-amps are involved, $R_{(\max)}$ should be calculated using the specified $I_{B(\max)}$ for that particular op-amp.

$$R_{(\max)} \approx \frac{0.1\ V_{BE}}{I_{B(\max)}} \tag{3-1}$$

Potential Divider Bias

Figure 3-2 shows a potential divider (R_1 and R_2) employed to derive a terminal bias voltage from the supply voltages. Potential divider bias is commonly used with op-amps and in transistor circuitry. The potential divider current (I_2) should be much larger than the op-amp maximum input bias current. This is to ensure that I_B, and

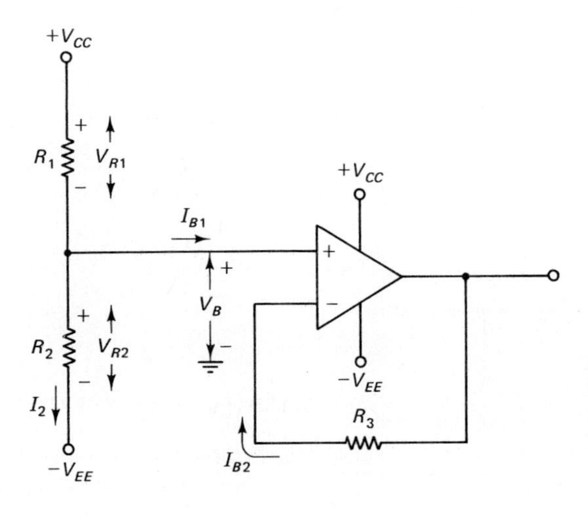

Figure 3-2 Potential divider bias for an op-amp input. The potential divider current (I_2) must be selected to be much greater than the maximum level of the op-amp input bias current (I_B).

any bias current variation, has a negligible effect upon the bias voltage level. Usually I_2 is made 100 or more times I_B. Then

$$R_1 = \frac{V_{R1}}{I_2} \quad \text{and} \quad R_2 = \frac{V_{R2}}{I_2}$$

Typically, $I_{B(max)}$ is 500 nA for a 741 op-amp, which gives

$$I_2 = 100 \times 500 \text{ nA} = 50 \ \mu\text{A}$$

This is a minimum level for I_2 when using a 741 and it would be quite satisfactory to use a current of 1 mA, for example. However, *all electronic circuit currents are normally selected as low as possible in order to minimize the current demand from the power supply*.

In Fig. 3-2, V_{R1} and V_{R2} would usually be selected to give V_B close to ground level, or half way between $+V_{CC}$ and $-V_{EE}$. But V_B can be above or below ground level so long as it is within the specified input voltage range for the op-amp (see Section 2-1). From Appendix 1-1, the input voltage range is minimum of ± 12 V for a 741 op amp with a ± 15 V supply.

The resistance "seen" when "looking out" of the noninverting input terminal in Fig. 3-2 is $R_1 \| R_2$. To equalize the voltage drops at the input terminals,

$$I_{B2} R_3 = I_{B1}(R_1 \| R_2)$$

or

$$R_3 \approx R_1 \| R_2$$

A single polarity supply voltage can be employed with an operational amplifier. For example, a 741 could use a +30 V supply as illustrated in Fig. 3-3. In this case the input terminal bias voltage should be approximately half the supply voltage (+15 V for a 30 V supply). Alternatively, V_B might be within the input voltage range (± 12 V for a 741 using a ± 15 V supply) of the half-way point between $+V_{CC}$ and ground. Since the circuit shown in Fig. 3-3 is a voltage follower, the dc output voltage will be equal to the bias voltage level.

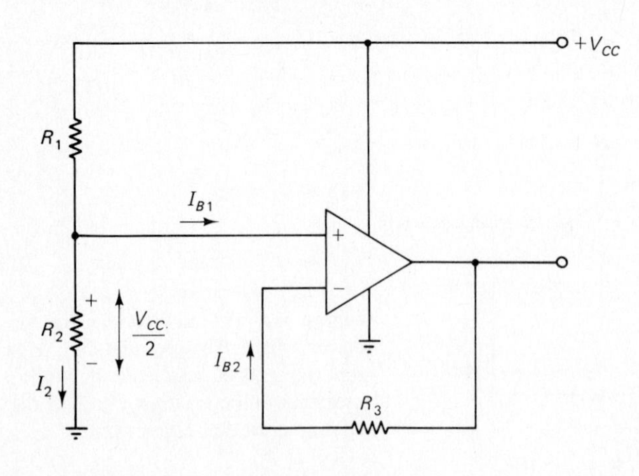

Figure 3-3 When using a single polarity supply with an op-amp, the input terminals should usually be biased at approximately half the supply voltage.

Biasing BIFET Op-Amps

BIFET Op-amps are operational amplifiers with FET input stages (see Fig. 3-4). They draw very low levels of input bias current; 50 pA is not unusual. In this case, the usual design approach of selecting resistor currents one hundred times $I_{B(\max)}$ would result in very high resistor values which are undesirable for several reasons. When the bias resistors at the gate terminal of a FET are extremely large, a charge can accumulate at the gate and this might take a relatively long time to discharge. Thus, the gate voltage would not be a stable quantity, and the op-amp bias conditions would be uncertain. Another reason for avoiding high resistance values with any op-amp circuit is that stray capacitance becomes more effective as resistance values increase, possibly resulting in unwanted circuit oscillations. This is further considered in Chapter 5.

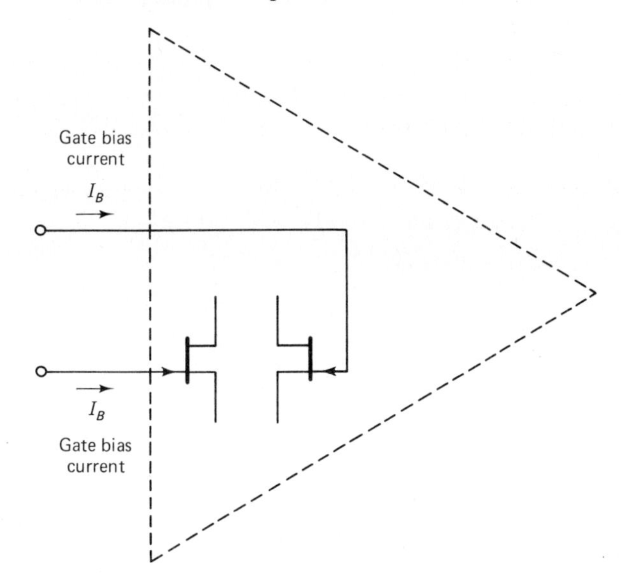

Figure 3-4 Op-amp with FET input stages have very low input bias currents. Bias resistor values should be determined by first selecting the largest resistor as 1 MΩ.

For statisfactory bias conditions when using BIFET op-amps, the resistance "seen" when "looking out" of either input terminal should normally not exceed 1 MΩ. A reasonable rule-of-thumb is to first select the largest resistor in a bias network as 1 MΩ, then calculate the other resistors accordingly. There are some op-amps which can operate with even larger resistors, notably the LM108 (not a BIFET op-amp) which can use signal source resistors as high as 10 MΩ.

3-2 DIRECT-COUPLED VOLTAGE FOLLOWERS

Design

As already discussed, an operational amplifier may be connected to function as a voltage follower with the use of any external components (Fig. 3-1(a)). However, as illustrated in Fig. 3-1(b), resistor R_1 is frequently included in series with the inverting input terminal to match the source resistance R_S in series with the noninverting input terminal.

Example 3-1

A voltage follower using a 741 op-amp is connected to a signal source via a 47 kΩ resistor, as in Fig. 3-1(b). Select a suitable value for resistor R_1. Also, calculate the maximum voltage drop across each resistor and the maximum input offset voltage produced by the input offset current.

Solution

$$R_1 = R_S = 47\ \text{k}\Omega$$

From the 741 data sheet,

$$I_{B(\text{max})} = 500\ \text{nA}, \quad \text{and} \quad I_{i(\text{offset})} = 20\ \text{nA}$$

$$I_{B(\text{max})1} \times R_S = I_{B(\text{max})2} \times R_1$$

$$= 500\ \text{nA} \times 47\ \text{k}\Omega$$

$$= 23.5\ \text{mV}$$

$$V_{i(\text{offset})} = I_{i(\text{offset})} \times (R_S \text{ or } R_1)$$

$$= 20\ \text{nA} \times 47\ \text{k}\Omega$$

$$= 0.94\ \text{mV}$$

Performance

From Equations 2-5 and 2-6, the input and output impedances of the voltage follower are

$$Z_{\text{in}} = (1 + M)Z_i \tag{3-2}$$

and

$$Z_{\text{out}} = \frac{Z_o}{1 + M} \tag{3-3}$$

These quantities are calculated exactly as in Examples 2-4 and 2-5.

As already explained, the voltage follower has a very high input impedance and a very low output impedance. Therefore, it is normally used to convert a high impedance source to a low output impedance. In this situation it is said to be employed as a *buffer* between the high impedance source and the low impedance load. Thus, it is termed a *buffer amplifer*.

Figure 3-5(a) illustrates the fact that a signal voltage is potentially divided across R_S and R_L when connected directly to a load. In Fig. 3-5(b), the voltage follower presents its very high input impedance to the signal source. Because Z_{in} is normally very much larger than R_S, there is virtually no signal loss at this point and effectively all of V_i appears at the op-amp input. From Eq. 1-4, the voltage follower output is

$$V_o = V_i\left(1 - \frac{1}{M}\right)$$

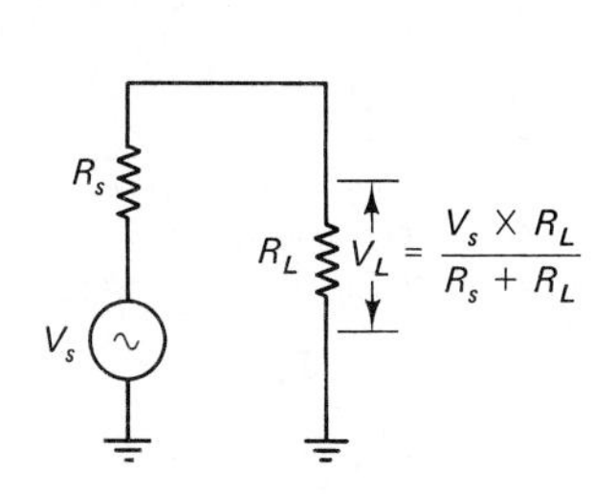

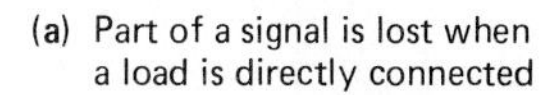
(a) Part of a signal is lost when a load is directly connected

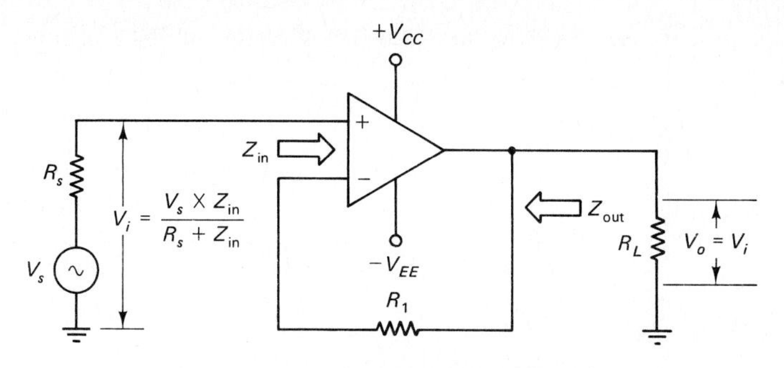

(b) The high input impedance of the voltage follower prevents any significant signal loss

Figure 3-5 The very high input impedance, very low output impedance, and gain of 1 of the voltage follower make it an ideal *buffer amplifier*, which passes the signal from source to load with no significant loss.

The output voltage can be thought of as being potentially divided across R_L and the voltage follower output impedance Z_{out}. But Z_{out} is normally much smaller than any load resistance that might be connected. So, once again, there is effectively no signal loss and all of V_i appears as V_o at the circuit output. The voltage follower performance is demonstrated by Example 3-2.

Example 3-2

The voltage follower in Example 3-1 has a 1 V signal and a 20 kΩ load. Calculate the load voltage (a) when the load is directly connected to the source, (b) when the voltage follower is between the load and the source.

Solution

(a)

$$V_L = \frac{V_S \times R_L}{R_S + R_L} = \frac{1\text{ V} \times 20\text{ k}\Omega}{47\text{ k}\Omega + 20\text{ k}\Omega}$$

$$= 298\text{ mV}$$

(b) Eq. 3-2,

$$Z_{in} = (1 + M)\,Z_i$$

Using typical values of M and Z_i for the 741,

$$Z_{in} = (1 + 200\,000)\ 2\text{ M}\Omega$$

$$= 4 \times 10^{11}\ \Omega$$

$$V_i = \frac{V_S \times Z_{in}}{R_S + Z_{in}} = \frac{1\text{ V} \times (4 \times 10^{11}\ \Omega)}{47\text{ k}\Omega + 4 \times 10^{11}\ \Omega}$$

$$= 1\text{ V (effectively)}$$

Eq. 1-4,

$$V_o = V_i\left(1 - \frac{1}{M}\right) = 1\text{ V}\left(1 - \frac{1}{200\,000}\right)$$

$$= 1\text{ V (effectively)}$$

Using typical values of M and Z_o for the 741,

Eq. 3-3,
$$Z_{out} = \frac{Z_o}{1 + M} = \frac{75\ \Omega}{1 + 200\ 000}$$
$$= 37 \times 10^{-5}\ \Omega$$
$$V_L = \frac{V_o \times R_L}{R_L + Z_{out}} = \frac{1\ \text{V} \times 20\ \text{k}\Omega}{20\ \text{k}\Omega + (37 \times 10^{-5}\ \Omega)}$$
$$= 1\ \text{V (effectively)}$$

Figure 3-6 illustrates the use of a voltage follower with a potential divider to produce a low impedance dc voltage source. In Fig. 3-6(a), load resistor R_L is shown directly connected in series with resistor R_1 to derive a voltage V_L from the supply V_{CC}. This simple arrangement has the disadvantage that the load voltage varies if the load resistance changes. In Fig. 3-6(b), the presence of the voltage follower maintains the load voltage constant regardless of the load resistor value.

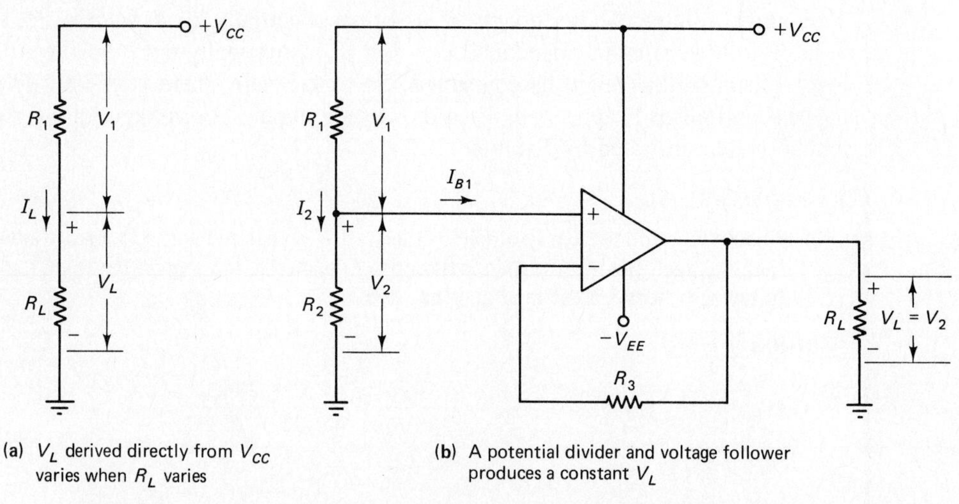

(a) V_L derived directly from V_{CC} varies when R_L varies

(b) A potential divider and voltage follower produces a constant V_L

Figure 3-6 A voltage follower together with a potential divider can be used to provide a voltage source that remains constant even when the load resistance varies.

Example 3-3

A 1 kΩ load is to have 5 V developed across it from a 15 V source. Design suitable circuits as in Fig. 3-6(a) and (b) and calculate the load voltage variation in each case when the load resistance varies by −10%. Use a 741 op-amp.

Solution

Circuit (a)
$$I_L = \frac{V_L}{R_L} = \frac{5\ \text{V}}{1\ \text{k}\Omega}$$
$$= 5\ \text{mA}$$

$$V_1 = V_{CC} - V_L = 15\text{ V} - 5\text{ V}$$

$$= 10\text{ V}$$

$$R_1 = \frac{V_1}{I_L} = \frac{10\text{ V}}{5\text{ mA}}$$

$$= 2\text{ k}\Omega$$

When R_L changes by -10%,

$$V_L = \frac{V_{CC} \times (R_L - 10\%)}{R_1 + (R_L - 10\%)}$$

$$= \frac{15\text{ V} \times (1\text{ k}\Omega - 10\%)}{2\text{ k}\Omega + (1\text{ k}\Omega - 10\%)}$$

$$= 4.66\text{ V}$$

Circuit (b),

$$V_2 = V_L = 5\text{ V}$$

$$V_1 = V_{CC} - V_L$$

$$= 10\text{ V}$$

For the 741

$$I_{B(max)} = 500\text{ nA}$$

let

$$I_2 = 100\ I_{B(\max)} = 100 \times 500\text{ nA}$$

$$= 50\ \mu\text{A}$$

$$R_2 = \frac{V_2}{I_2} = \frac{5\text{ V}}{50\ \mu\text{A}}$$

$$= 100\text{ k}\Omega$$

$$R_1 = \frac{V_1}{I_2} = \frac{10\text{ V}}{50\ \mu\text{A}}$$

$$= 200\text{ k}\Omega$$

When R_L changes by -10%,

$$V_L = V_2 = 5\text{ V (effectively)}$$

Voltage Follower Compared to an Emitter Follower

Both the voltage follower and the emitter follower are buffer amplifiers. However, the voltage follower has a much higher input impedance and much lower output impedance than the emitter follower. The most obvious disadvantage of the emitter follower is the dc voltage loss due to the transistor base-emitter voltage drop (see Fig. 3-7(a)). But the voltage follower dc loss of V_i/M is insignificant [see Eq. 1-4 and Fig. 3-7(b)]. There can also be a greater loss of ac signal voltage in the emitter follower than in the voltage follower because of the lower input impedance and higher output impedance with the emitter follower.

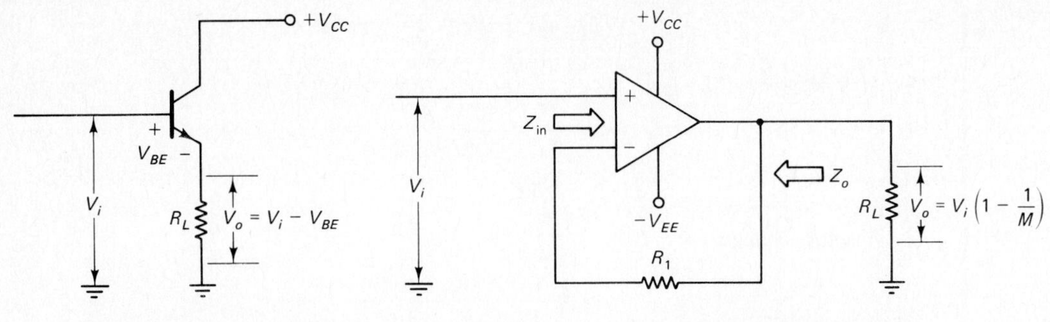

Figure 3-7 The voltage follower has a much higher input impedance and lower output impedance than the emitter follower. Also, the emitter follower has a base-emitter dc voltage drop between input and output (V_{BE}) that does not occur with the voltage follower.

3-3 DIRECT-COUPLED NONINVERTING AMPLIFIERS

Design

In Section 1-5 it is shown that the voltage gain of a noninverting amplifier, as reproduced in Fig. 3-8, is

Eq. 1-5, $$A_v = \frac{R_2 + R_3}{R_3}$$

As always with a bipolar op-amp, design commences by selecting the potential divider current (I_2) very much larger than the maximum input bias current ($I_{B(\max)}$). For a BIFET op-amp, the largest-value resistor is first selected as 1 MΩ. In each case, potential divider resistor values are then determined using V_i, V_o, and I_2.

Because $V_{R3} = V_i$, $$R_3 = \frac{V_i}{I_2} \tag{3-4}$$

and V_o appears across $(R_2 + R_3)$,

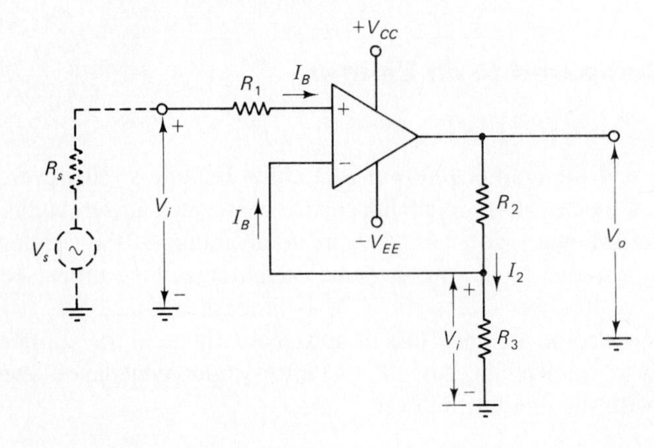

Figure 3-8 Design of a noninverting amplifier (using a bipolar op-amp) starts by selecting the potential divider current (I_2) to be much greater than the maximum level of the input bias current (I_B). Then, $R_3 = V_i/I_2$, and $(R_2 + R_3) = V_o/I_2$.

so
$$R_2 + R_3 = \frac{V_o}{I_2} \tag{3-5}$$

Finally, to equalize the $I_B R$ voltage drops at the op-amp inputs, R_1 is calculated as

$$R_1 \approx R_2 \| R_3 \tag{3-6}$$

If R_1, as determined from Eq. 3-6 is not very much larger than the source resistance, then R_1 should be determined from,

$$R_S + R_1 \approx R_2 \| R_3$$

Example 3-4

Using a 741 op-amp, design a noninverting amplifier to have a voltage gain of approximately 66. The signal amplitude is to be 15 mV.

Solution

For the 741, $I_{B(\max)} = 500 \text{ nA}$

let $I_2 = 100 \times I_{B(\max)}$

$= 50\ \mu\text{A}$

Eq. 3-4, $R_3 = \frac{V_i}{I_2} = \frac{15 \text{ mV}}{50\ \mu\text{A}}$

$= 300\ \Omega$ (Use a 270 Ω standard value resistor for R_3, see Appendix 2-1. This will give a level of I_2 slightly greater than the selected 50 μA)

I_2 becomes, $I_2 = \frac{V_i}{R_3} = \frac{15 \text{ mV}}{270\ \Omega}$

$= 55.6\ \mu\text{A}$

$V_o = A_v \times V_i = 66 \times 15 \text{ mV}$

$= 990 \text{ mV}$

Eq. 3-5, $R_2 + R_3 = \frac{V_o}{I_2} = \frac{990 \text{ mV}}{55.6\ \mu\text{A}}$

$= 17.8 \text{ k}\Omega$

$R_2 = (R_2 + R_3) - R_3 = 17.8 \text{ k}\Omega - 270\ \Omega$

$= 17.53 \text{ k}\Omega$ (Use 18 kΩ standard value resistor to give $A_v > 66$. Alternatively, use 15 kΩ in series with 2.7 kΩ)

Eq. 3-6, $R_1 \approx R_2 \| R_3 = 270\ \Omega \| 18 \text{ k}\Omega$

$\approx 270\ \Omega$ (use 270 Ω standard value)

The op-amp supply voltage should normally be ±9 V to ±18 V, or within whatever range is specified on the data sheet.

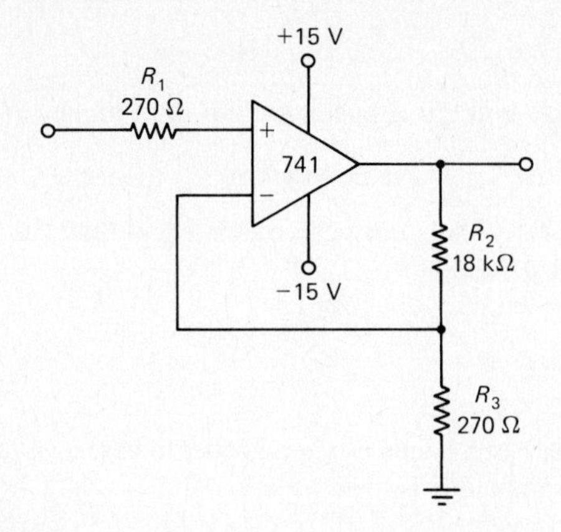

Figure 3-9 Noninverting amplifier with $A_v \approx 66$ designed in Example 3-4.

Example 3-5

Redesign the noninverting amplifier in Example 3-4 using a LF353 BIFET op-amp (see Appendix 1-3).

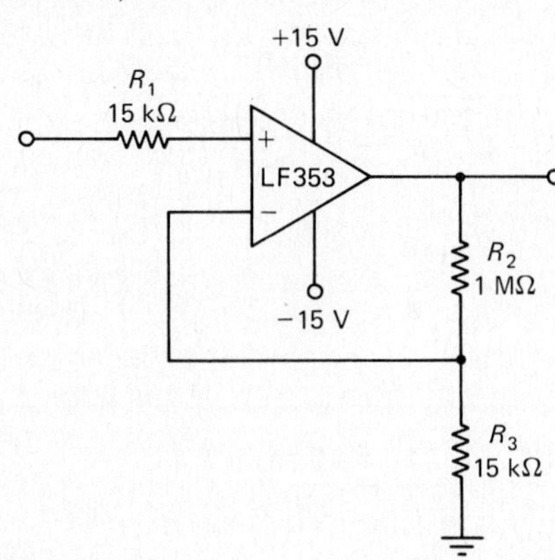

Figure 3-10 When using a BIFET op-amp as a noninverting amplifier, design begins by first selecting the largest potential divider resistor as 1MΩ. The other resistor is then calculated from $A_v = (R_2 + R_3)/R_3$.

Solution

For the LF353, $$I_{B(max)} = 200 \text{ pA}$$

As explained, using $I_2 = 100\, I_{B(max)}$ in this case would result in very large resistor values. Therefore,

let $$R_2 = 1 \text{ M}\Omega$$

Eq. 1-5, $$A_v = \frac{R_2 + R_3}{R_3} = \frac{R_2}{R_3} + 1$$

or, $$R_3 = \frac{R_2}{A_v - 1} = \frac{1 \text{ M}\Omega}{66 - 1}$$

$$= 15.38 \text{ k}\Omega$$ (using a 15 kΩ standard value resistor will give A_v slightly greater than 66)

Eq. 3-6, $$R_1 \approx R_2 \| R_3 = 1 \text{ M}\Omega \| 15 \text{ k}\Omega$$

$$\approx 15 \text{ k}\Omega \text{ (use 15 k}\Omega\text{ standard value)}$$

Performance

From Eq. 2-5, the input impedance of an op-amp circuit is,

$$Z_{in} = (1 + M\beta)Z_i$$

For a noninverting amplifier, the feedback factor is

$$\beta = \frac{R_3}{R_2 + R_3} = \frac{1}{A_v}$$

Therefore, for a noninverting amplifier

$$Z_{in} = \left(1 + \frac{M}{A_v}\right)Z_i \tag{3-7}$$

Referring to Fig. 3-11, the input impedance given by Eq. 3-7 is the impedance "seen" when "looking into" the noninverting input terminal; it does not include R_1. Thus, the impedance seen from the signal source is

$$Z'_{in} = R_1 + Z_{in} \tag{3-8}$$

Since Z_{in} is always much larger than R_1 in a noninverting amplifier, the inclusion of R_1 normally makes no significant difference. As will be explained, it can be important in other circuits.

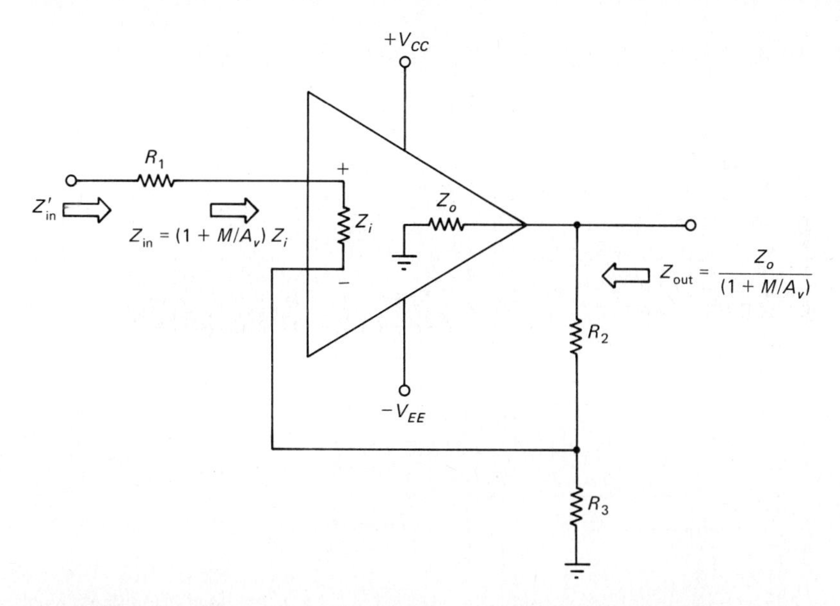

Figure 3-11 The noninverting amplifier has an input impedance of $(R_1 + Z_{in})$, where $Z_{in} = (1 + M/A_v)Z_i$. The output impedance is $Z_{out} = Z_o/(1 + M/A_v)$.

Equation 2-6 gives the output impedance of an op-amp circuit as

$$Z_{out} = \frac{Z_o}{(1 + M\beta)}$$

Since $\beta = 1/A_v$ this becomes,

$$Z_{out} = \frac{Z_o}{(1 + M/A_v)} \tag{3-9}$$

Example 3-6

Calculate the input impedance of the noninverting amplifier designed in Example 3-5. Use the typical parameters for the LF 353 op-amp.

Solution

From the LF353 data sheet in Appendix 1-3, the typical parameters are, $M = 100\ 000$, $Z_i = 10^{12}\ \Omega$.

Eq. 3-7, $$Z_{in} = \left(1 + \frac{M}{A_v}\right)Z_i = \left(1 + \frac{100\ 000}{66}\right) \times 10^{12}\ \Omega$$

$$= 1.5 \times 10^{15}\ \Omega$$

Eq. 3-8, $$Z'_{in} = R_1 + Z_{in} = 15\ \text{k}\Omega + 1.5 \times 10^{15}\ \Omega$$

$$= 1.5 \times 10^{15}\ \Omega \text{ (effectively)}$$

3-4 DIRECT-COUPLED INVERTING AMPLIFIERS

Design

The inverting amplifier circuit in Fig. 3-12 includes resistor R_3 at the noninverting terminal to equalize the dc voltage drops due to the input bias currents. As already discussed for other circuits, approximately equal resistances should be "seen" when

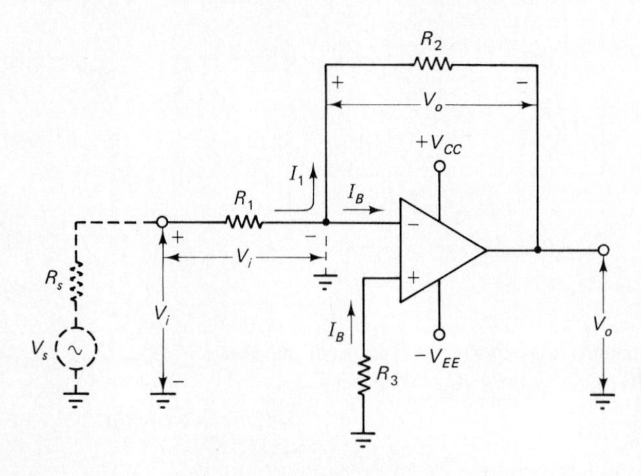

Figure 3-12 Design of an inverting amplifier (using a bipolar op-amp) starts by selecting the potential divider current (I_1) to be much greater than the maximum level of the input bias current (I_B). Then, $R_1 = V_i/I_1$, and $R_2 = V_o/I_1$.

"looking out" from each input terminal of the op-amp. Therefore,

$$R_3 \approx R_1 \| R_2 \tag{3-10}$$

And if R_1 is not very much larger than the source resistance, then

$$R_3 \approx (R_1 + R_S) \| R_2$$

As with other bipolar op-amp circuits, the resistor current (I_1) is first selected very much larger than the maximum input bias current ($I_{B(\text{max})}$). When using a BIFET op-amp, the largest-value resistor is first selected as 1 MΩ. Then

$$R_1 = \frac{V_i}{I_1} \tag{3-11}$$

and

$$R_2 = \frac{V_o}{I_1} \tag{3-12}$$

Example 3-7

Design an inverting amplifier using a 741 op-amp. The voltage gain is to be 50 and the output voltage amplitude is to be 2.5 V

Solution

For the 741,

$$I_{B(\text{max})} = 500 \text{ nA}$$

let

$$I_1 = 100 \times I_{B(\text{max})}$$

$$= 50 \ \mu\text{A}$$

$$V_1 = \frac{V_o}{A_v} = \frac{2.5 \text{ V}}{50}$$

$$= 50 \text{ mV}$$

Eq. 3-11,

$$R_1 = \frac{V_i}{I_1} = \frac{50 \text{ mV}}{50 \ \mu\text{A}}$$

$$= 1 \text{ k}\Omega \text{ (standard value)}$$

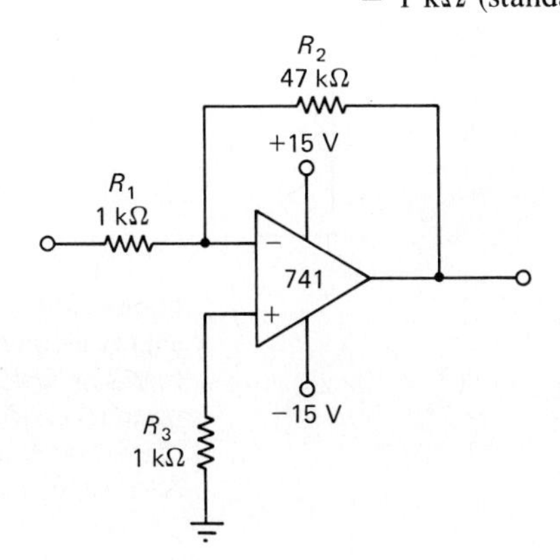

Figure 3-13 Inverting amplifier designed in Example 3-7.

Eq. 3-12, $$R_2 = \frac{V_o}{I_1} = \frac{2.5\text{ V}}{50\ \mu\text{A}}$$

$$= 50\text{ k}\Omega$$ (Use a 47 kΩ standard value resistor for R_2 to give a gain slightly lower than 50; or a 56 kΩ resistor to give a slightly higher gain. Alternatively, use 47 kΩ and 3.3 kΩ in series.)

$$R_3 = R_1 \| R_2 = 1\text{ k}\Omega \| 50\text{ k}\Omega$$

$$\approx 1\text{ k}\Omega$$

Example 3-8

Redesign the inverting amplifier in Example 3-7 using an LF353 BIFET op-amp (see Appendix 1-3).

Solution

Select the largest resistor as 1 MΩ.

let $$R_2 = 1\text{ M}\Omega$$

From Eq. 1-6, $$R_1 = \frac{R_2}{A_v} = \frac{1\text{ M}\Omega}{50}$$

$$= 200\text{ k}\Omega$$ (use a 180 kΩ standard value resistor for R_1 to give a gain slightly higher than 50; or a 220 kΩ resistor to give a slightly lower gain. Alternatively, use 180 kΩ and 22 kΩ in series.)

$$R_3 \approx R_1 \| R_2 = 200\text{ k}\Omega \| 1\text{ M}\Omega$$

$$\approx 200\text{ k}\Omega \text{ (use 180 k}\Omega\text{ standard value)}$$

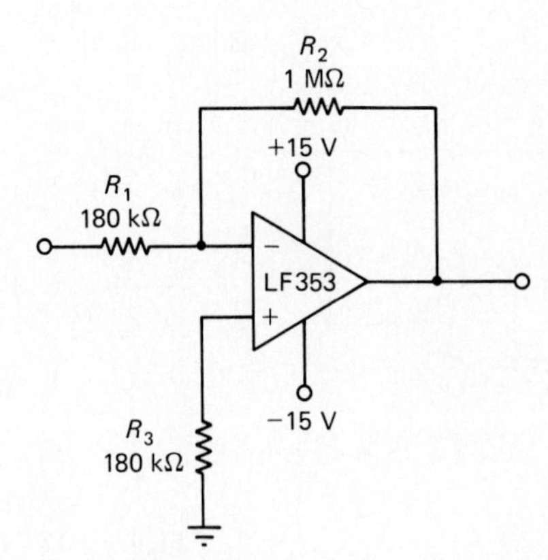

Figure 3-14 When using a BIFET op-amp as an inverting amplifier, design begins by first selecting the largest potential divider resistor as 1 MΩ. The other resistor is then calculated from $A_v = R_2/R_1$.

Performance

In the case of the inverting amplifier, the input impedance cannot be determined by the use of Eq. 2-5. Look at the inverting amplifier circuit in Fig. 3-15 and recall that the op-amp inverting input terminal always remains close to ground potential. This means the junction of R_1 and R_2 is always close to ground level. Consequently, "looking into" the inverting amplifier from the signal source, the resistor R_1 is "seen" with its other end at ground level. So,

$$Z_{in} = R_1 \tag{3-13}$$

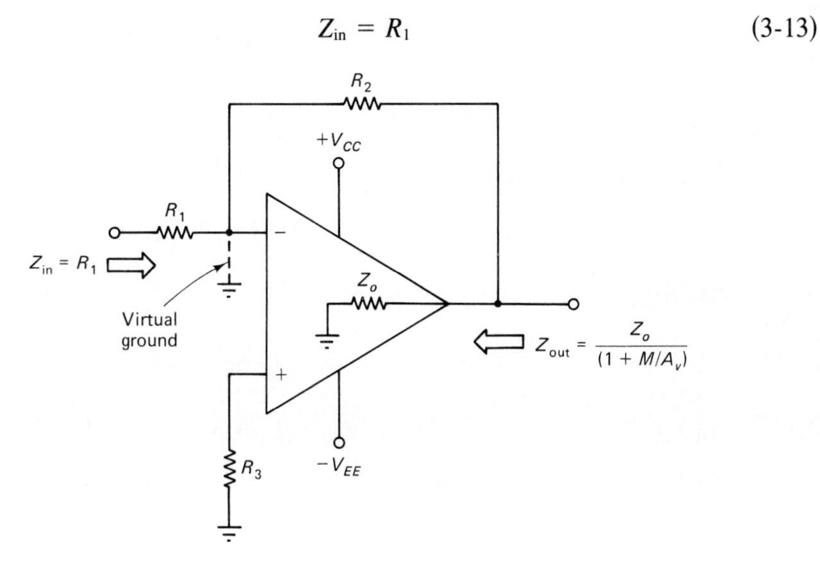

Figure 3-15 The inverting amplifier has an input impedance of R_1 and an output impedance of $Z_{out} \approx Z_o/(1 + M/A_v)$.

The output impedance of the inverting amplifier is determined exactly as for any other op-amp circuit.

Eq. 2-6,
$$Z_{out} = \frac{Z_o}{(1 + M\beta)}$$

For an inverting amplifier,

$$\beta = \frac{R_1}{R_1 + R_2}$$

Therefore
$$Z_{out} = \frac{Z_o}{[1 + MR_1/(R_1 + R_2)]} \tag{3-14}$$

When $R_2 \gg R_1$,
$$Z_{out} \approx \frac{Z_o}{(1 + M/A_v)}$$

as in the case of the noninverting amplifier.

3-5 SUMMING AMPLIFIERS

Inverting Summing Circuit

Figure 3-16 shows a circuit that amplifies the sum of two or more inputs. This is essentially an inverting amplifier with two input terminals and two input resistors. As with other inverting amplifiers, the inverting input terminal of the op-amp behaves as a virtual ground. So,

$$I_1 = \frac{v_1}{R_1} \quad \text{and} \quad I_2 = \frac{v_2}{R_2}$$

All of $(I_1 + I_2)$ flows through resistor R_3, giving

$$v_o = -(I_1 + I_2)R_3$$

$$= -\left(\frac{v_1}{R_1} + \frac{v_2}{R_2}\right)R_3$$

With $R_1 = R_2$,

$$v_o = -\frac{R_3}{R_1}(v_1 + v_2) \tag{3-15}$$

or

$$v_o = A_v(v_1 + v_2)$$

When the summing circuit in Fig. 3-16 has $R_3 = R_1 = R_2$,

$$A_v = -1$$

and the output voltage is the *direct sum* of the input voltages (inverted).

When $R_3 > R_1$ and R_2, $\quad |A_v| > 1$

The circuit output voltage is now *the sum* of the input voltages multiplied by R_3/R_1.

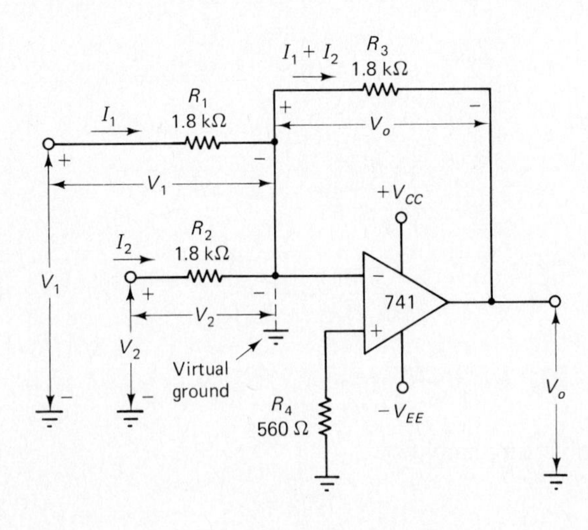

Figure 3-16 An op-amp summing amplifier amplifies the sum of two or more inputs. With $R_1 = R_2$, $v_o = -R_3(v_1 + v_2)/R_1$.

If R_3 is made less than R_1 and R_2, the output is the sum of the inputs divided by the factor R_1/R_3.

Suppose there are three inputs to a summing circuit as in Fig. 3-17. And suppose that

$$R_1 = R_2 = R_3$$

and

$$R_4 = \frac{R_1}{3}$$

In this case, the output voltage is

$$v_o = -\frac{1}{3}(v_1 + v_2 + v_3)$$

The output is the *average* of the inputs. Thus, the summing amplifier can be designed to be an *averaging circuit*.

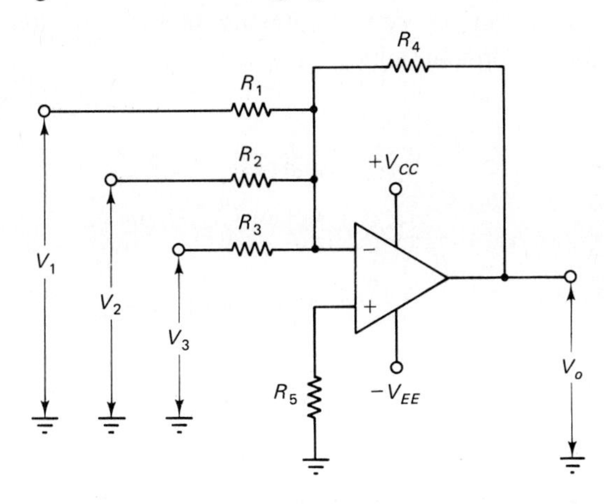

Figure 3-17 A summing amplifier can be used to determine the average of two or more inputs. For a three-input circuit with $R_1 = R_2 = R_3$ and $R_4/R_1 = 1/3$, $v_o = -(v_1 + v_2 + v_3)/3$.

The summing amplifier is generally designed the same way an ordinary inverting amplifier.

Example 3-9

Design a summing amplifier as in Fig. 3-16 to give the direct sum of two inputs which each range from 0.1 V to 1 V. Use a 741 op-amp.

Solution

let

$$I_{1(\min)} = 100 \times I_{B(\max)} = 100 \times 500 \text{ nA}$$

$$= 50 \ \mu\text{A}$$

$$R_1 = \frac{V_{s(\min)}}{I_{1(\min)}} = \frac{0.1 \text{ V}}{50 \ \mu\text{A}}$$

$$= 2 \text{ k}\Omega \text{ (use 1.8 k}\Omega \text{ standard value)}$$

$$R_2 = R_1 = 1.8 \text{ k}\Omega$$

and for $A_v = 1$, $R_3 = R_1 = 1.8\ \text{k}\Omega$

$$R_4 \approx R_1 \| R_2 \| R_3 = 1.8\ \text{k}\Omega \| 1.8\ \text{k}\Omega \| 1.8\ \text{k}\Omega$$

$$\approx 600\ \Omega \text{ (use 560 } \Omega \text{ standard value)}$$

The summing amplifier can function as a multichannel *audio mixer* for several audio channels. No interference (feedback from one channel to the input of another channel) occurs because each signal source is applied via one resistor with its opposite end at ground potential. Making the input resistors adjustable allows the output volume from each channel to be separately controlled.

Noninverting Summing Circuit

A noninverting amplifier can be employed as a summing circuit as illustrated in Fig. 3-18. This gives the direct sum of the inputs instead of the inverted sum. The equation for the output voltage can be derived by first applying the superposition theorem to determine the voltage v_i at the op-amp noninverting terminal.

With $v_2 = 0$, and $R_1 = R_2 = R$,

$$v_{i1} = \frac{v_1 R}{R + R} = \frac{v_1}{2}$$

and with $v_1 = 0$,

$$v_{i2} = \frac{v_2}{2}$$

giving

$$v_i = \frac{v_1}{2} + \frac{v_2}{2} = \frac{v_1 + v_2}{2}$$

For the noninverting amplifer,

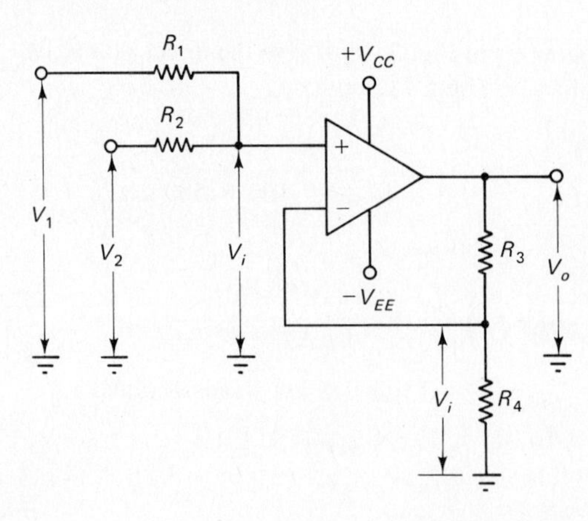

Figure 3-18 Two-input noninverting summing circuit. With all resistors equal, $v_i = (v_1 + v_2)/2$, and $v_o = 2v_i = (v_1 + v_2)$.

$$A_v = \frac{R_3 + R_4}{R_4}$$

and

$$v_o = A_v v_i$$

or

$$v_o = \frac{R_3 + R_4}{R_4} \times \frac{v_1 + v_2}{2} \tag{3-16}$$

with $R_3 = R_4$,

$$v_o = v_1 + v_2$$

For the three-input noninverting summing circuit in Fig. 3-19, the voltage at the op-amp noninverting input when all three input resistors are equal is

$$v_i = \frac{v_1 + v_2 + v_3}{3}$$

And, with $A_v = 3$,

$$v_o = v_1 + v_2 + v_3$$

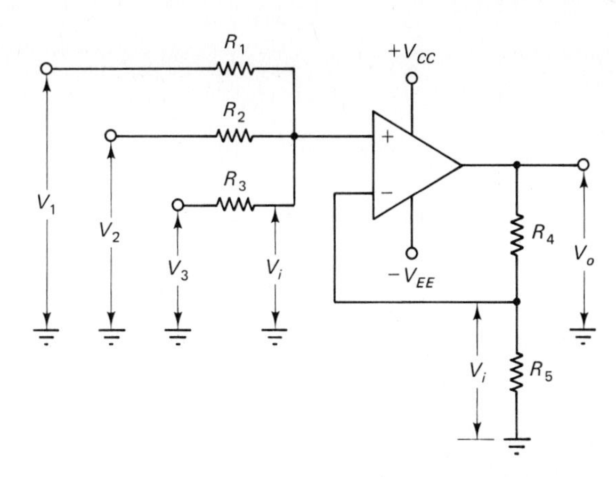

Figure 3-19 Three-input noninverting summing circuit. In this case, the amplifier must have a gain of 3 in order to give $v_o = (v_1 + v_2 + v_3)$.

Design of a noninverting summing circuit is approached by first designing the noninverting amplifier to have the required voltage gain. Then the input resistors are usually selected as large as possible to suit the type of op-amp used.

3-6 DIFFERENCE AMPLIFIER

Circuit Operation

A *difference amplifier*, or *differential amplifier*, amplifies the difference between two input signals. An IC operational amplifier is a difference amplifier; it has two (inverting and noninverting) inputs. But the open-loop voltage gain of operational amplifiers is too great for one to be used without feedback. So, like other op-amp circuits, a practical difference amplifier must have negative feedback.

The difference amplifier-circuit illustrated in Fig. 3-20(a) is a combination of inverting and noninverting amplifiers. If input terminal 2 is grounded, the circuit operates as an inverting amplifier and input v_1 is amplified by $-R_2/R_1$. With terminal 1 grounded, R_2 and R_1 function as the feedback components of a noninverting am-

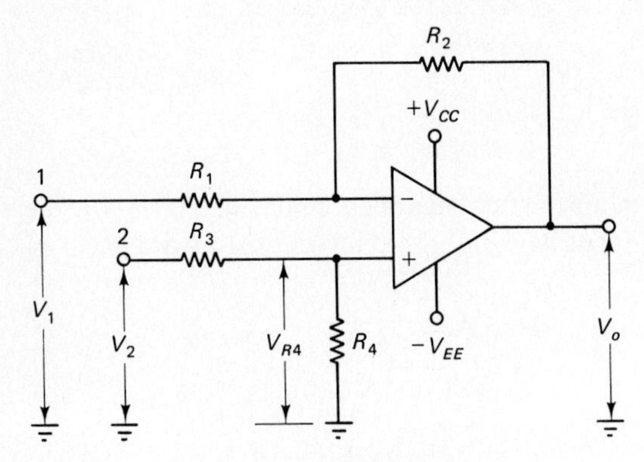

(a) Difference amplifier circuit

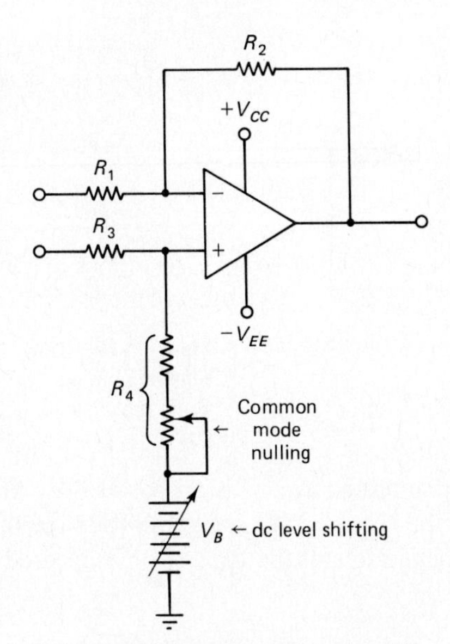

(b) Modifications for common mode nulling and dc output voltage level shifting

Figure 3-20 Op-amp difference amplifier. With $R_4/R_3 = R_2/R_1$, the output voltage is $v_o = R_2(v_2 - v_1)/R_1$. Common-mode output nulling is provided by making part of R_4 adjustable. Adjustable voltage source V_B provides dc output voltage level shifting.

plifier. Input v_2 is potentially divided across resistors R_3 and R_4 to give v_{R4}, and then v_{R4} is amplified by $(R_2 + R_1)/R_1$.

With $v_2 = 0$,

$$v_{o1} = \frac{-R_2}{R_1} \times v_1$$

With $v_1 = 0$,
$$v_{R4} = \frac{R_4}{R_3 + R_4} \times v_2$$

and
$$v_{o2} = \frac{R_1 + R_2}{R_1} \times v_{R4}$$

$$= \frac{R_1 + R_2}{R_1} \times \frac{R_4}{R_3 + R_4} \times v_2$$

With $R_3 = R_1$ and $R_4 = R_2$,

$$v_{o2} = \frac{R_2}{R_1} \times v_2$$

With both signals present,

$$v_o = v_{o2} + v_{o1}$$

$$= \frac{R_2}{R_1} v_2 - \frac{R_2}{R_1} v_1$$

giving,
$$v_o = \frac{R_2}{R_1}(v_2 - v_1) \qquad (3\text{-}17)$$

When R_2 and R_1 are equal value resistors, the output is the direct difference of the two inputs. By selecting R_2 greater than R_1, the output can be made an amplified version of the input difference.

Input Resistances

One problem with selecting the difference amplifier resistors as $R_3 = R_1$ and $R_4 = R_2$ is that the two input resistances are unequal. The input resistance for voltage v_1 in Fig. 3-20(a) is R_1, as in the case of an inverting amplifier. At the op-amp noninverting input terminal, the input resistance is very high, as it is for a noninverting amplifier. So, for voltage v_2, the input resistance is $R_2 + R_4$.

Returning to the derivation of Eq. 3-17, it can be shown that the same result would be obtained if the ratio R_4/R_3 is the same as R_2/R_1 instead of making $R_3 = R_1$ and $R_4 = R_2$. Therefore, when the resistance of R_1 has been determined, $R_3 + R_4$ can be made equal to R_1, as long as the resistor ratio is correct. This will give equal input resistances at the two input terminals of the circuit.

The input resistance difference above (if $R_3 = R_1$ and $R_4 = R_2$) will not present any problem if the signal source resistances are much smaller than the input resistances. Also, it is usually desirable to have $R_3 = R_1$ and $R_4 = R_2$, in order to equalize the resistance seen "looking out" of each input ($R_1 \| R_2 = R_3 \| R_4$) to minimize input offset voltages (see Section 2-3). Two more items to consider are the *differential input resistance* ($R_{i(\text{dif})}$) and the *common mode input resistance* ($R_{i(\text{cm})}$). The differential input resistance is the resistance offered to a signal source which is connected directly across the input terminals. It is the sum of the two input resistances.

$$R_{i(\text{dif})} = R_1 + R_3 + R_4 \tag{3-18}$$

The common mode input resistance is the resistance offered to a signal source which is connected between ground and both input terminals, that is, the parallel combination of the two input resistances.

$$R_{i(\text{cm})} = R_1 \| (R_3 + R_4) \tag{3-19}$$

Common Mode Voltages

As already discussed, Eq. 3-17 shows that the output voltage is the amplified difference of the two input voltages. A common mode input voltage v_n would give inputs of $(v_1 + v_n)$ and $(v_2 + v_n)$ which would result in v_n being cancelled out when the difference of the two is amplified. Now recall that to derive Eq. 3-17 the resistor ratios must be equal $(R_4/R_3 = R_2/R_1)$. If these ratios are not exactly equal, one input voltage will be amplified by a greater amount than the other. Also, the common mode voltage at one input will be amplified by a greater amount than that at the other input. Consequently, common mode voltages will not be completely cancelled. Because it is impossible to perfectly match the resistor ratios, there is likely to be some common mode output voltage. One way of minimizing the common mode output from of a difference amplifier is illustrated in Fig. 3-20(b). Resistor R_4 is made up of a fixed-value resistor and a much smaller adjustable resistor. This allows the ratio R_4/R_3 to be adjusted to closely match R_2/R_1 in order to null the common mode output voltage to zero.

Output Level Shifting

In Fig. 3-20(b) R_4 is connected to a bias voltage V_B instead of being grounded in the usual way. To understand the effect of V_B, assume that both input voltages are zero and that V_B is 1 V. The voltage at the op-amp noninverting input terminal will be

$$v_+ = \frac{v_B \times R_3}{R_3 + R_4}$$

Consequently, the output voltage will move to the level that gives

$$v_- = v_+$$

and the output will be

$$v_o = \frac{v_+(R_1 + R_2)}{R_1}$$

Substituting for v_+ and using the resistor relationships $(R_4/R_3 = R_2/R_1)$, the output voltage is

$$v_o = v_B$$

Therefore if v_B is adjustable, the dc output voltage level can be shifted as desired.

Circuit Design

The design of an op-amp difference amplifier is quite simple. Resistors R_1 and R_2 are determined exactly as for an inverting amplifier. Then the other resistors are selected to give R_4/R_3 equal to R_2/R_1 and usually, $R_3 = R_1$ and $R_4 = R_2$, depending upon input resistance requirements.

Example 3-10

The difference of two input signals is to be amplified by a factor of 37. Each input has an amplitude of approximately 50 mV. Using an LF353 op-amp, design a suitable circuit and calculate the differential and common mode input resistances.

Solution

With the LF353, let

$$R_2 = 1\ \text{M}\Omega$$

$$A_v = \frac{R_2}{R_1}$$

or

$$R_1 = \frac{R_2}{A_v} = \frac{1\ \text{M}\Omega}{37}$$

$$\approx 27\ \text{k}\Omega \text{ (standard value)}$$

$$R_3 = R_1 = 27\ \text{k}\Omega$$

and

$$R_4 = R_2 = 1\ \text{M}\Omega$$

$$R_{i(\text{dif})} = R_1 + (R_3 + R_4) = 27\ \text{k}\Omega + 27\ \text{k}\Omega + 1\ \text{M}\Omega$$

$$\approx 1\ \text{M}\Omega$$

$$R_{i(\text{cm})} = R_1 \| (R_3 + R_4) = 27\ \text{k}\Omega \| (27\ \text{k}\Omega + 1\ \text{M}\Omega)$$

$$= 26.3\ \text{k}\Omega$$

Example 3-11

Modify the circuit designed in Example 3-10 to give approximately equal input resistance at the two input terminals and to provide output voltage nulling.

Solution

$$R_1 = 27\ \text{k}\Omega \quad \text{and} \quad R_2 = 1\ \text{M}\Omega$$

$$R_3 + R_4 = R_1 = 27\ \text{k}\Omega$$

and

$$\frac{R_4}{R_3} = \frac{R_2}{R_1} = 37$$

giving

$$R_4 = 37R_3$$

$$R_3 + 37R_3 = 27\ \text{k}\Omega$$

$$R_3 = \frac{27\ \text{k}\Omega}{1 + 37} = 710\ \Omega \qquad \text{(use 680 } \Omega \text{ standard value)}$$

$$R_4 = 37R_3 = 37 \times 680\ \Omega$$

$$= 25.2\ \text{k}\Omega$$

Allowing a $\pm$ 10% adjustment of R_4, the total resistance of R_4 [as in Fig. 3-20(b)] is

$$R_4 = 25.2\ \text{k}\Omega + 10\% \approx 27.7\ \text{k}\Omega$$

and the variable portion is

$$R_v = 20\%\ \text{of}\ R_4 = 20\%\ \text{of}\ 27.7\ \text{k}\Omega$$

$$\approx 5\ \text{k}\Omega\ \text{(standard variable resistance value)}$$

Fixed portion of R_4 is

$$R_f = 27.7\ \text{k}\Omega - 5\ \text{k}\Omega$$

$$= 22.7\ \text{k}\Omega \qquad \text{(use 22 k}\Omega\text{ standard value)}$$

REVIEW QUESTIONS

3-1. Explain the need for uninterrupted current paths at each input terminal of an IC operational amplifier.

3-2. Discuss the effect of using resistors which are too large at the input terminals of a bipolar operational amplifier. Write an equation for calculating a suitable maximum resistance value.

3-3. Explain why the resistances at the two input terminals of an operational amplifier should be approximately equal in value.

3-4. Sketch an op-amp voltage follower circuit with a potential divider providing bias to one input terminal of the operational amplifier. Show the various currents and explain the preferred relationship between the potential divider current and the op-amp input bias current.

3-5. Write equations for each resistor value in the circuit drawn for Question 3-4.

3-6. Sketch a circuit to show how an op-amp could be connected to use a single-polarity supply. Explain. Also, discuss the limits of the op-amp bias voltage.

3-7. Discuss the approach used to determine suitable resistor values for a BIFET op-amp circuit. Explain.

3-8. Sketch the circuit of a direct-coupled voltage follower and explain how the value of the feedback resistor should be determined.

3-9. Write equations for input impedance, output impedance, and output voltage for a voltage follower.

3-10. Sketch a circuit to show how a voltage follower may be used to provide a constant voltage across a load resistor. Explain.

3-11. Briefly explain the advantages of a voltage follower compared to an emitter follower.

3-12. Sketch the complete circuit of an op-amp noninverting amplifier. Write equations for determining suitable values for each resistor (a) using a bipolar op-amp and (b) using a BIFET op-amp.

3-13. Write equations for input impedance, output impedance, and voltage gain for a noninverting amplifier.

3-14. Sketch the complete circuit of an op-amp inverting amplifier. Write equations for determining suitable values for each resistor (a) using a bipolar op-amp and (b) using a BIFET op-amp.

3-15. Write equations for input impedance, output impedance, and voltage gain for an inverting amplifier.

3-16. Sketch the circuit of a two-input inverting summing amplifier. Explain the operation of the circuit and derive an equation for the output voltage.

3-17. Write equations for determining suitable values for each resistor in the summing amplifier in Question 3-16, (a) using a bipolar op-amp and (b) using a BIFET op-amp.

3-18. Sketch the circuit of a three-input inverting summing amplifier and explain how it may be used as an averaging circuit.

3-19. Sketch a two-input noninverting summing circuit. Explain the operation of the circuit and derive an equation for the output voltage.

3-20. Sketch a three-input noninverting summing circuit and derive an equation for the output voltage.

3-21. Sketch an op-amp difference amplifier circuit. Explain the operation of the circuit and derive an equation for the output voltage.

3-22. Discuss the various aspects of differential amplifier input resistance.

3-23. Discuss common mode rejection ratio for differential amplifiers and show how it can be improved.

3-24. Show how output dc voltage level shifting can be provided for differential amplifiers. Briefly explain.

PROBLEMS

3-1. A voltage follower using a 741 op-amp is to be connected to a signal source with a resistance of 22 kΩ. Sketch the circuit and select suitable components. Also, determine the maximum input offset voltage that could be produced by the input offset current.

3-2. Substitute a LMI08 op-amp (see Appendix 1-2) in place of the 741 in Problem 3-1 and recalculate the input offset voltage.

3-3. If the voltage follower in Problem 3-2 has a 300 mV input signal and a 10 kΩ load, calculate the load voltage (a) when the load is directly connected to the source and (b) when the voltage follower is used.

3-4. A signal applied to a voltage follower is to be developed across a load with a maximum signal loss of 0.005%. Determine the minimum open-loop gain of a suitable op-amp.

3-5. A voltage follower is to be used to provide a constant 3 V to a load with an approximate resistance of 12 kΩ. The available supply is ±12 V. Design a suitable circuit using a 741 op-amp.

3-6. Redesign the circuit for Problem 3-5 substituting a LF353 op-amp (see Appendix 1-3) in place of the 741.

3-7. Redesign the circuit for Problem 3-5 substituting a LM108 op-amp (see Appendix 1-2) in place of the 741.

3-8. Redesign the circuit for Problem 3-5 using the 741 and using +24 V instead of the ±12 V supply.

3-9. A noninverting amplifier is to amplify a 100 mV signal to a level of 3 V. Using a 741 op-amp, design a suitable circuit.

3-10. Redesign the circuit for Problem 3-9 substituting a LF353 op-amp in place of the 741.

3-11. Calculate the typical input and output impedances for the noninverting amplifier designed in Problem 3-9.

3-12. A noninverting amplifier with a ±15 V supply is to produce maximum possible output voltage and is to have a voltage gain of 25. Using an LF353 op-amp, design a suitable circuit.

3-13. An inverting amplifier is to amplify a 50 mV signal to a level of 4 V. Using a LF353 op-amp, design a suitable circuit.

3-14. Redesign the circuit for Problem 3-13 substituting a 741 op-amp in place of the LF353.

3-15. Determine the input impedance for each of the inverting amplifiers designed in Problem 3-13 and 3-14.

3-16. An inverting amplifier with a ±12 V supply is to produce maximum possible output voltage and is to have a voltage gain of 33. Using an LF353 op-amp, design a suitable circuit.

3-17. Two signals which each range from 0.1 V to 1 V are to be summed. Using a 741 op-amp, design a suitable inverting summing circuit.

3-18. Two signals as in Problem 3-17 are to be summed and amplified by a factor of 5. Using a LF353 op-amp, design a suitable inverting summing circuit.

3-19. Three signals which each range from 0.1 V to 1 V are to be averaged. Using an LM108 op-amp, design a suitable inverting averaging circuit.

3-20. Design a noninverting summing circuit to meet the requirement of Problem 3-17.

3-21. Design a noninverting summing circuit to meet the requirment of Problem 3-18.

3-22. The difference of two signals which each range from 0.1 V to 1 V is to be determined. Using a 741 op-amp, design a suitable difference amplifier. Determine the input resistance at each input, the differential input resistance, and the common mode input resistance.

3-23. Redesign the circuit for Problem 3-22 to amplify the input voltage difference by a factor of 15 and to provide a common mode nulling facility.

3-24. Redesign the circuit for Problem 3-22 to use an LF353 op-amp, to multiply the difference by a factor of 10, and to approximately equalize the input resistances at the two input terminals.

COMPUTER PROBLEMS

3-25. Write a computer program to design a direct-coupled noninverting amplifier using a bipolar op-amp. Given: $I_{B(\max)}$, A_v, and v_i.

3-26. Write a computer program to design a direct-coupled noninverting amplifier using a BIFET op-amp. Given: A_v and v_i.

3-27. Write a computer program to analyze a noninverting amplifier for A_v, Z_{in}, and Z_{out}. Given: resistor values and op-amp parameters.

3-28. Write a computer program to design a direct-coupled inverting amplifier using a bipolar op-amp. Given: $I_{B(max)}$, A_v, and v_i.

3-29. Write a computer program to analyze an inverting amplifier, for A_v, Z_{in}, and Z_{out}. Given: resistor values and op-amp parameters.

LABORATORY EXERCISES

3-1 Direct-Coupled Noninverting Amplifier

1. Construct the 741 noninverting amplifier circuit shown in Fig. 3-9 as designed in Example 3-4.
2. Apply a ± 15 mV, 1 kHz sinusoidal signal. Monitor the input and output waveforms on an oscilloscope and measure the output amplitude.
3. Calculate the voltage gain and compare it to the designed gain.
4. Connect a 1 MΩ resistor in series with the amplifier input. Check that the output voltage is unaffected to demonstrate that $Z_{in} \gg 1$ MΩ.
5. Connect a 100 Ω resistor in parallel with the output. Check that the output voltage is unaffected to demonstrate that $Z_{out} \ll 100$ Ω.
6. Construct the LF353 noninverting amplifier circuit in Fig. 3-10 as designed in Example 3-5.
7. Repeat Procedures 2 through 5.

3-2 Direct-Coupled Inverting Amplifier

1. Construct the 741 inverting amplifier circuit shown in Fig. 3-13 as designed in Example 3-7.
2. Apply a 1 kHz sinusoidal signal and adjust the signal amplitude to give an output peak amplitude of 2.5 V as displayed on an oscilloscope.
3. Measure the input peak amplitude, calculate the voltage gain, and compare it to the designed gain.
4. Connect a 1 kΩ resistor in series with the amplifier input. Note the output voltage change and calculate Z_{in}.
5. Adjust the signal to give a peak output of 1 V. Then, connect a 100 Ω resistor in parallel with the output. Check that the output voltage is unaffected to demonstrate that $Z_{out} \ll 100$ Ω.
6. Construct the LF353 inverting amplifier circuit in Fig. 3-14 as designed in Example 3-8.
7. Repeat Procedures 2 through 5.

3-3 Direct-Coupled Summing Circuits

1. Construct the inverting summing circuit in Fig. 3-16. Use a 741 op-amp, a supply of ±15 V, and 1.8 kΩ resistors throughout (as calculated in Example 3-9).
2. Connect adjustable dc voltage sources to each input and connect dc voltmeters to monitor the levels of V_1, V_2, and V_o.

3. Set each input to several voltage levels beween 0.1 V and 1 V and measure the output level for each case to check the accuracy of Eq. 3-15.
4. Change R_3 to 18 kΩ and repeat Procedure 3 using maximum input levels of 0.5 V.

3-4 Direct-Coupled Difference Amplifier

1. Construct the difference amplifier circuit shown in Fig. 3-20(a). Use a LF353 op-amp with a supply of ±15 V, and resistor values $R_1 = R_3 = 27$ kΩ, $R_2 = R_4 = 1$ MΩ as calculated in Example 3-10.
2. Connect adjustable dc voltage sources at each input terminal and voltmeters to monitor V_1, V_2, and V_o.
3. Set V_1 and V_2 to several different levels from zero to +50 mV and −50 mV. Note the level of V_o at each setting of the inputs, and check the accuracy of Eq. 3-17.
4. Connect another adjustable dc voltage source between R_4 and ground as illustrated in Fig. 3-20(b). Investigate its effect as an output level shifter.
5. Ground R_4 once again, connect the two inputs terminals together, and apply a single adjustable dc voltage source at the inputs.
6. Adjust the input to 10 V, measure the output voltage level, and calculate the common mode gain for the circuit.

CHAPTER 4

OP-AMPS as AC Amplifiers

Objectives

- Sketch the following capacitor-coupled op-amp circuits, explain the operation of each circuit, and specify the input impedance of each: *voltage follower, high-input-impedance voltage follower, noninverting amplifier, high-input-impedance noninverting amplifier, inverting amplifier, difference amplifier.*
- *Easily* design each of the above circuits using bipolar and BIFET op-amps.
- Analyze each of the above circuits to determine input impedance, output impedance, voltage gain, lower cutoff frequency, and other performance criteria.
- Show how the upper cutoff frequency for an op-amp circuit can be set to any desired frequency.
- Show how various capacitor-coupled amplifier circuits can use a single-polarity supply; design such circuits.

INTRODUCTION

Operational amplifier circuits can readily be capacitor-coupled at the input and output so that they operate only as ac amplifiers. Capacitors must not be allowed to interrupt the bias current paths to the op-amp input terminals. This sometimes requires additional bias resistors which can affect the circuit input impedance. Since capacitors have their highest impedances at the lowest signal frequency, all coupling capacitor values must be calculated at the desired lower cutoff frequency (f_1). The impedance of coupling capacitors at f_1 is usually determined as one-tenth of the resistance in series with them. The largest capacitor in the circuit is normally selected to determine f_1 and in this case the capacitive impedance is made equal to the series-connected resistance.

4-1 CAPACITOR-COUPLED VOLTAGE FOLLOWER

When a voltage follower is to have its input and output capacitor-coupled, the noninverting input terminal must be grounded via a resistor (R_1 in Fig. 4-1(a)). As explained in Section 3-1, the resistor is required to pass bias current to the amplifier noninverting input terminal. A resistor equal to R_1 might be included in series with the inverting terminal to equalize the $I_B R_B$ voltage drop and thus minimize the output offset voltage. However, in the case of a circuit with its output capacitor-coupled, small dc offset output voltages are unimportant because they are blocked by the capacitor.

Design of a capacitor-coupled voltage follower as in Fig. 4-1 involves calcula-

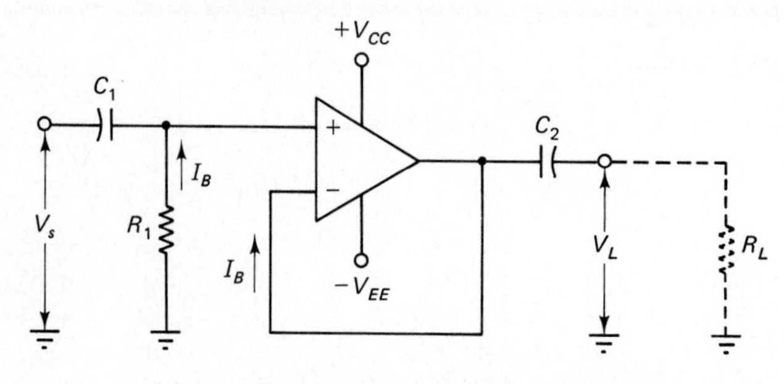

(a) Capacitor-coupled voltage follower circuit

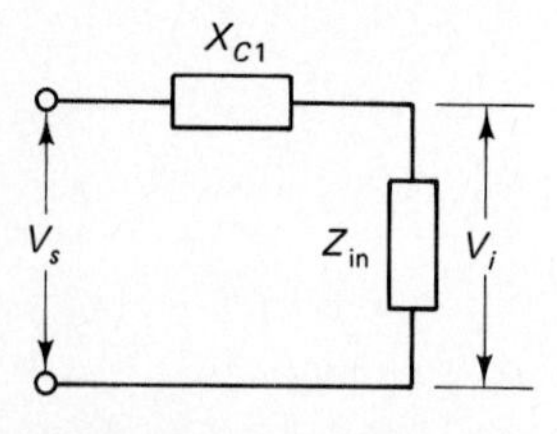

(b) The signal voltage is divided across X_{C1} and Z_{in}

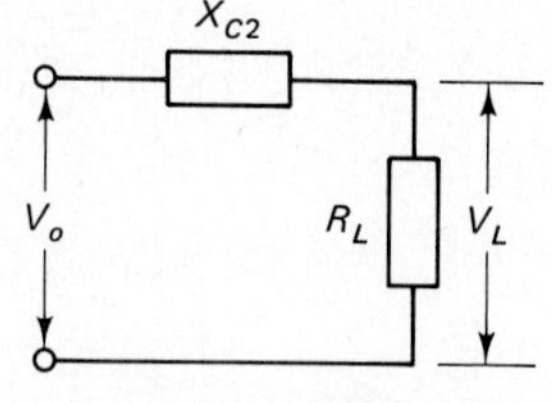

(c) The output voltage is divided across X_{C2} and R_L

Figure 4-1 Capacitor-coupled voltage follower. The op-amp noninverting terminal must be grounded via a resistor to provide a path for input bias current. The capacitor values are determined by making $X_{C1} = Z_{in}/10$ at f_1, and $X_{C2} = R_L$ at f_1.

tion of R_1, C_1 and C_2. As always, the largest possible resistor values are normally selected to ensure minimum circuit power dissipation and minimum current demand from the power supply. The smallest possible capacitor values are normally used for their small physical size and low cost. From Eq. 3-1, a maximum value for R_1 for a bipolar op-amp is determined as $(0.1\ V_{BE})/I_B$.

The circuit input impedance is $R_1 \| Z_i'$, where Z_i' is the input impedance at the op-amp noninverting input terminal, $[Z_i' = Z_i(1 + M\beta)]$. But, with the 100% negative feedback employed in a voltage follower circuit, Z_i' is always much larger than R_1. Consequently, for the circuit in Fig. 4-1,

$$Z_{in} = R_1 \tag{4-1}$$

Load resistor R_L normally has a lower resistance than R_1. Therefore, because each capacitor value is inversely proportional to the resistance in series with it, C_2 is usually larger than C_1. At the circuit low 3 dB frequency (f_1), the impedance of C_1 should be much smaller than Z_{in}, so that there is no significant division of the signal across X_{C1} and Z_{in}, (see Fig. 4-1(b)). In this case, C_1 will have no effect on the circuit low 3 dB frequency. Thus, C_1 is calculated from

$$X_{C1} = (Z_{in}/10) \text{ at } f_1 \tag{4-2}$$

giving

$$C_1 = \frac{1}{2\pi f_1(R_1/10)}$$

As illustrated in Fig. 4-1(c), the circuit output voltage v_o is divided across X_{C2} and R_L to give the load voltage v_L. The equation for v_L is

$$V_L = \frac{v_o \times R_L}{\sqrt{(R_L^2 + X_{C2}^2)}}$$

When $X_{C2} = R_L$,

$$V_L = \frac{v_o \times R_L}{\sqrt{(2R_L^2)}} = \frac{v_o}{\sqrt{2}}$$

$$= 0.707\ v_o = v_o - 3 \text{ dB}$$

Or, the circuit low 3 dB frequency (f_1) occurs when $X_{C2} = R_L$. Therefore, C_2 is calculated from

$$X_{C2} = R_L \text{ at } f_1 \tag{4-3}$$

which gives,

$$C_2 = \frac{1}{2\pi f_1 R_L}$$

The above design approach gives the smallest possible capacitor values. When selecting standard value components, the next larger standard size should be chosen to give capacitive impedances slightly smaller than calculated.

In the unusual situation where R_1 is smaller than R_L, C_1 would be a larger capacitor than C_2. To give the smallest capacitor values in this case, it is best to let C_1 determine the low 3 dB frequency, by making $X_{C1} = R_1$ at f_1. Then, $X_{C2} = R_L/10$ at f_1.

Example 4-1

Design a capacitor-coupled voltage follower using a 741 operational amplifier. The lower cutoff frequency for the circuit is to be 50 Hz and the load resistance is R_L = 3.9 kΩ.

Solution

Eq. 3-1, $$R_{1(max)} = \frac{0.1\ V_{BE}}{I_{B(max)}} = \frac{0.1 \times 0.7\ \text{V}}{500\ \text{nA}}$$

$$\approx 140\ \text{k}\Omega \qquad \text{(use 120 k}\Omega\text{ (next lower) standard value)}$$

Eq. 4-2, $$X_{C1} = R_1/10 \text{ at } f_1$$

or, $$C_1 = \frac{1}{2\pi f_1 (R_1/10)} = \frac{1}{2\pi \times 50\ \text{Hz} \times (120\ \text{k}\Omega/10)}$$

$$= 0.27\ \mu\text{F} \qquad \text{(standard value–see Appendix 2-2)}$$

Eq. 4-3, $$X_{C2} = R_L \text{ at } f_1$$

or $$C_2 = \frac{1}{2\pi f_1 R_L} = \frac{1}{2\pi \times 50\ \text{Hz} \times 3.9\ \text{k}\Omega}$$

$$\approx 0.82\ \mu\text{F} \qquad \text{(standard value)}$$

The circuit supply voltages should normally be ±9 V to ±18 V or within whatever range is specified on the op-amp data sheet.

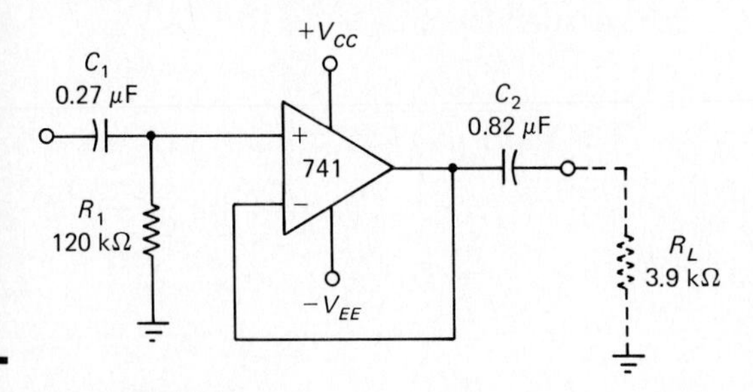

Figure 4-2 Capacitor-coupled voltage follower (using a bipolar op-amp) designed in Example 4-1.

Example 4-2

Redesign the capacitor-coupled voltage follower circuit in Example 4-1 to use a LF353 BIFET op-amp.

Solution

$$R_{1(max)} = 1\ \text{M}\Omega$$ (largest resistance to be connected at the input of a BIFET op-amp—see Section 3-1)

$$C_2 = 0.82\ \mu\text{F}$$ (as an Example 4-1)

From Eq. 4-2, $$C_1 = \frac{1}{2\pi f_1 (R_1/10)} = \frac{1}{2\pi \times 50\ \text{Hz} \times (1\ \text{M}\Omega/10)}$$

$$= 0.032\ \mu\text{F}$$ [use 0.033 μF standard value, to give $X_{C1} < (R_1/10)$ at f_1]

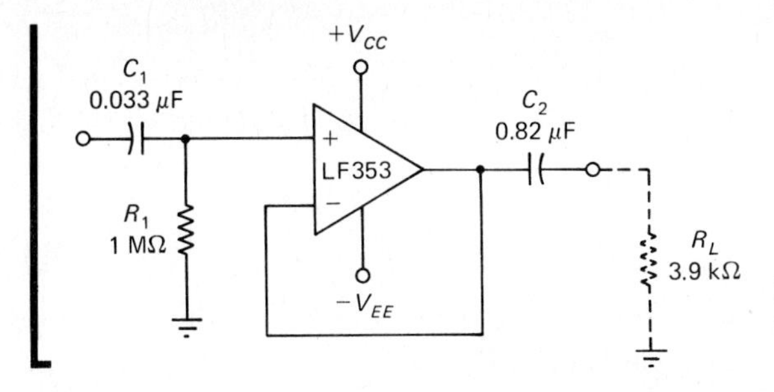

Figure 4-3 Capacitor-coupled voltage follower (using a BIFET op-amp) designed in Example 4-2.

4-2 HIGH Z_{in} CAPACITOR-COUPLED VOLTAGE FOLLOWER

The input impedance of the capacitor-coupled voltage follower discussed in Section 4-1 is set by the value of resistor R_1 in Fig. 4-1(a). This gives a much smaller input impedance than the direct-coupled voltage follower. Fig. 4-4 shows a method by which the input impedance of the capacitor-coupled voltage follower can be substantially increased.

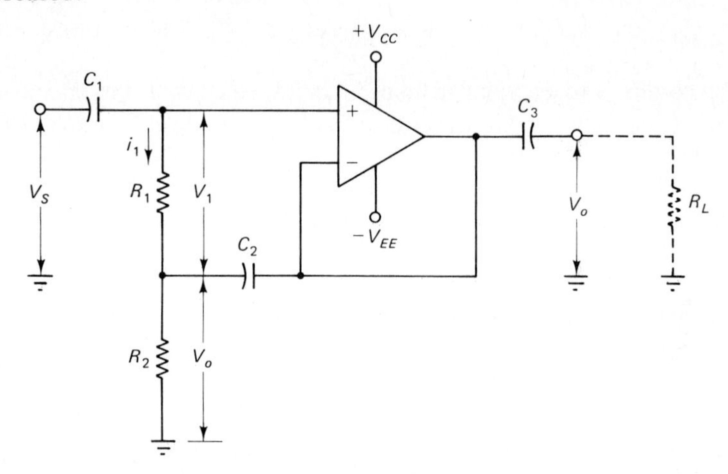

Figure 4-4 High input impedance capacitor-coupled voltage follower. Feedback via C_2 to the junction of R_1 and R_2 gives an input impedance of $Z_{in} = (1 + M)R_1$.

Capacitor C_2 in Fig. 4-4 couples the circuit output voltage to the junction of resistors R_1 and R_2. C_2 behaves as an ac short circuit so that v_o is developed across R_2. The voltage developed across R_1 is

$$v_1 = v_s - v_o$$

$$= v_s - Mv_1$$

giving

$$v_1(1 + M) = v_s$$

or

$$v_1 = \frac{v_s}{(1 + M)}$$

and $$i_1 = \frac{v_1}{R_1} = \frac{v_s}{(1 + M)\,R_1}$$

input resistance $$Z_{in} = \frac{v_s}{i_1}$$

or $$Z_{in} = (1 + M)R_1 \tag{4-4}$$

Equation 4-4 shows that this circuit does indeed have a very high input impedance. For example, with an open loop gain of 200 000 and a 47 kΩ resistor for R_1, the circuit input impedance would be

$$Z_{in} \approx 200\,000 \times 47\ \text{k}\Omega$$
$$= 9.4 \times 10^9\ \Omega$$

This extremely high input impedance is unrealistic when it is remembered that some stray capacitance is always present. If the stray capacitance (C_s) between the circuit input and ground is 3 pF (which is quite possible), the impedance of C_s at 1 kHz is

$$X_{CS} = \frac{1}{2\pi \times 1\ \text{kHz} \times 3\ \text{pF}}$$
$$= 53\ \text{M}\Omega$$

Since this is much smaller than Z_{in} as just calculated, the effective input impedance of a *high* Z_{in} voltage follower is normally much lower than that determined from Eq. 4-4.

To design a high input impedance capacitor-coupled voltage follower, resistors R_1 and R_2 are first calculated as a single resistor $R_{1(max)}$ using Eq. 3-1. Then, $R_{1(max)}$ is split into two equal resistors R_1 and R_2. To ensure that the feedback voltage remains close to 100% of the output voltage at the lowest operating frequency, the required feedback capacitor (C_2) is determined from

$$X_{C2} = \frac{R_2}{10} \text{ at } f_1 \tag{4-5}$$

As in the case of the voltage follower discussed in Section 4-1, the output capacitor can be used to set the lower 3 dB frequency for the circuit. From Eq. 4-3, $X_{C3} = R_L$ at f_1.

The impedance of input capacitor C_1 should theoretically be determined from Eq. 4-2 as $X_{C1} = Z_{in}/10$ at f_1. However, as the previous discussion of input impedance shows, the actual input impedance is affected by stray capacitance. Therefore, instead of calculating C_1 in terms of Z_{in}, C_1 should be selected much larger than any stray capacitance at the circuit input. A reasonable rule-of-thumb in the case of any very high input impedance circuit, is to select C_1 as a minimum of 1000 pF.

A resistor could be included in series with the op-amp inverting input terminal in the circuit of Fig. 4-4 to equalize the $I_B R_B$ voltage drops, but, as already explained, this is not usually necessary when the output is capacitor-coupled. If such a resistor is to be used, it should be equal to $(R_1 + R_2)$ and it should be connected between the inverting input terminal and the junction of C_2 and the op-amp output. In other words, it should *not* be in series with C_2.

Example 4-3

Modify the circuit designed in Example 4-1 to make it a high input impedance voltage follower. Also, determine the minimum theoretical input impedance of the circuit.

Solution

$$R_1 + R_2 = R_{1(max)} = 140\ \text{k}\Omega \ \text{(as in Example 4-1)}$$

$$R_1 = R_2 = \frac{140\ \text{k}\Omega}{2}$$

$$= 70\ \text{k}\Omega \qquad \text{(use 68 k}\Omega\text{ standard value resistors)}$$

From Eq. 4-3,

$$C_3 = \frac{1}{2\pi f_1 R_L} = 0.82\ \mu\text{F} \qquad \text{(as for } C_2 \text{ in Example 4-1)}$$

From Eq. 4-5,

$$C_2 = \frac{1}{2\pi f_1 (R_2/10)} = \frac{1}{2\pi \times 50\ \text{Hz} \times (68\ \text{k}\Omega/10)}$$

$$\approx 0.5\ \mu\text{F} \qquad \text{(standard value)}$$

$$C_1 = 1000\ \text{pF} \qquad \text{(to be much greater than stray input capacitance)}$$

Eq. 4-4

$$Z_{in} = (1 + M)R_1$$

From the 741 data sheet,

$$M_{(min)} = 50\ 000$$

so,

$$Z_{in(min)} = (1 + 50\ 000)\ 68\ \text{k}\Omega$$

$$= 3400\ \text{M}\Omega$$

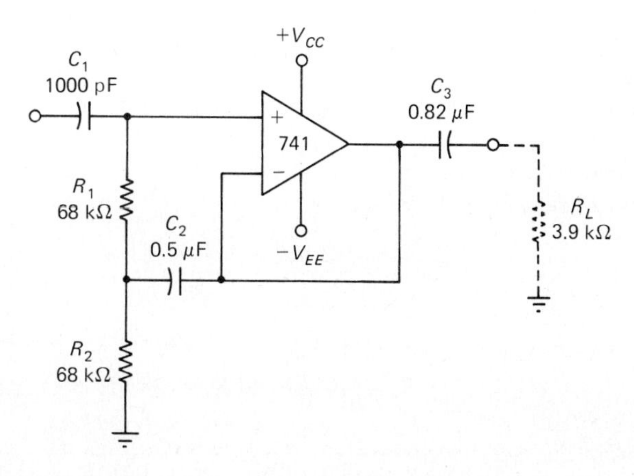

Figure 4-5 High input impedance capacitor-coupled voltage follower designed in Example 4-3.

4-3 CAPACITOR-COUPLED NONINVERTING AMPLIFIER

When the input of a noninverting amplifier is to be capacitor-coupled, the noninverting input terminal must be grounded via a resistor to provide a path for the input bias current. Fig. 4-6 shows the arrangement. R_1 may be made equal to $R_2 \| R_3$ as in the direct-coupled case. However, as already explained, dc offset is unimportant when the output is capacitor-coupled, so any reasonable value of R_1 can be chosen within the limit set by Eq. 3-1.

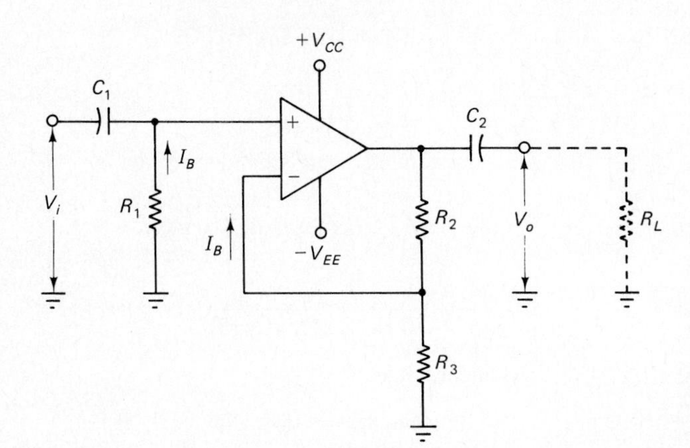

Figure 4-6 Capacitor-coupled noninverting amplifier. The op-amp noninverting terminal is grounded via a resistor to provide a path for input bias current. The capacitor values are determined by making $X_{C1} = Z_{in}/10$ at f_1, and $X_{C2} = R_L$ at f_1.

As in the case of the capacitor-coupled voltage follower, $Z_{in} = R_1$ for a capacitor-coupled noninverting amplifier. Resistors R_2 and R_3 in Fig. 4-6 are calculated in the same manner as for a direct-coupled circuit. The capacitors are determined in the same way as for the capacitor-coupled voltage follower.

Example 4-4

The noninverting amplifier designed in Example 3-4 is to be capacitor-coupled at input and output. The load resistor is 2.2 kΩ and the lower cut-off frequency is to be 120 Hz. Make the necessary modifications to give the highest input impedance and determine the required capacitor values.

Solution

The circuit is rearranged as in Fig. 4-6.
For highest Z_{in},

Eq. 3-1,
$$R_{1(max)} = \frac{0.1\, V_{BE}}{I_{B(max)}} = \frac{0.1 \times 0.7\text{ V}}{500\text{ nA}}$$

$$\approx 140\text{ k}\Omega \qquad \text{(use 120 k}\Omega\text{ standard value)}$$

$$R_2 = 18\text{ k}\Omega \text{ and } R_3 = 270\ \Omega \text{ as in Example 3-4}$$

From Eq. 4-2,
$$C_1 = \frac{1}{2\pi f_1 (R_1/10)} = \frac{1}{2\pi \times 120\text{ Hz} \times (120\text{ k}\Omega/10)}$$

$$= 0.11\ \mu\text{F} \qquad [\text{use } 0.12\ \mu\text{F standard value}]$$

From Eq.4-3, $C_2 = \frac{1}{2\pi f_1 R_L} = \frac{1}{2\pi \times 120 \text{ Hz} \times 2.2 \text{ k}\Omega}$

$\approx 0.6\ \mu\text{F}$ (use 0.68 μF standard value)

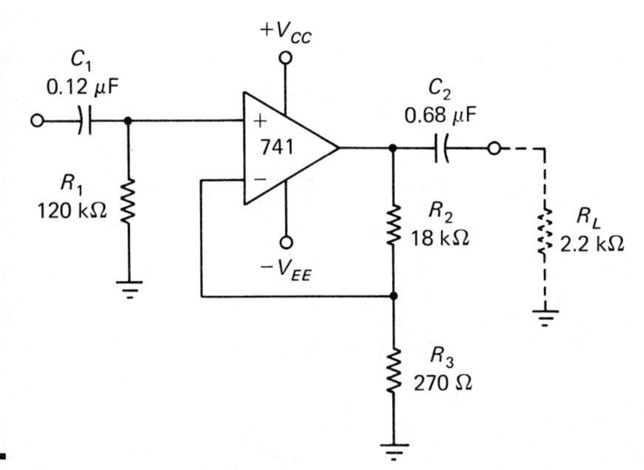

Figure 4-7 Capacitor-coupled noninverting amplifier designed in Example 4-4.

4-4 HIGH Z_{in} CAPACITOR-COUPLED NONINVERTING AMPLIFIER

The input impedance of the noninverting amplifier in Fig. 4-8 is improved in the same way as for the high Z_{in} voltage follower discussed in Section 4-2. In this case, the voltage fed back from the output to the input via R_2, C_2, and R_3 is not 100%, but is potential divided by a factor β, where

$$\beta = R_3/(R_2 + R_3)$$

Substituting this quantity into the analysis for Eq. 4-4 gives

$$Z_{in} = (1 + M\beta)R_1 \tag{4-6}$$

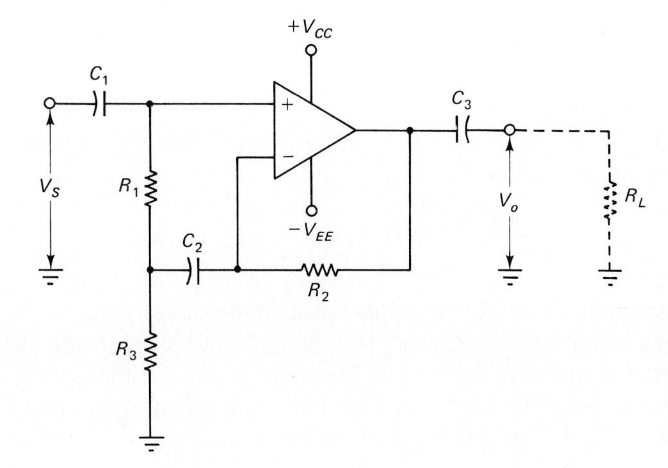

Figure 4-8 High input impedance capacitor-coupled noninverting amplifier. Feedback via C_2 to the junction of R_1 and R_3 gives an input impedance of $Z_{in} = (1 + M\beta)R_1$, where $\beta = 1/A_v$.

The values of resistors R_2 and R_3 for the high Z_{in} circuit in Fig. 4-8 are determined exactly as for a direct-coupled noninverting amplifier. Then, for equal $I_B R_B$ voltage drops,

$$R_1 + R_3 = R_2$$

which usually gives

$$R_1 \approx R_2$$

Alternatively, for the highest input impedance (without equalizing the $I_B R_B$ voltage drops),

$$R_1 + R_3 = R_{(max)}$$

$$= \frac{0.1\ V_{BE}}{I_{B(max)}} \quad \text{as determined by Eq. 3-1.}$$

Equation 3-1 applies only in the case of a bipolar op-amp. For a BIFET circuit, the maximum resistor values normally are

$$R_1 + R_3 = 1\ \text{M}\Omega$$

The capacitor values can be calculated exactly as for the high Z_{in} voltage follower. An alternative to this for a circuit which might have a variable load is to use C_2 to determine the low 3 dB frequency for the circuit. With C_2 in the circuit, the voltage gain is

$$A_v = \frac{R_2 + R_3 - jX_{C2}}{R_3 - jX_{C2}}$$

If $X_{C2} \ll (R_2 + R_3)$,

$$A_v \approx \frac{R_2 + R_3}{\sqrt{(R_3^2 + X_{C2}^2)}}$$

and when $X_{C2} = R_3$,

$$A_v = \frac{R_2 + R_3}{R_3} \times \frac{1}{\sqrt{2}}$$

Or, when $X_{C2} = R_3$, A_v is 3 dB below the normal mid-frequency gain of $(R_2 + R_3)/R_3$. Thus, for C_2 to determine f_1,

$$X_{C2} = R_3 \text{ at } f_1 \tag{4-7}$$

Note that the assumption of $X_{C2} \ll (R_2 + R_3)$ is reasonable when $X_{C2} = R_3$ and $R_3 \ll R_2$, which is the case only when the circuit has substantial voltage gain.

When C_2 is to set the lower cutoff frequency, C_3 should have no significant effect on the low 3 dB frequency of the circuit. The capacitance of C_3 is then determined, like other coupling capacitors, to have an impedance at f_1 equal to one-tenth of the minimum resistance in series with C_3.

Example 4-5

Using a LF353 BIFET op-amp, design a high Z_{in} capacitor-coupled noninverting amplifier to have a low cutoff frequency of 200 Hz. The input and output voltages are to be 15 mV and 3 V respectively, and the minimum load resistance is 12 kΩ.

Solution

$$A_v = \frac{V_o}{V_i} = \frac{3\text{ V}}{15\text{ mV}} = 200$$

For a BIFET op-amp, let $R_2 = 1\text{ M}\Omega$

$$A_v = \frac{R_2 + R_3}{R_3}$$

or,

$$R_3 = \frac{R_2}{A_v - 1} = \frac{1\text{ M}\Omega}{200 - 1}$$

$$\approx 5\text{ k}\Omega \quad \text{(use 4.7 k}\Omega\text{ to give a slightly larger voltage gain)}$$

$$R_1 = 1\text{ M}\Omega - R_3 \approx 1\text{ M}\Omega$$

From Eq. 4-7,

$$C_2 = \frac{1}{2\pi f_1 R_3} = \frac{1}{2\pi \times 200\text{ Hz} \times 4.7\text{ k}\Omega}$$

$$= 0.17\ \mu\text{F} \quad \text{(use 0.18 }\mu\text{F standard value)}$$

$$C_1 = 1000\text{ pF} \quad \text{(to be much larger than stray capacitance)}$$

$$C_3 = \frac{1}{2\pi f_1 (R_L/10)} = \frac{1}{2\pi \times 200\text{ Hz} \times (12\text{ k}\Omega/10)}$$

$$= 0.66\ \mu\text{F} \quad \text{(use 0.68 }\mu\text{F standard value)}$$

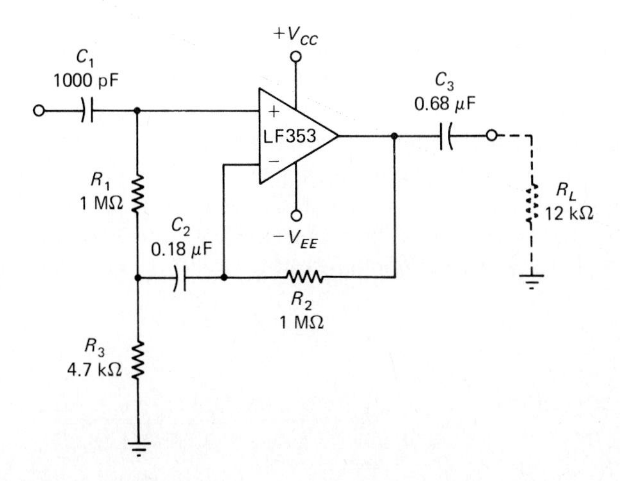

Figure 4-9 High input impedance capacitor-coupled noninverting amplifier (using a BIFET op-amp) designed in Example 4-5.

4-5 CAPACITOR-COUPLED INVERTING AMPLIFIER

A capacitor-coupled inverting amplifier is shown in Fig. 4-10. In this case, bias current to the op-amp inverting input terminal flows via resistor R_2, so coupling-capacitor C_1 does not interrupt the input bias current. No resistor is included in series with the noninverting input terminal, because a small dc offset is unimportant with a capacitor-coupled output. If it is desired to equalize the $I_B R_B$ voltage drops, the resistance in series with the noninverting input should equal R_2 because R_1 is not part of the bias current path at the inverting input terminal.

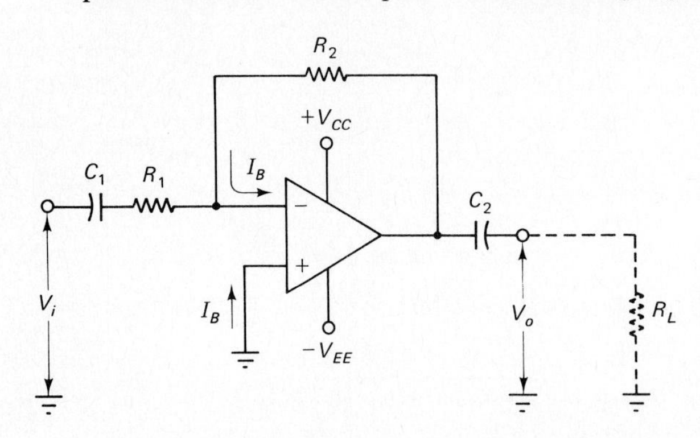

Figure 4-10 Capacitor-coupled inverting amplifier. No additional bias resistors are required because neither of the op-amp input bias current paths is interrupted by a capacitor.

The resistor values are determined as for a direct-coupled inverting amplifier circuit. Then C_1 and C_2 are calculated to give $X_{C1} = R_1/10$ at f_1, and $X_{C2} = R_L$ at f_1, as in the case of a capacitor-coupled noninverting amplifier.

4-6 SETTING THE UPPER CUTOFF FREQUENCY

The highest signal frequency that can be processed by an op-amp circuit depends on the op-amp selected. This is further considered in Section 5-5. In some circumstances, the upper cutoff frequency may be higher than desired. One example of this is where very low frequency signals are to be amplified and unwanted higher frequency noise voltages are to be excluded. In this case, the circuit voltage gain should be made to fall off just above the highest desired signal frequency. The usual method of doing this is to connect a feedback capacitor C_f from the op-amp output to its inverting input terminal as illustrated in Fig. 4-11(a) and (b).

For the inverting amplifier in Fig. 4-11(a), the voltage gain is

$$A_v = \frac{R_2 \| X_{Cf}}{R_1} \tag{4-8}$$

or

$$A_v = \frac{1}{R_1\sqrt{(1/R_2)^2 + (1/X_{Cf})^2}}$$

when $X_{Cf} = R_2$,

$$A_v = \frac{1}{\sqrt{2}}\left(\frac{R_2}{R_1}\right)$$

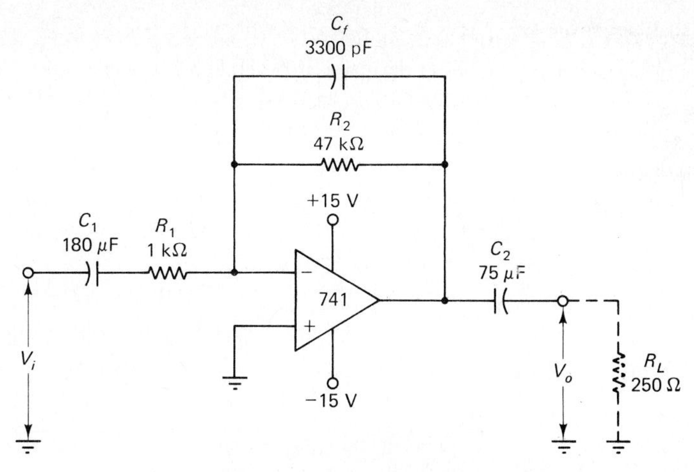

(a) Inverting amplifier with feedback capacitor C_f to set the upper cutoff frequency

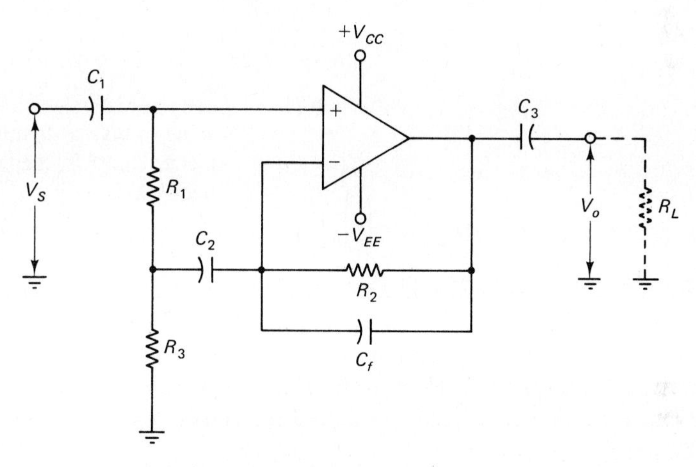

(b) Noninverting amplifier with feedback capacitor C_f to set the upper cutoff frequency

Figure 4-11 The upper cutoff frequency of an amplifier can be selected by including capacitor C_f across the feedback resistor. This gives f_2 at $X_{Cf} = R_2$.

or A_v is 3 dB below the normal voltage gain of R_2/R_1. Thus, the upper cutoff frequency for the circuit can be set at the desired frequency (f_2) by making

$$X_{Cf} = R_2 \text{ at } f_2 \tag{4-9}$$

An analysis of the noninverting amplifier gives a similar result.

It must be emphasized that this method of setting the circuit upper cutoff frequency is applicable only where the op-amp has a much higher cutoff frequency. In fact, as explained in Section 5-5, the op-amp cutoff frequency must be greater than the circuit cutoff frequency multiplied by the amplifier closed-loop gain.

Example 4-6

The inverting amplifier designed in Example 3-7 is to be capacitor-coupled and to have a signal frequency range of 10 Hz to 1 kHz. If the load resistance is 250 Ω, calculate the required capacitor values [as in Fig. 4-11(a)].

Solution

From Eq. 4-2,

$$C_1 = \frac{1}{2\pi f_1(R_1/10)} = \frac{1}{2\pi \times 10\ \text{Hz} \times (1\ \text{k}\Omega/10)}$$

$$= 159\ \mu\text{F} \qquad \text{(use 180 } \mu\text{F standard value)}$$

From Eq. 4-3,

$$C_2 = \frac{1}{2\pi f_1 R_L} = \frac{1}{2\pi \times 10\ \text{Hz} \times 250\ \Omega}$$

$$= 64\ \mu\text{F} \qquad \text{(use 75 } \mu\text{F standard value)}$$

From Eq. 4-9,

$$C_f = \frac{1}{2\pi f_2 R_2} = \frac{1}{2\pi \times 1\ \text{kHz} \times 47\ \text{k}\Omega}$$

$$= 3386\ \text{pF}$$ (use 3300 pF standard value to give a slightly higher cutoff frequency than 1 kHz)

4-7 CAPACITOR-COUPLED DIFFERENCE AMPLIFIER

The difference amplifier circuit discussed in Section 3-6 can be capacitor-coupled as illustrated in Fig. 4-12. Here again, the resistor values can be calculated in the same way as for a direct-coupled circuit and the capacitors can be determined in the usual way. At f_1

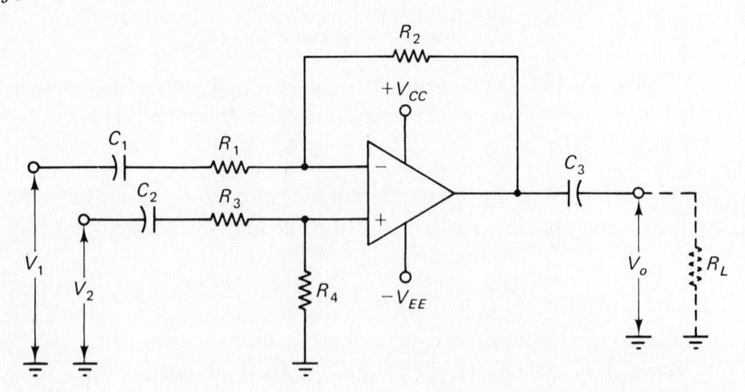

Figure 4-12 Capacitor-coupled difference amplifier. At f_1, input coupling-capacitors C_1 and C_2 are calculated in terms of the impedance at each input terminal, and X_{C3} is again made equal to R_L.

$$X_{C1} = R_1/10, \qquad X_{C2} = (R_3 + R_4)/10, \qquad \text{and } X_{C3} = R_L$$

The voltage gain of this circuit could be rolled off at a desired upper cutoff frequency as discussed in Section 4-6. For the difference amplifier, capacitors should be placed across resistors R_2 and R_4 with each one calculated at the desired cutoff frequency as X_C = (resistance in parallel).

4-8 USE OF A SINGLE-POLARITY SUPPLY

Voltage Follower

Capacitor-coupled op-amp circuits can be easily adapted to use a single-polarity supply voltage because the capacitors block the dc bias voltages at input and output. A capacitor-coupled voltage follower circuit using a single-polarity supply is illustrated in Fig. 4-13(a). If the op-amp data sheet lists the minimum supply voltage as ±9 V, then a minimum of 18 V should be used in a single-polarity supply situation. Similarly, the specified maximum supply voltage must not be exceeded. The potential divider (R_1 and R_2) sets the bias voltage at the noninverting input terminal as approximately $V_{CC}/2$. This means that the dc levels of the output terminal and the inverting input are also at $V_{CC}/2$. Thus, with an 18 V supply, the positive supply terminal is +9 V with respect to the bias level at the input and output terminals, and the negative supply terminal is −9 V with respect to those terminals.

The potential divider resistors are determined in the usual way, by choosing a resistor current (I_2) very much larger than the op-amp input bias current. The voltage drop across each resistor is usually selected as $V_{CC}/2$; although it could be above or below this point within the specified input voltage range for the op-amp. The input impedance of the circuit becomes

$$Z_{in} = R_1 \| R_2$$

so as usual,

$$X_{C1} = (R_1 \| R_2)/10 \text{ at } f_1$$

and

$$X_{C2} = R_L \text{ at } f_1$$

Figure 4-13(b) shows a high input impedance voltage follower using a single-polarity supply. Potential divider R_1 and R_2 again serve to set the bias voltage at approximately $V_{CC}/2$. Resistor R_3 is now included so that its bottom end may be pushed up and down by feedback via C_2 and thus offer an input impedance of $(1 + M)R_3$. In this circuit, the resistance in series with C_2 is $R_1 \| R_2$ because the top of R_1 is ac grounded via V_{CC}. So, C_2 is calculated from $X_{C2} = (R_1 \| R_2)/10$ at f_1.

Noninverting Amplifier

Figure 4-14(a) shows the circuit of a capacitor-coupled noninverting amplifier using a single-polarity supply. Here again, the potential divider constituted by R_1 and R_2 biases the op-amp noninverting input terminal at $V_{CC}/2$. The bottom of resistor R_4 is capacitor-coupled to ground via capacitor C_3. If this point was directly grounded, the

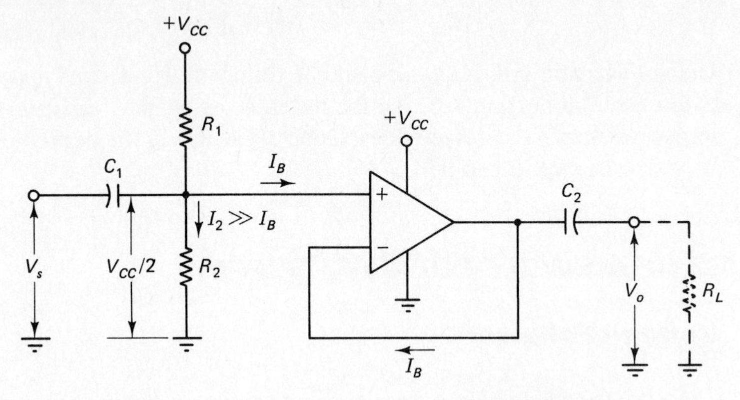

(a) Capacitor-coupled voltage follower using a single-polarity supply

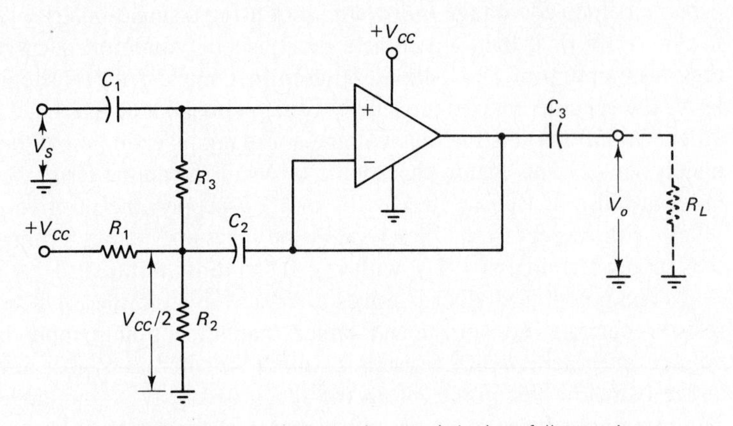

(b) High input impedance capacitor-coupled voltage follower using a single-polarity supply

Figure 4-13 Capacitor-coupled voltage follower and high input impedance voltage follower using single-polarity supplies. The potential divider (R_1 and R_2) is required to bias the op-amp noninverting input terminal at $V_{CC}/2$.

dc voltage at the op-amp output terminal would tend toward $A_v \times$ (bias level at the noninverting input), or $A_v \times V_{CC}/2$. This would saturate the output at approximately $V_{CC} - 1$ V. With C_3 in the circuit as shown, and R_3 connecting the inverting input terminal to the output, the op-amp behaves as a dc voltage follower. The dc voltage level at the op-amp output terminal is then the same as that at the noninverting input terminal ($V_{CC}/2$). For ac voltages, C_3 behaves as a short circuit, so that the ac voltage gain is $(R_3 + R_4)/R_4$.

Components C_1, C_2, R_1, and R_2 are calculated exactly as for the voltage follower discussed above, and R_3 and R_4 are determined in the usual way for a noninverting amplifier. Capacitor C_3 should be selected to have an impedance very much smaller than R_4 at the low 3 dB frequency of the circuit. As usual, let $X_{C3} = R_4/10$ at f_1. An alternative is to have C_3 determine the lower cutoff frequency for the circuit. This might be desirable for a circuit that has a variable load. In this case, $X_{C3} = R_4$ at f_1, and $X_{C2} = R_{L(\min)}/10$ at f_1.

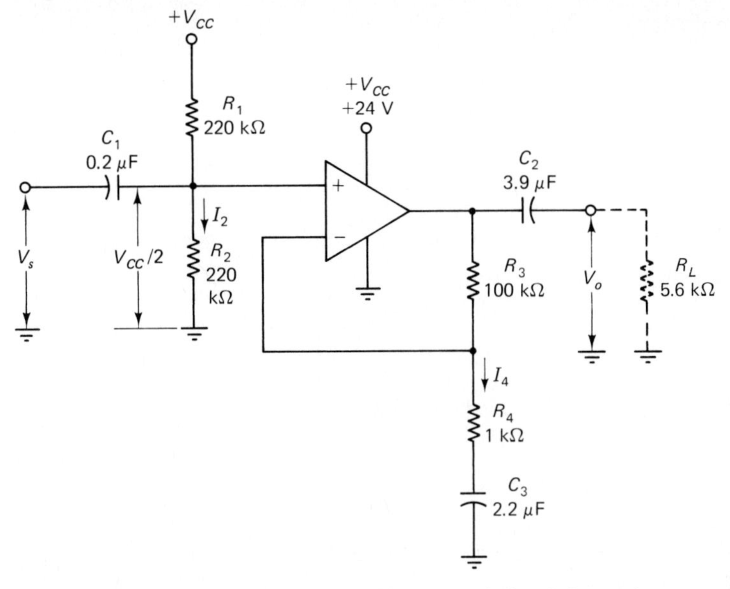

(a) Capacitor-coupled noninverting amplifier using a single-polarity supply

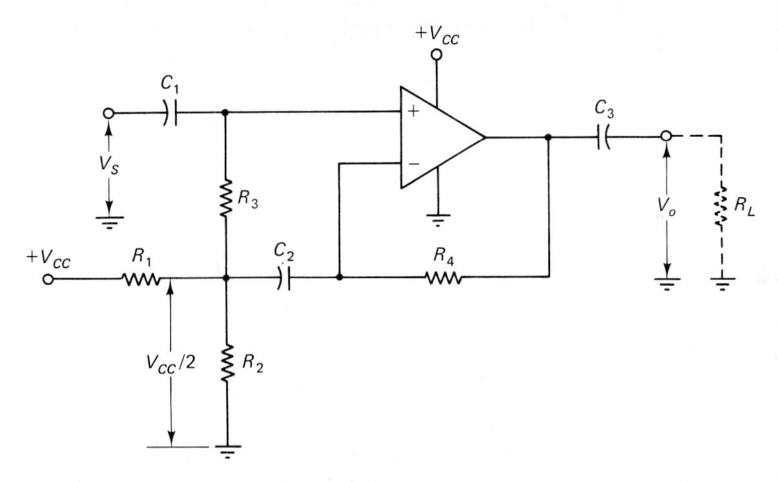

(b) High input impedance capacitor-coupled noninverting amplifier using a single-polarity supply

Figure 4-14 Capacitor-coupled noninverting amplifier and high input impedance noninverting amplifier using single-polarity supplies. R_1 and R_2 bias the op-amp noninverting input terminal at $V_{CC}/2$.

Example 4-7

A capacitor-coupled noninverting amplifier is to have a +24 V supply, a voltage gain of 100, an output amplitude of 5 V, a lower cutoff frequency of 75 Hz, and a minimum load resistance of 5.6 kΩ. Using a 741 op-amp, design a suitable circuit.

Solution

$$I_2 \gg I_{B(\max)}$$

Let $\quad I_2 = 100 \times I_{B(\max)} = 50\ \mu\text{A}$

$$R_1 = R_2 = \frac{V_{CC}/2}{I_2} = \frac{24\ \text{V}/2}{50\ \mu\text{A}}$$

$$= 240\ \text{k}\Omega \qquad \text{(use 220 k}\Omega\text{ standard value)}$$

$$V_i = \frac{V_o}{A_v} = \frac{5\ \text{V}}{100}$$

$$= 50\ \text{mV}$$

$$I_4 \gg I_{B(\max)}$$

Let $\quad I_4 = 100 \times I_{B(\max)} = 100 \times 500\ \text{nA}$

$$= 50\ \mu\text{A}$$

$$R_4 = \frac{V_i}{I_4} = \frac{50\ \text{mV}}{50\ \mu\text{A}}$$

$$= 1\ \text{k}\Omega \qquad \text{(standard value)}$$

$$R_3 + R_4 = \frac{V_o}{I_4} = \frac{5\ \text{V}}{50\ \mu\text{A}}$$

$$= 100\ \text{k}\Omega$$

$$R_3 = (R_3 + R_4) - R_4 = 100\ \text{k}\Omega - 1\ \text{k}\Omega$$

$$= 99\ \text{k}\Omega \qquad \text{(use 100 k}\Omega\text{ standard value)}$$

From Eq. 4-2, $\quad C_1 = \dfrac{1}{2\pi f_1 (R_1 \| R_2)/10} = \dfrac{1}{2\pi \times 75\ \text{Hz} \times (220\ \text{k}\Omega \| 220\ \text{k}\Omega)/10}$

$$\approx 0.2\ \mu\text{F} \qquad \text{(standard value)}$$

Allowing C_3 to determine the lower cutoff frequency gives the smallest possible capacitor values. Therefore,

$$C_2 = \frac{1}{2\pi f_1 R_L/10} = \frac{1}{2\pi \times 75\ \text{Hz} \times (5.6\ \text{k}\Omega/10)}$$

$$\approx 3.8\ \mu\text{F} \qquad \text{(use 3.9 }\mu\text{F standard value)}$$

$$C_3 = \frac{1}{2\pi f_1 R_4} = \frac{1}{2\pi \times 75\ \text{Hz} \times 1\ \text{k}\Omega}$$

$$= 2.12\ \mu\text{F} \qquad \text{(use 2.2 }\mu\text{F standard value)}$$

The high input impedance noninverting amplifier with a single polarity supply in Fig. 4-14(b) is very similar to the high input impedance voltage follower in Fig. 4-13(b). The difference is that resistor R_4 is included in the noninverting amplifier circuit to give a voltage gain greater than 1. The voltage gain is

$$A_v = \frac{R_4 + (R_1 \| R_2)}{R_1 \| R_2} \tag{4-10}$$

To design the circuit, R_1 and R_2 should be determined in the usual way. Then R_4 should be calculated from Eq. 4-10. The capacitors should be determined as discussed for the high input impedance noninverting amplifier in Section 4-4, bearing in mind that (for ac) $R_1 \| R_2$ is in series with C_2.

Inverting Amplifier

The circuit of an inverting amplifier using a single-polarity supply is illustrated in Fig. 4-15. In this case, a potential divider (R_3 and R_4) is used to set the noninverting input terminal at $V_{CC}/2$. The dc voltage level of the output and the inverting input terminal will then also be $V_{CC}/2$. As always, the potential divider is designed by first selecting a current (I_4) which is much greater than the current flowing out of the potential divider (I_B in this case). Then

$$R_3 = R_4 = \frac{V_{CC}/2}{I_4}$$

Other components are determined exactly as discussed in Section 4-5.

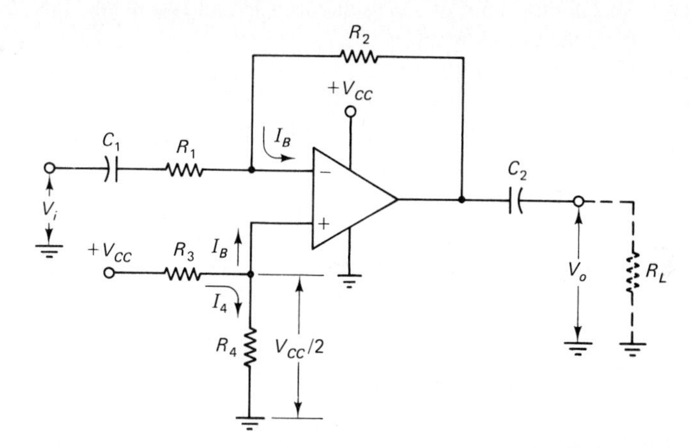

Figure 4-15 Capacitor-coupled inverting amplifier using a single-polarity supply. The potential divider (R_3 and R_4) biases the op-amp noninverting input terminal at $V_{CC}/2$.

REVIEW QUESTIONS

4-1. Sketch the circuit of a capacitor-coupled voltage follower. Briefly explain. Define the input impedance of the circuit and show how a portion of the signal can be lost across the coupling capacitors.

4-2. Write the equations for calculating the capacitance values for a capacitor-coupled voltage follower.

4-3. Sketch the circuit of a high input impedance capacitor-coupled voltage follower. Briefly explain the circuit operation.

4-4. Develop the equation for Z_{in} for a high input impedance capacitor-coupled voltage follower.

4-5. Explain how to determine the capacitor values for a high input impedance capacitor-coupled voltage follower.

4-6. Sketch the circuit of a capacitor-coupled noninverting amplifier. Briefly explain the circuit operation and define its input impedance.

4-7. Sketch the circuit of a high input impedance capacitor-coupled noninverting amplifier. Briefly explain the circuit operation.

4-8. Develop the equation for Z_{in} for a high input impedance capacitor-coupled noninverting amplifier.

4-9. Explain how to determine the capacitor values for a high input impedance capacitor-coupled noninverting amplifier.

4-10. Sketch the circuit of a capacitor-coupled inverting amplifier. Briefly explain the circuit operation, and define its input impedance.

4-11. Write the equations for calculating the the capacitance values for a capacitor-coupled inverting amplifier.

4-12. Briefly discuss the upper cutoff frequency of an op-amp circuit and show how the cutoff frequency can be set for inverting, noninverting, and difference amplifiers.

4-13. Sketch the circuit of a capacitor-coupled difference amplifier. Explain how to determine the capacitor values.

4-14. Draw a sketch to show how a capacitor-coupled voltage follower should be used with a single-polarity supply. Briefly explain.

4-15. Sketch the circuit of a high input impedance capacitor-coupled voltage follower using a single-polarity supply. Briefly explain.

4-16. Draw a sketch to show how a capacitor-coupled noninverting amplifier should be connected to use a single-polarity supply. Briefly explain.

4-17. Sketch the circuit of a high input impedance capacitor-coupled noninverting amplifier using a single-polarity supply. Briefly explain.

4-18. Sketch the circuit of a capacitor-coupled inverting amplifier using a single-polarity supply. Briefly explain.

PROBLEMS

4-1. A capacitor-coupled voltage follower is to be designed to have a lower cutoff frequency of 120 Hz. The load resistance is 8.2 kΩ and the op-amp used has a maximum input bias current of 600 nA. Design a suitable circuit.

4-2. Redesign the circuit for Problem 4-1 using a BIFET op-amp.

4-3. Calculate the new lower cutoff frequency for the circuit in Example 4-1 when the load resistance is changed to 4.7 kΩ.

4-4. A capacitor-coupled voltage follower using a 741 op-amp is to have an input impedance of 12 kΩ. The lower cutoff frequency is to be a maximum of 100 Hz and the load resistance can vary from 3.3 kΩ to 15 kΩ. Design a suitable circuit.

4-5. A capacitor-coupled voltage follower as in Fig. 4-1(a) has the following components:

$R_1 = 68$ kΩ, $R_L = 2.7$ kΩ, $C_1 = 0.15$ μF, $C_2 = 0.33$ μF. Determine the circuit input impedance and the lower cutoff frequency, and calculate the impedance of C_1 at f_1.

4-6. Modify the circuit designed for Problem 4-4 to make it a high input impedance capacitor-coupled voltage follower using a 709 op-amp. Calculate the minimum theoretical input impedance of the circuit.

4-7. Using a BIFET op-amp design a high input impedance capacitor-coupled voltage follower to meet the specification in Problem 4-1.

4-8. A voltage follower as in Fig. 4-4 has the following components: $R_1 = 47$ kΩ, $R_2 = 47$ kΩ, $R_L = 1.5$ kΩ, $C_1 = 1000$ pF, $C_2 = 0.18$ μF, $C_3 = 0.5$ μF. Calculate the circuit lower cutoff frequency, the impedance of C_1 and C_2 at f_1, and the theoretical input impedance. Assume $M = 30\,000$.

4-9. A capacitor-coupled noninverting amplifier as in Fig. 4-6 is to have $A_v = 90$ and $V_o = 3$ V. The load resistance is 10 kΩ, and the lower cutoff frequency is to be 70 Hz. Design a suitable circuit using a BIFET op-amp.

4-10. Modify the circuit designed for Problem 4-9 to set its upper cutoff frequency at 3 kHz.

4-11. Redesign the circuit for Problem 4-9 to use a 741 op-amp.

4-12. Modify the circuit designed for Problem 4-9 to have an input impedance of approximately 20 kΩ.

4-13. A noninverting amplifier as in Fig. 4-6 has the following components: $R_1 = 33$ kΩ, $R_2 = 150$ kΩ, $R_3 = 1.5$ kΩ, $R_L = 4.7$ kΩ, $C_1 = 0.39$ μF, $C_2 = 0.27$ μF. Determine the circuit voltage gain, input impedance, lower cutoff frequency, and the impedance of C_1 at f_1.

4-14. Modify the circuit designed for Problem 4-9 to make it a high input impedance capacitor-coupled noninverting amplifier. Calculate the theoretical input impedance. Assume $M = 50\,000$.

4-15. A high input impedance capacitor-coupled noninverting amplifier is to be designed to have $A_v = 120$ and $f_1 = 100$ Hz. The input signal is 50 mV, and the load resistance ranges from 2.7 kΩ to 27 kΩ. Design a suitable circuit using a 741 op-amp.

4-16. Modify the circuit designed for Problem 4-15 to set its upper cutoff frequency at 3 kHz.

4-17. A capacitor-coupled inverting amplifier with an input signal of 30 mV and a load resistance of 2.2 kΩ is to have $A_v = 150$ and $f_1 = 80$ Hz. Design a suitable circuit using a BIFET op-amp.

4-18. Redesign the circuit in Problem 4-17 to use a 741 op-amp.

4-19. A capacitor-coupled inverting amplifier as in Fig. 4-10 has the following components: $R_1 = 2.7$ kΩ, $R_2 = 100$ kΩ, $R_L = 1.5$ kΩ, $C_1 = 3.9$ μF, $C_2 = 0.68$ μF. Determine the circuit voltage gain, input impedance, lower cutoff frequency, and the impedance of C_1 at f_1.

4-20. The difference amplifier designed for Problem 3-22 is to be capacitor-coupled at input and output. The load resistance is 3.3 kΩ and the lower cutoff frequency is to be 200 Hz. Determine suitable capacitor values.

4-21. A capacitor-coupled difference amplifier is to have two input voltages which each range from 200 mV to 2 V. The lowest signal frequency is 300 Hz, the load resistance is 3.3 kΩ, and the voltage gain is to be 5. Design a suitable circuit using a BIFET op-amp.

4-22. Redesign the circuit in Problem 4-21 to use a 741 op-amp.

4-23. A difference amplifier as in Fig. 4-12 has the following components: $R_1 = 6.8\ \text{k}\Omega$, $R_2 = 68\ \text{k}\Omega$, $R_3 = 6.8\ \text{k}\Omega$, $R_4 = 68\ \text{k}\Omega$, $R_L = 1.5\ \text{k}\Omega$, $C_1 = 2.2\ \mu\text{F}$, $C_2 = 0.2\ \mu\text{F}$, and $C_3 = 1\ \mu\text{F}$. Determine the circuit lower cutoff frequency and the maximum differential input voltage if the output amplitude is not to exceed 5 V.

4-24. Modify the voltage follower designed for Problem 4-1 to use a +30 V power supply.

4-25. Redesign the voltage follower circuit for Problem 4-4 to use a +25 V supply.

4-26. The noninverting amplifier designed for Problem 4-9 is to be modified to use a +30 V power supply. Determine suitable components.

4-27. Using a 741 op-amp, design a high Z_{in} noninverting amplifier to operate with a +36 V power supply. The load resistance is 12 kΩ, the lower cutoff frequency is to be 150 Hz, and the voltage gain is to be 7.

4-28. The inverting amplifier circuit designed for Problem 4-17 is to be modified to use a +24 V power supply. Redesign the circuit as necessary.

4-29. Design a capacitor-coupled inverting amplifier to operate with a +20 V supply. The minimum input signal level is 50 mV, the voltage gain is to be 68, the load resistance is 500 Ω, and the lower cutoff frequency is to be 200 Hz. Use a 741 op-amp.

COMPUTER PROBLEMS

4-30. Write a computer program to design a capacitor-coupled, high Z_{in}, voltage follower using a BIFET op-amp. Given: v_i, f_1, and R_L.

4-31. Write a computer program to design a capacitor-coupled, noninverting amplifier using a bipolar op-amp. Given: $I_{B(max)}$, A_v, V_i, f_1, and R_L.

4-32. Write a computer program to analyze a capacitor-coupled, noninverting amplifier, for A_v, Z_{in}, Z_{out}, and f_1. Given: component values and op-amp parameters.

4-33. Write a computer program to design a capacitor-coupled, high Z_{in}, noninverting amplifier using a bipolar op-amp. Given: $I_{B(max)}$, A_v, V_i, f_1, and R_L.

LABORATORY EXERCISES

4-1 Capacitor-Coupled Voltage Follower

1. Construct the capacitor-coupled voltage follower circuit shown in Fig. 4-2, as designed in Example 4-1. Use a supply of ±9 V to ±15 V.
2. Apply a ±1 V, 1 kHz sinusoidal signal input. Monitor the input and output voltages on an oscilloscope.
3. Maintaining the input voltage (v_i) constant, reduce the signal frequency until the output is approximately 0.707 v_i. Note the lower cutoff frequency (f_1) and compare it to the designed cutoff frequency.
4. Return the signal to 1 kHz. Connect a 120 kΩ resistor in series with the amplifier input. Check the effect on the output voltage and calculate Z_{in}.

4-2 High Z_{in} Capacitor-Coupled Voltage Follower

1. Construct the high input impedance capacitor-coupled voltage follower circuit shown in Fig. 4-5, as designed in Example 4-3. Use a supply of ±9 V to ±15 V.

2–3. Same as Procedures 2 through 3 in Laboratory Exercise 4-1.

4. Return the signal frequency to 1 kHz. Connect a 1 MΩ resistor in series with the amplifier input. Check that the output voltage is unaffected to demonstrate that $Z_{in} \gg$ 1 MΩ.

4-3 Capacitor-Coupled Noninverting Amplifier

1. Construct the capacitor-coupled noninverting amplifier circuit shown in Fig. 4-7, as designed in Examples 3-4 and 4-4. Use a supply of ±9 V to ±15 V.
2. Apply a ±50 mV, 1 kHz sinusoidal signal input. Using an osciloscope, measure the peak amplitude of the output voltage, calculate the circuit voltage gain and compare it to the designed gain.
3. Maintaining the input voltage (v_i) constant, reduce the signal frequency until the output is approximately 0.707 times the v_o level at 1 kHz. Note the lower cutoff frequency (f_1) and compare it to the designed cutoff frequency.
4. Return the signal frequency to 1 kHz. Connect a 120 kΩ resistor in series with the amplifier input. Check the effect on the output voltage, and calculate Z_{in}.

4-4 Capacitor-Coupled Inverting Amplifier

1. Construct the capacitor-coupled inverting amplifier circuit shown in Fig. 4-11(a), as designed in Examples 3-7 and 4-6. Use a supply of ±9 V to ±15 V.

2–3. Same as Procedures 2 through 3 for Laboratory Experience 4-3.

4. Still maintaining the input voltage (v_i) constant, increase the signal frequency until the output is approximately 0.707 times the V_o level at 1 kHz. Note the upper cutoff frequency (f_2), and compare it to the designed cutoff frequency.
5. Modify the circuit to use a single-polarity supply voltage, as in Fig. 4-15. Use a supply of +18 V to +30 V, and $R_3 = R_4 = 100$ kΩ.
6. Repeat Procedure 2.

CHAPTER 5

Op-Amp Frequency Response and Compensation

Objectives

- Show how feedback in op-amp circuits can produce instability and state the conditions that produce oscillations.
- Define *loop phase shift, loop gain, open-loop gain, closed-loop gain.*
- Sketch typical open-loop *gain/frequency response* and *phase/frequency response* graphs for an operational amplifier.
- Sketch circuits for *lag compensation, lead compensation,* and *Miller-effect compensation.* Explain the operation of each circuit and how each affects the op-amp frequency response.
- Select compensating components for inverting and noninverting amplifiers from the manufacturer's recommended components.
- Determine the *upper cutoff frequency* for various op-amp circuits directly from the op-amp open-loop gain/frequency response.
- Define the *gain-bandwidth product* for an op-amp and use it to calculate the upper cutoff frequency for an op-amp circuit.
- Calculate the *slew rate-limited frequency* and the *slew rate-limited rise time* for a given op-amp output amplitude.
- Calculate the *slew rate-limited amplitude* at the circuit cutoff frequency or rise time.
- Explain the effects of *stray capacitance* and *load capacitance* on op-amp circuit stability and discuss how the effects can be minimized.
- Sketch compensating circuits for stray and load capacitances, and calculate the values of compensating components.
- Sketch a circuit to show the Z_{in} *mod* method of frequency compensation. Explain how it can be employed to extend the bandwidth of an op-amp circuit.
- List precautions that should be observed for operational amplifier circuit stability.

INTRODUCTION

Operational amplifiers have internal phase shifts from input to output. The phase shifts are greatest at high frequencies and at some particular frequency the total phase shift can add up to 360°. When this happens, the op-amp circuit is likely to become unstable enough to break into oscillation. For oscillation to occur, the loop voltage gain (from the inverting input terminal to the output and back to the input via the feedback network) must be equal to, or greater than, unity when the phase shift approaches 360°. To combat instability, compensating capacitors and resistors are used either to reduce the voltage gain or to minimize the phase shift.

The upper cutoff frequency of an operational amplifier circuit depends upon the particular op-amp employed and upon the compensating components. The frequency response is also limited by the op-amp slew rate. Some IC operational amplifiers (notably the 741) have internal compensation, which makes them very easy to use but limits the frequency response.

5-1 OP-AMP CIRCUIT STABILITY

Consider the inverting amplifier circuit and waveforms shown in Fig. 5-1(a). The signal voltage v_s is amplified by a factor R_2/R_1 (see Section 1-6) and phase shifted through $-180°$. The circuit is redrawn in Fig. 5-1(b) to show that output voltage v_o is divided by the feedback network to produce feedback voltage v_f. Whether or not an input signal is present, any ac voltage at the op-amp inverting input terminal is amplified by the open-loop gain M to produce an output voltage v_o and then divided to produce a feedback voltage v_f.

Suppose an ac voltage v_f occurs at the inverting input terminal. The amplified v_f is $v_o = M v_f$ and the divided v_o is $v_f = \beta v_o$. Neglecting signal source resistance R_s,

$$\beta = R_1/(R_1 + R_2) \qquad (5\text{-}1)$$

If βv_o is exactly equal to and in phase with the original v_f, the circuit is supplying its own ac input and a state of continuous oscillation exists.

Now look at the noninverting amplifier in Fig. 5-2(a). When this circuit is redrawn in Fig. 5-2(b), the circuit is similar to the inverting amplifier in Fig. 5-1(b) except for the position of R_s. Once again, output voltage v_o is divided to produce a feedback voltage v_f which is applied to the amplifier inverting input terminal. Oscillation of the circuit is quite possible.

Because of the feedback network, virtually every operational amplifier circuit is a potential oscillator. Measures taken to prevent oscillation are referred to as *frequency compensation.* Some operational amplifiers, such as the 741, are internally compensated. These amplifiers have limited frequency response and are unsuitable for many applications. When an uncompensated amplifier is employed, it is usually necessary to connect external components to stablize the circuit.

There are two conditions, known as the *Barkhausen criteria,* that have to be

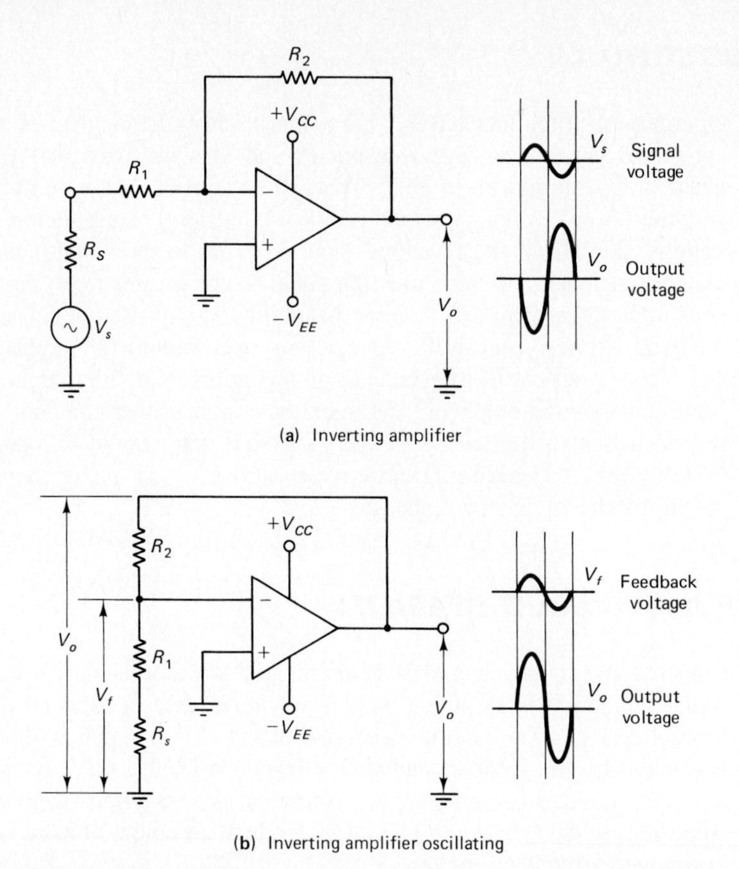

Figure 5-1 Because an inverting amplifier consists of an operational amplifier and a feedback network, the circuit can supply its own ac input (v_f derived from v_o) and a state of continuous oscillations can occur.

fulfilled for a circuit to oscillate. The *loop gain* should be equal to or greater than one and the *loop phase shift* should equal 360°. The loop gain is the gain around the loop from the inverting input terminal to the amplifier output and back to the inverting input via the feedback network. The gain from the inverting input terminal to the output is the op-amp *open-loop gain M*. For the feedback network, the gain from the amplifier output back to the input is actually an attenuation. Therefore,

$$\text{loop gain} = (\text{amplifier gain}) \times (\text{feedback network attenuation})$$
$$= M\beta$$

The *loop phase shift* is the total phase shift around the loop from the inverting input terminal to the amplifier output and back to the inverting input via the feedback network. Assuming that it is purely resistive, the feedback network has no phase shift, so the loop phase shift is essentially the amplifier phase shift. The phase shift from the inverting input terminal to the output is normally −180°. The output goes negative when the input goes positive, and vice versa. At high frequencies there

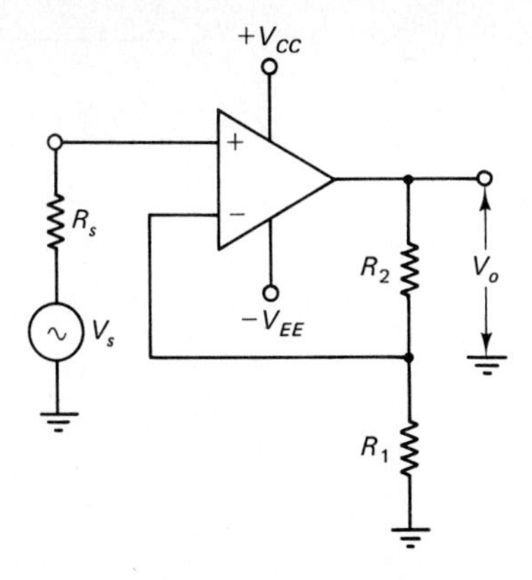

(a) Noninverting amplifier

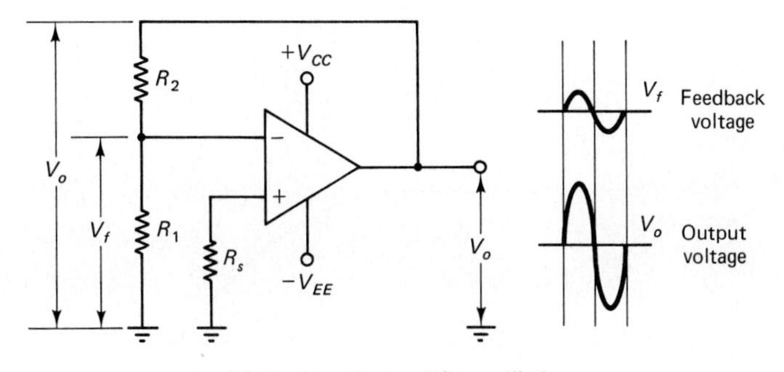

(b) Noninverting amplifier oscillating

Figure 5-2 The feedback network employed with a noninverting amplifier behaves exactly the same as that with an inverting amplifier (Fig. 5-1(b)). A state of continuous oscillations can occur.

is additional phase shift (explained in Section 5-2) and the total can approach −360°.

5-2 FREQUENCY AND PHASE RESPONSE

Single-Stage Amplifier Response

The voltage gain of a single-stage transistor amplifier commences to fall off at some high frequency. This may be due to the construction of the individual transistor or to stray capacitance in the circuit. Figure 5-3(a) shows a single-stage transistor amplifier circuit while Fig. 5-3(b) shows the high frequency ends of the gain/frequency and phase/frequency responses of the circuit. Note that frequency is plotted to a logarithmic base. The voltage gain falls off at a rate of 6 dB for each doubling of frequency (−6 *dB per octave*). This can also be stated as a fall of 20 dB for each ten

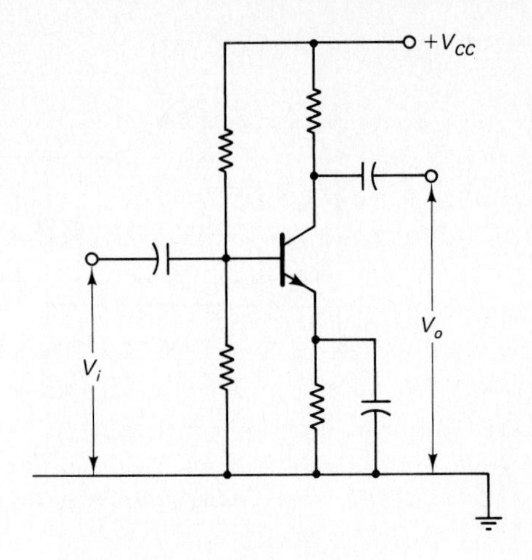

(a) Single-stage transistor amplifier

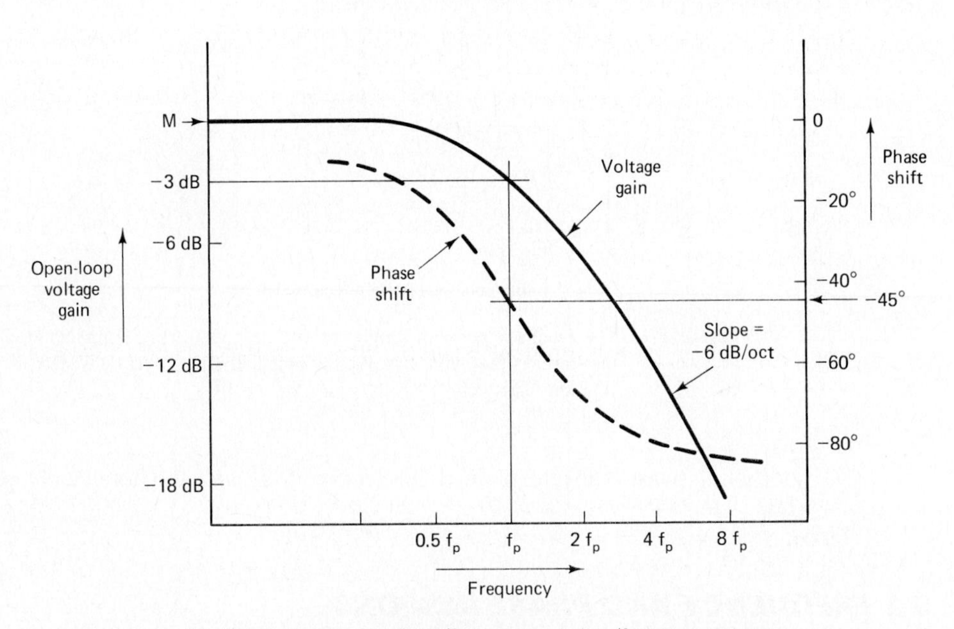

(b) High frequency ends of gain/frequency and phase/frequency responses

Figure 5-3 The open-loop gain of a single-stage transistor amplifier falls off at a rate of 6 dB/octave at high frequencies. The phase shift increases to −45° at the pole frequency (f_p) and continues to increase to a maximum of −90°.

times increase in frequency (−20 *dB/decade*). The *pole frequency* (f_p) is the frequency at which the gain is down by 3 dB from the mid-band gain.

From the graph of phase-shift versus frequency [broken line in Fig. 5-3(b)], it is seen that the phase lag increases from zero to −45° at the pole frequency. As the frequency increases, the lag continues to increase to a maximum of −90°.

Operational Amplifier Response

Operational amplifiers generally have a differential input stage, an intermediate amplification stage, and a low impedance output stage [see Fig. 5-4(a)]. Each of these three stages has its own gain/frequency and phase/frequency response. Usually the pole frequency (f_{p2}) of Stage 2 is higher than the pole frequency (f_{p1}) of Stage 1; and the pole frequency (f_{p3}) of Stage 3 is higher still. A straight line approximation of the gain/frequency response graph for a typical operational amplifier is shown in Fig. 5-4(b). This is known as a *Bode plot*. Note that the overall voltage gain (M) initially falls off at 6 dB/octave (−20 dB/decade) from f_{p1} when only the gain of Stage 1 is decreasing. At f_{p2}, Stage 2 gain is decreasing at 6 dB/octave, so the total rate of decline is 12 dB/octave (−40 dB/decade). Finally, when the frequency reaches f_{p3}, the gain of Stage 3 is decreasing at 6 db/octave and the overall rate of decline of M is 18 dB/octave (−60 dB/decade).

The phase shifts of each stage also add together to give the total phase shift for the op-amp. A typical phase shift versus frequency (θ/f) response is shown as a broken line in Fig. 5-4(b). At f_{p1} the Stage 1 phase shift is −45° while the phase shift for each of the other two stages is negligibly small, so the total phase shift is −45°. At f_{p2}, Stage 2 adds another −45°. But at this point the Stage 1 phase shift is at its maximum of −90°. Consequently, the total phase shift at f_{p2} is

$$\theta = -45^\circ - 90^\circ = -135^\circ$$

When the frequency is f_{p3}, Stage 1 and Stage 2 are each contributing −90° of phase shift and Stage 3 adds a further −45°. Thus, the total phase shift at f_{p3} is

$$\theta = -90^\circ - 90^\circ - 45^\circ = -225^\circ$$

This open-loop phase shift is additional to the −180° phase shift that normally occurs from the op-amp inverting input terminal to the output.

Phase Margin

As already discussed, oscillations occur in an op-amp circuit when the loop gain $M\beta$ is equal to or greater than 1 and the loop phase shift is −360°. In fact, the phase shift does not have to be exactly −360° for oscillation to occur. A loop phase shift of −330° at $M\beta \geq 1$ makes the circuit unstable. To avoid oscillations the total loop phase shift must not be greater than −315° when $M\beta = 1$. The difference between 360° and the actual phase shift at $M\beta = 1$ is referred to as the *phase margin*. So for stability,

$$\text{minimum phase margin} = 360^\circ - 315^\circ = 45^\circ$$

High Gain Amplifier Stability

For an op-amp with the frequency response shown in Fig. 5-4(b), the maximum gain is 100 dβ which is equivalent to a voltage gain of 100 000. From Eqs. 1-5, 1-6, and 5-1, the overall voltage gain of an amplifier with negative feedback is

$$A_v \approx 1/\beta$$

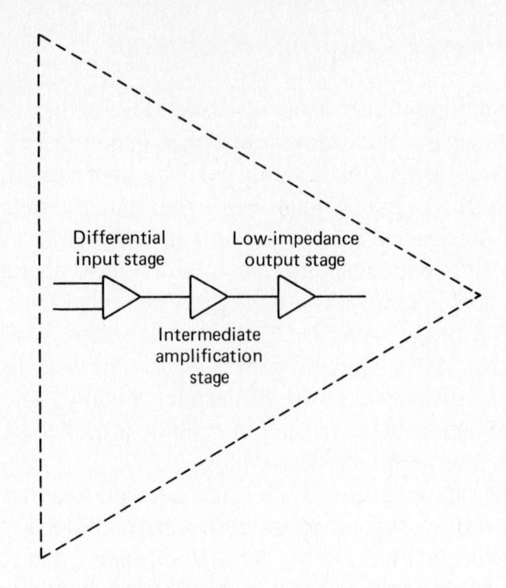

(a) An operational amplifier has three stages of amplification

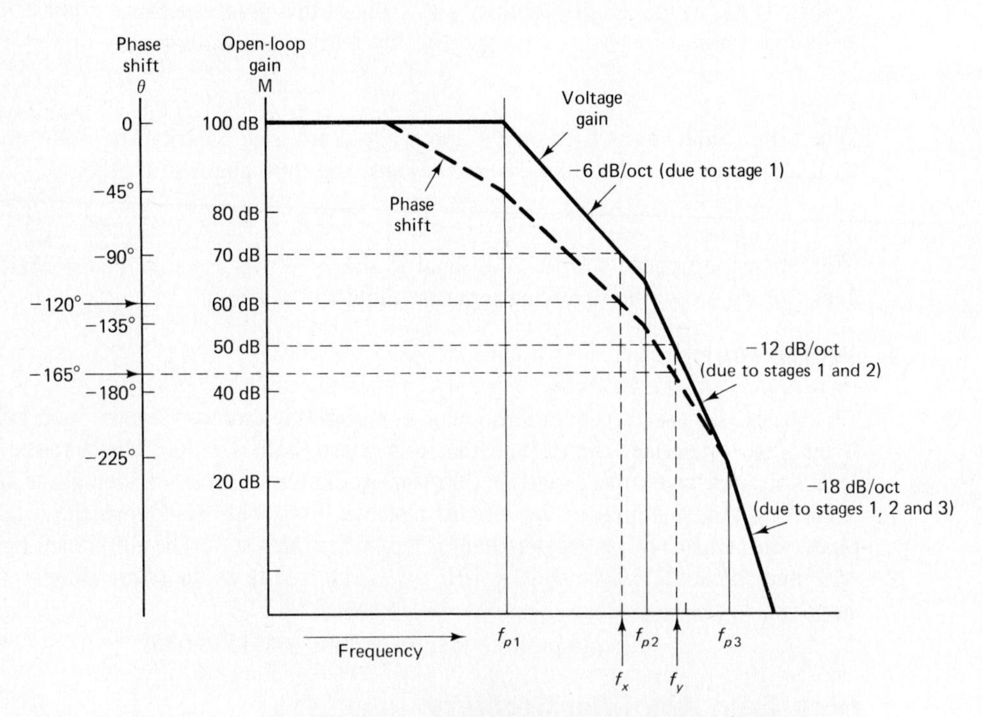

(b) Straight-line approximation of op-amp open-loop gain/frequency and phase/frequency responses

Figure 5-4 Each of the three stages of an operational amplifier has its own particular upper cutoff frequency with a fall-off rate of 6 dB/octave, and each has a maximum phase shift of −90°. The three responses are combined to give a straight line approximations of gain/frequency and phase/frequency response for the op-amp.

This gives $$\beta \approx 1/A_v$$

and loop gain $$M\beta \approx M/A_v$$

Suppose an op-amp with a frequency response as in Fig. 5-4(b) is used with feedback to produce an amplifier with a closed loop gain of 70 dB. A horizontal line drawn on the frequency response graph at A_v = 70 dB intersects the open-loop gain/frequency graph at frequency f_x. At this point, A_v = 70 dB and M = 70 dB. Consequently, the loop gain is

$$M\beta \approx M/A_v = 70 \text{ dB}/70 \text{ dB}$$
$$= 1$$

Thus, the frequency f_x, at which the closed-loop gain A_v equals the open-loop gain M, is the frequency at which the loop gain is $M\beta = 1$. This is one of the conditions required for oscillation.

Now estimate the loop phase shift at frequency f_x. From the intersection of the vertical line drawn from f_x and the θ/f graph, $\theta \approx -120°$. Because of the $-180°$ phase shift between the op-amp inverting input terminal and the output, the total loop phase shift (ϕ) is

$$\phi = -180° + \theta = -180° - 120°$$
$$= -300°$$

and $$\text{phase margin} = 360° - \phi = 360° - 300°$$
$$= 60°$$

At $M\beta = 1$ the phase margin is 60°. Since the minimum phase margin for stability is 45°, this particular circuit (with A_v = 70 dB) is likely to be stable. In other words, it is unlikely to oscillate.

The circuit will not oscillate at the frequency at which $M\beta = 1$ because only one of the two conditions required for oscillation ($M\beta \geq 1$ and $\phi = -330°$ to $-360°$) is fulfilled. At frequencies lower than f_x, $M\beta$ is greater than 1 but the open-loop phase shift θ is smaller than $-120°$. Consequently, the closed-loop phase shift ϕ is smaller than $-300°$. So, if the conditions for oscillation did not occur at the frequency where $M\beta = 1$, they will not occur at lower frequencies.

At frequencies greater than f_x the open-loop phase shift increases to a phase angle greater than $-120°$. But M gets progressively smaller beyond f_x which makes $M\beta$ less than one. Although the condition $\phi = -330°$ to $-360°$ might occur at some frequency greater than f_x, the other condition required for oscillation ($M\beta = 1$) cannot be fulfilled. If the conditions for oscillation did not occur at $M\beta = 1$, they will not occur at any higher frequency.

Stability of Lower Gain Amplifiers

Now consider an amplifier with a gain of 50 dB that uses an op-amp with the frequency response in Fig. 5-4(b). A horizontal line drawn at 50 dB gives the frequency f_y at which $M\beta = 1$. The vertical line from f_y intersects the θ/f graph at $\theta \approx -165°$, giving a loop phase shift

$$\phi = -180° - 165°$$

$$= -345°$$

and

$$\text{phase margin} = 360° - 345° = 15°$$

With $A_v = 50$ dB at $M\beta = 1$, the phase margin is 15°. This phase margin is not adequate for stability, so the circuit is likely to oscillate. A means of increasing the phase margin (reducing ϕ) must be found to stabilize this circuit. Alternatively, a way of moving the frequency at which $M\beta = 1$ to some other stable point on the θ/f graph can be employed.

Note that the amplifier with $A_v = 70$ dB was found to be stable and the amplifier with $A_v = 50$ dB is unstable. That is, the amplifier with the higher closed-loop gain is stable and the one with the lower closed-loop gain is unstable. This is because for $A_v = 70$ dB the feedback network has an attenuation of $\beta \approx 70$ dB, while for $A_v = 50$ dB the network attenuation is $\beta \approx 50$ dB. The lower attenuation of the feedback network gives a greater amount of feedback and thus makes oscillations more likely.

The lesson from this is that an op-amp circuit with a low closed-loop gain is more difficult to stabilize than one with a high closed-loop gain. The voltage follower, with a gain of 1, is one of the most difficult circuits to stabilize.

5-3 FREQUENCY COMPENSATING METHODS

Phase-Lag Compensation

Lag compensation and *lead compensation* are two methods often employed to stabilize op-amp circuits. In each case a resistance-capacitance network is connected into the circuit so that it is part of the loop. Usually, as recommended by the device manufacturer, the network is connected to specified points within the op-amp. For this purpose, external terminals are provided for access to points within the IC.

Referring to the phase lag network in Fig. 5-5(a), it can be shown that at frequencies where $X_{C1} \gg R_2$ the voltage v_2 lags v_1. A phase lag as great as $-90°$ might be introduced. However, at higher frequencies where $X_{C1} \ll R_2$ the network is largely resistive and no significant phase lag occurs. At these frequencies the lag network merely introduces some attenuation.

The components of the lag network in Fig. 5-5(a) are calculated to introduce additional phase lag at some low frequency where the op-amp open-loop phase shift is still so small that additional phase shift has no effect on the circuit stability. Then, at higher frequencies there is only the attenuation without any additional phase lag. The network attenuation alters the op-amp gain/frequency response as illustrated in Fig. 5-5(b).

The effect of the lag network attenuation is to move the frequency f_{x1} at which $M\beta = 1$ for a given closed-loop gain A_v, to a lower frequency f_{x2}. Because f_{x2} is less

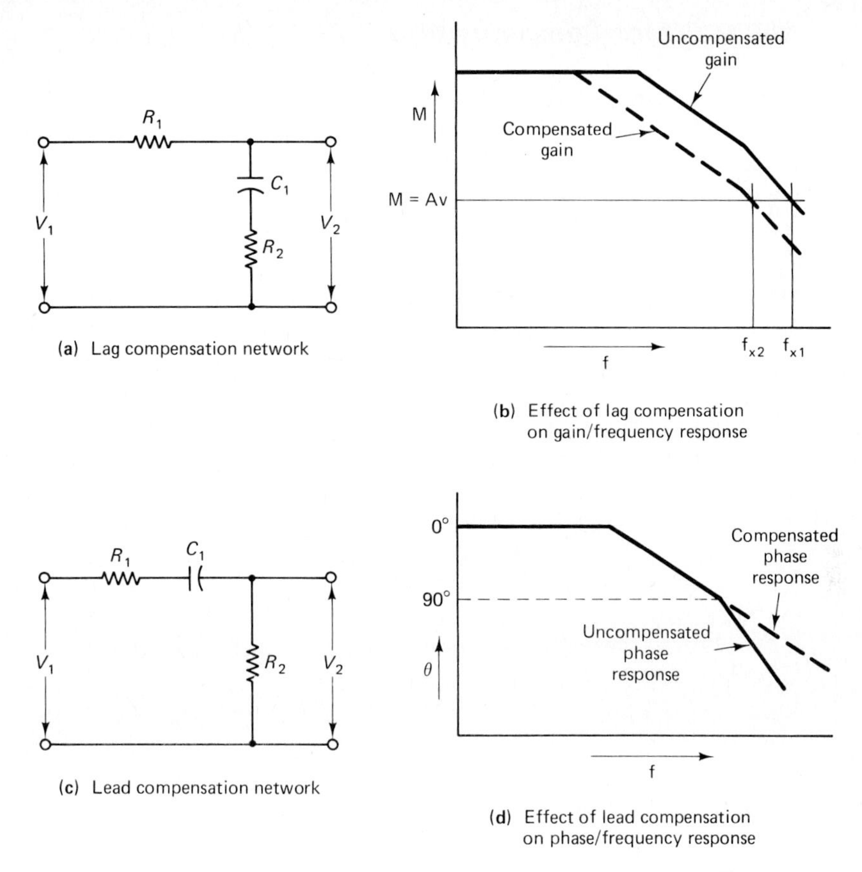

Figure 5-5 A phase-lag network stabilizes an amplifier by reducing the open-loop gain, so that the frequency at which $M\beta = 1$ occurs when the phase shift is too small to cause instability. A phase-lead network cancels some of the phase lag.

than f_{x1}, the op-amp phase shift at f_{x2} is less than that at f_{x1}, so the circuit is likely to be stable. Recall that no additional phase shift is introduced by the lag network at high frequencies.

As explained, despite its name, the lag network is not used to introduce a phase lag, (which would, increase the total phase lag in the loop). Instead, it is employed to attenuate the loop gain, so that $M\beta = 1$ occurs at a frequency at which the amplifier phase shift is too small to cause oscillations.

Phase-Lead Compensation

The CR network in Fig. 5-5(c) introduces a phase lead. When $X_{C1} \gg R_1$, v_2 leads v_1. Unlike the case of the lag network which is used only for attenuation, the phase lead network is actually employed to introduce a phase lead. The phase lead cancels some of the unwanted phase lag in the op-amp (Fig. 5-5(d)). Consequently, this increases the phase margin at $M\beta = 1$, and improves the circuit stability.

Miller Effect Compensation

Consider a common-emitter transistor amplifier stage with a capacitor connected between its collector and base terminals as illustrated in Fig. 5-6(a). Amplification and phase inversion occurs. When v_i goes positive, v_o moves in a negative direction by an amount $A_v v_i$.

$$v_o = -A_v v_i$$

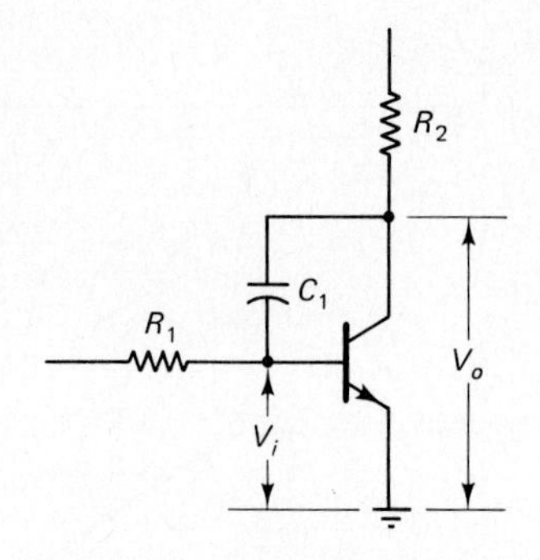

(a) Single-stage transistor amplifier with capacitor between input and output

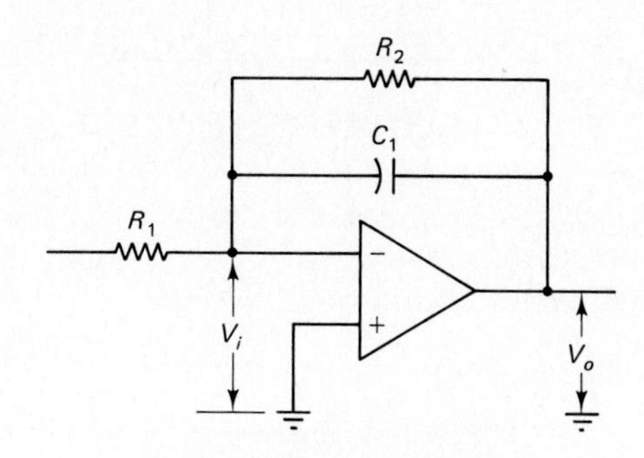

(b) Op-amp inverting amplifier with capacitor between input and output

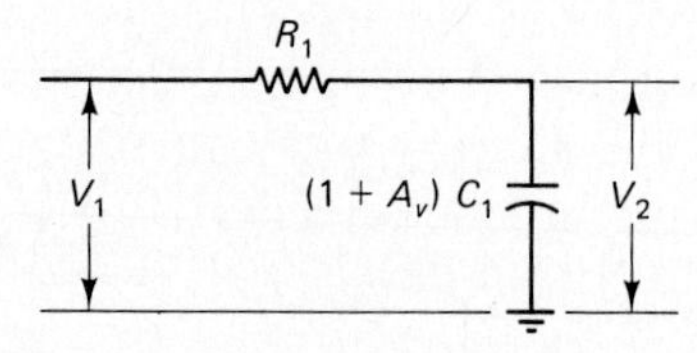

(c) R_1 and $(1 + A_v)\, C_1$ in (a) above constitutes a lag network for signal voltages

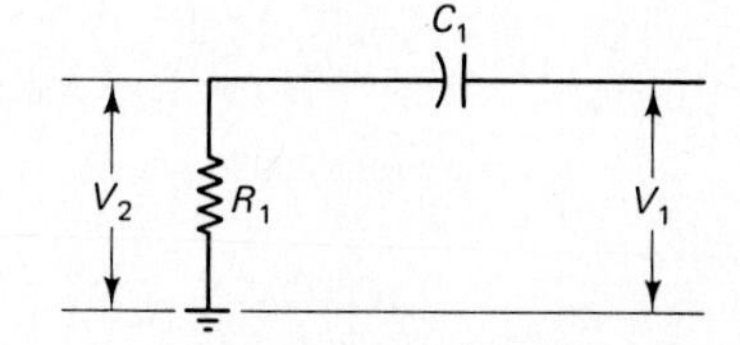

(d) R_1 and C_1 in (b) above constitutes a lead network for feedback voltages

Figure 5-6 A small capacitor connected between the input and output terminals of an inverting amplifier stage behaves as a much larger input capacitance. This can be used in phase lag or phase lead compensation for operational amplifiers.

When the base voltage goes up by v_i and the collector voltage goes down by $A_v v_i$, the voltage change across capacitor C_1 is,

$$\Delta v_{C1} = v_i + A_v v_i$$

$$= v_i(1 + A_v)$$

The charge supplied to the capacitor is,

$$Q = C_1 \times \Delta v_{C1}$$

$$= C_1 \times v_i(1 + A_v)$$

or

$$Q = C_1(1 + A_v)v_i \tag{5-2}$$

Therefore, when the transistor base voltage changes by v_i, capacitor C_1 is charged. The charge is *not* $Q = C_1 v_i$, as might be expected, but $Q = C_1(1 + A_v)v_i$. Thus, the capacitance appears to be amplified by a factor of $(1 + A_v)$. This is known as *Miller effect*.

The above reasoning can also be applied to the op-amp inverting amplifier circuit illustrated in Fig. 5-6(b) to show that C_1 is amplified by a factor of $(1 + A_v)$. In this case, A_v is the closed-loop gain of the amplifier.

A capacitor and resistor connected in the same way as C_1 and R_1 in Fig. 5-6(a) are used for frequency compensation inside the 741 op-amp. Fig. 5-6(c) shows that in this situation the circuit behaves as a lag network. Compare this to Fig. 5-5(a). The advantage here is that the capacitance is amplified by the Miller effect, so a very small value capacitor (30 pF for the 741) can be used.

In the case of the externally connected Miller-effect capacitor in Fig. 5-6(b), the combination of C_1 and R_1 behaves as a phase lead network within the feedback loop. This is illustrated in Fig. 5-6(d). The capacitance amplification is not a factor here because the direction of feedback is from the output to the input and there is no capacitance amplification in that direction. Thus, C_1 and R_1 in Fig. 5-6(d) introduce a phase lead to cancel some of the phase lag in the loop.

Feedforward Compensation

In *feedforward compensation* a capacitor is connected between the input and output terminals of the high-gain stage of the op-amp. The high-gain stage tends to introduce large phase shifts into high frequencies. Bypassing it with a capacitor reduces this effect.

5-4 MANUFACTURER'S RECOMMENDED COMPENSATION

Component Selection

Calculation of component values for an operational amplifier compensation network can be a tedious process. The frequency response graph is required for the particular op-amp used. To simplify the compensation process, IC manufacturers publish recommended compensation networks and component values on op-amp data sheets. The recommended component values for the amplifier closed-loop gain are read from the data sheet and the components are then connected to the op-amp as specified on the data sheet. Example 5-1 demonstrates the process.

Example 5-1

Using a 715 op-amp, design the inverting amplifier in Fig. 5-7(a) to have a maximum signal voltage of 0.5 V and a voltage gain of 10. Select suitable compensating components.

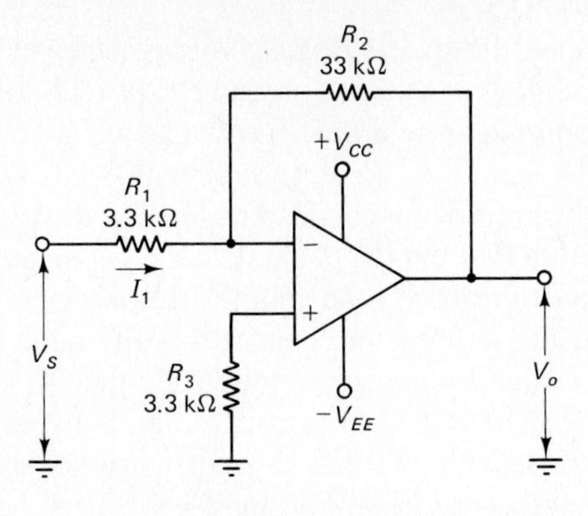

(a) Inverting amplifier

Noninverting Compensation Components Values

Closed loop gain	C_1	C_2	C_3
1000	10 pF	——	——
100	50 pF	——	250 pF
10	100 pF	500 pF	1000 pF
1	500 pF	2000 pF	1000 pF

Frequency Compensation Circuit

C_1 100 pF

1, 3, 9, 6, 715, 4, 10, 7

C_2 500 pF

C_3 1000 pF

(b) Compensating components list and connection diagram reproduced from 715 operational data sheet. (Reprinted with permission of National Semiconductor Corporation)

Figure 5-7 Op-amp circuits can be compensated by using the manufacturers recommended method and components. The component values are selected for each closed-loop gain.

Solution

From the 715 data sheet in Appendix 1-4,

$$I_{B(\max)} = 1.5\ \mu\text{A}$$

Let

$$I_1 = 100\ I_{B(\max)} = 100 \times 1.5\ \mu\text{A}$$

$$= 150\ \mu\text{A}$$

$$R_1 = \frac{V_s}{I_1} = \frac{0.5\ \text{V}}{150\ \mu\text{A}}$$

$$= 3.3\ \text{k}\Omega\ \text{(standard value)}$$

$$R_2 = A_v R_1 = 10 \times 3.3\ \text{k}\Omega$$

$$= 33 \text{ k}\Omega \text{ (standard value)}$$

$$R_3 = R_1 \| R_2 \approx R_1$$

$$= 3.3 \text{ k}\Omega \text{ (standard value)}$$

Refer to the portion of the 715 data sheet reproduced in Fig. 5-7(b). For $A_v = 10$: $C_1 = 100$ pF, $C_2 = 500$ pF, $C_3 = 1000$ pF. Connect C_1, C_2 and C_3 as illustrated in Fig. 5-7(b).

Inverting and Noninverting Amplifiers

Note that on the 715 data sheet portion in Fig. 5-7(b), the compensating components are listed as for noninverting amplifiers. These can be taken as applying equally to high-gain inverting amplifiers. If the circuit in Fig. 5-7(a) was a noninverting amplifier instead of an inverting amplifier, the voltage gain would be $(R_1 + R_2)/R_1$ instead of R_2/R_1. Clearly, there is little difference in these two quantities for voltage gains of 10 or greater. For an inverting amplifier with a voltage gain less than 10, A_v should normally be recalculated as $(R_1 + R_2)/R_1$ before selecting the compensating components.

Example 5-2

A 715 op-amp is to be used as a voltage follower. Select suitable compensating components.

Solution

$$A_v = 1$$

From Fig. 5-7(b):

$$C_1 = 500 \text{ pF}, \qquad C_2 = 2000 \text{ pF}, \qquad C_3 = 1000 \text{ pF (see Fig. 5-8)}$$

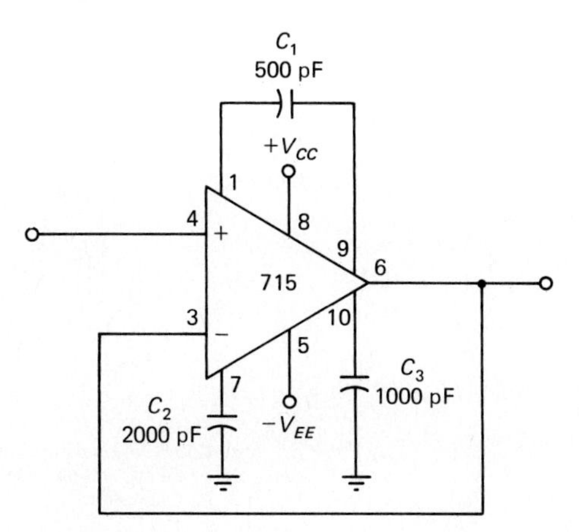

Figure 5-8 Frequency compensation components for a 715 op-amp employed as a voltage follower.

LM108 Op-Amp

The LM108 op-amp uses a Miller-effect lag network with internal resistors and an externally connected capacitor. This is shown in Fig. 5-9(a), which is reproduced from the manufacturer's data sheet. The compensating capacitor (C_f), connected between terminals 1 and 8, introduces a lag network which gives the circuit good stability and a frequency response roughly similar to that of the 741. The capacitor value is calculated as

$$C_f = \frac{30\ \text{pF} \times R_1}{R_1 + R_2} \tag{5-3}$$

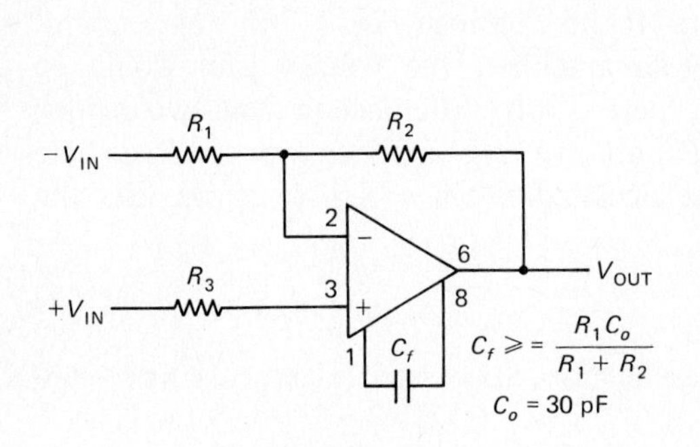

(a) Phase lag compensation

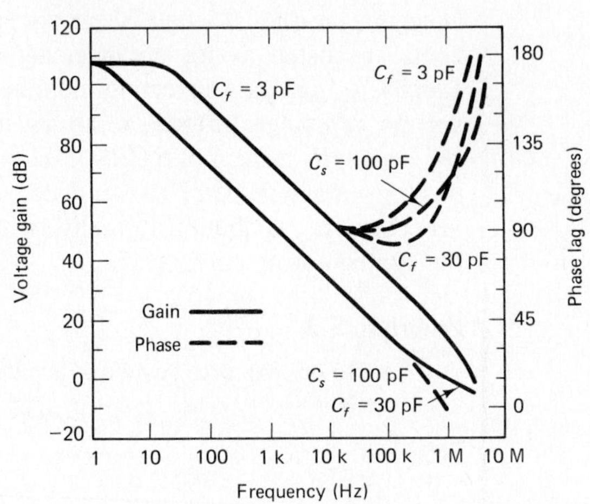

(b) Gain/frequency and phase/frequency responses

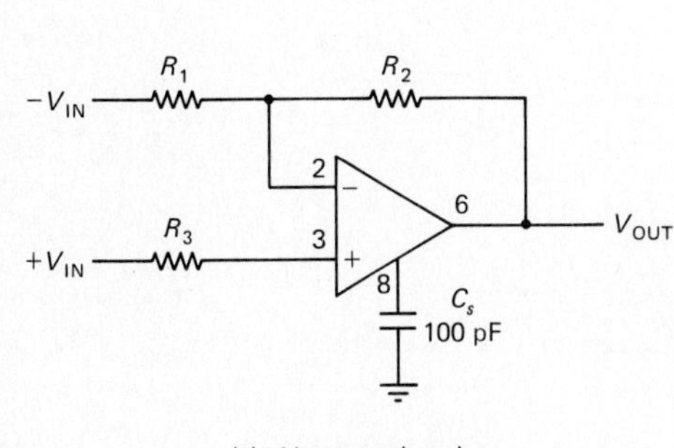

(c) Alternate phase lag compensation

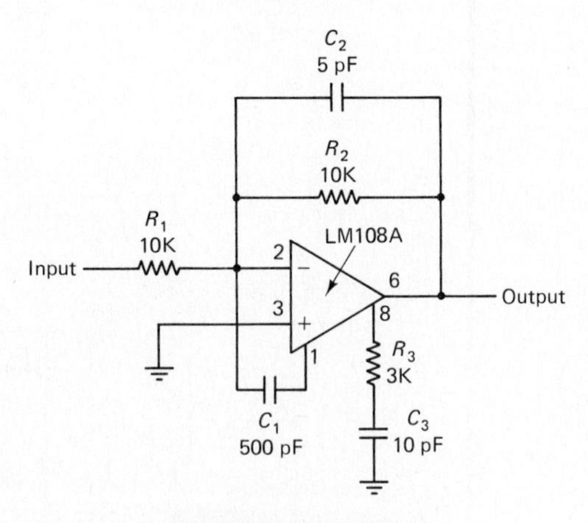

(d) Combination of feedforward, phase lag, and phase lead compensation

Figure 5-9 Manufacturer's recommended compensation method for an LM 108 op-amp. (Reprinted with permission of National Semiconductor Corp.)

So, the maximum capacitance of C_f is 30 pF. This can be reduced to a minimum of 3 pF as the closed-loop gain of the amplifier is increased. The gain/frequency graph in Fig. 5-9(b) shows that the circuit frequency response is improved by the use of a small capacitance value for C_f.

Two other compensating circuits for the LM108 are illustrated in Fig. 5-9. Figure 5-9(c) shows an alternate type of lag compensation using a larger (non-Miller-effect) capacitor. Figure 5-9(d) has a combination of feedforward compensation (C_1), Miller-effect phase lead compensation (C_2), and phase lag compensation (C_3).

Example 5-3

Using an LM108 op-amp, design an inverting amplifier as in Fig. 5-9(a) to amplify a 100 mV signal by a factor of 3. Select suitable frequency compensation.

Solution

From the LM108 data sheet in Appendix 1-2,

$$I_{B(max)} = 2 \text{ nA}$$

Let

$$I_1 = 100\, I_{B(max)} = 100 \times 2 \text{ nA}$$

$$= 200 \text{ nA}$$

$$R_1 = \frac{v_1}{I_1} = \frac{100 \text{ mV}}{200 \text{ nA}}$$

$$= 500 \text{ k}\Omega \text{ (use 470 k}\Omega \text{ standard value)}$$

$$R_2 = A_v R_1 = 3 \times 470 \text{ k}\Omega$$

$$= 1.41 \text{ M}\Omega \text{ (use 1.5 M}\Omega \text{ standard value)}$$

$$R_3 = R_1 \| R_2 \approx R_1$$

$$= 470 \text{ k}\Omega \text{ (standard value)}$$

Eq. 5-3,

$$C_f = \frac{30 \text{ pF} \times R_1}{R_1 + R_2} = \frac{30 \text{ pF} \times 470 \text{ k}\Omega}{470 \text{ k}\Omega + 1.5 \text{ M}\Omega}$$

$$= 7.2 \text{ pF (use 10 pF standard value)}$$

Connect C_f between terminals 1 and 8.

Over-Compensation

When the voltage gain of an amplifier falls between gain values for which compensating components are specified, components should be selected for the next *lower* gain. This provision of larger components than actually required, is termed *over-compensation*. Over-compensation results in better amplifier stability. It also produces an upper cutoff frequency which is not as high as could be obtained with smaller compensating components. When compensating components are selected for an inverting amplifier from those recommended for a noninverting amplifier, the circuit is actually being over-compensated because the inverting gain is always lower than the noninverting gain.

Example 5-4

Using a 709 op-amp, design an inverting amplifier as in Fig. 5-7(a) to have $A_v = 100$, and $V_s = 50$ mV.

Solution

From the 709 data sheet in Appendix 1-5

$$I_{B(\max)} = 200 \text{ nA}$$

Let

$$I_1 = 100 \times 200 \text{ nA}$$

$$= 20\ \mu\text{A}$$

$$R_1 = \frac{V_s}{I_1} = \frac{50 \text{ mV}}{20\ \mu\text{A}}$$

$$= 2.5 \text{ k}\Omega \text{ (use 2.2 k}\Omega \text{ standard value)}$$

$$R_2 = A_v R_1 = 100 \times 2.2 \text{ k}\Omega$$

$$= 220 \text{ k}\Omega \text{ (standard value)}$$

$$R_3 = R_1 \| R_2 \approx R_1$$

$$= 2.2 \text{ k}\Omega \text{ (standard value)}$$

$$A_v = 100 = 40 \text{ dB}$$

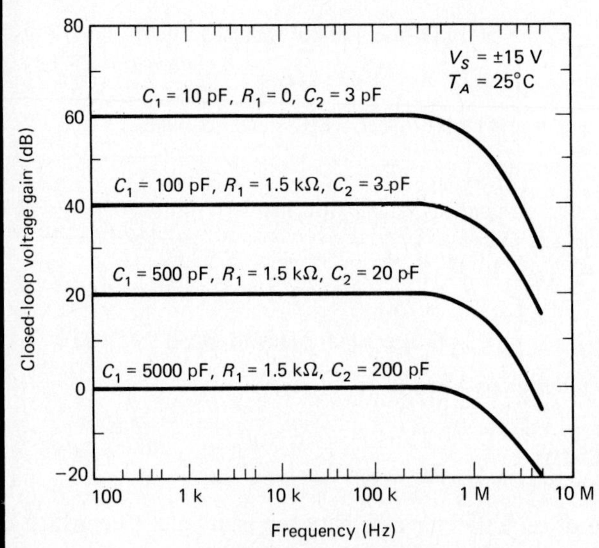

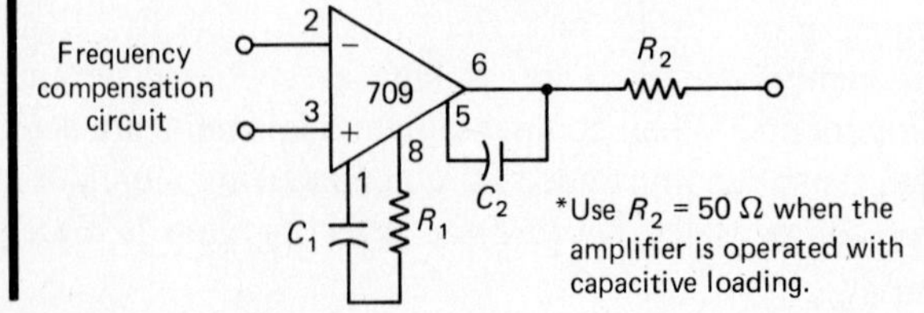

Figure 5-10 Manufacturer's recommended components and compensating method for the 709 operational amplifier. (Reprinted with permission of National Semiconductor Corp.)

Refer to the portion of the 709 data sheet reproduced in Fig. 5-10.

For $A_v = 40$ dB: $C_1 = 100$ pF, $R_1' = 1.5$ kΩ, $C_2 = 3$ pF

Connect C_1, R_1' and C_2 as illustrated in Fig. 5-10.

The complete circuit is shown in Fig. 5-11.

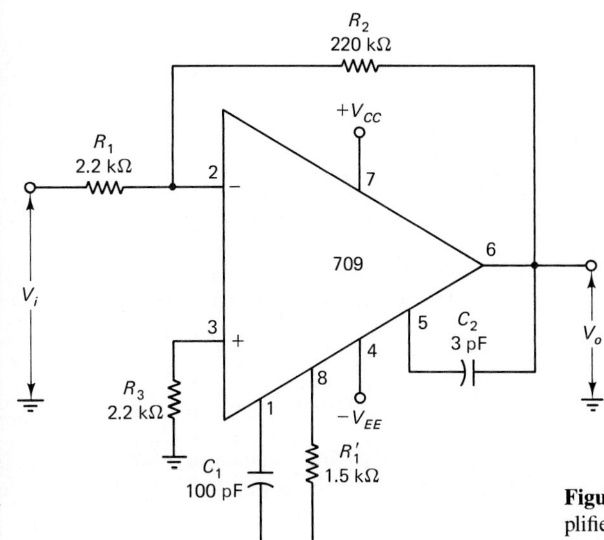

Figure 5-11 Circuit of inverting amplifier with frequency compensation, as determined in Example 5-4.

Example 5-5

A noninverting amplifier using a 709 op-amp is designed to have a voltage gain of 50. Select suitable compensating components.

Solution

$$A_v = 50 \approx 34 \text{ dB}$$

Compensation components are listed in Fig. 5-10 for $A_v = 20$ dB and for $A_v = 40$ dB. There is no listing for $A_v = 34$ dB.

For over-compensation, use the components for $A_v = 20$ dB:
$C_1 = 500$ pF, $R_1 = 1.5$ kΩ, $C_2 = 20$ pF

5-5 OP-AMP CIRCUIT BANDWIDTH

Cutoff Frequencies

Operational amplifiers are directly coupled internally. When the signal and load are also direct-coupled, the circuit lower cutoff frequency is zero. In capacitor-coupled circuits, the lower cutoff frequency is determined by the selection of capacitors (see

Chapter 4). The circuit upper cutoff frequency is dependent upon the frequency response of the op-amp, the compensating components, and the circuit voltage gain.

Upper Cutoff Frequency Determination

For any operational amplifier circuit, the frequency at which $M\beta = 1$ may be found simply by drawing a horizontal line at A_v on the op-amp frequency response graph, as explained in Section 5-2. This is the frequency at which the circuit might oscillate. As will be explained, it is also the upper cutoff frequency for the amplifier.

The upper cutoff frequency (f_2) is the frequency at which the closed-loop voltage gain falls to 3 dB below the normal mid-frequency gain. The voltage gain of an amplifier with negative feedback is

$$A_v = \frac{M}{1 + M\beta} \tag{5-4}$$

Refer once again to Fig. 5-4(b) and recall that the open-loop phase shift of the op-amp progressively increases with increase in signal frequency. So, Eq. 5-4 is more correctly expressed as

$$A_v \approx \frac{M\angle\theta}{1 + (M\angle\theta\,\beta)}$$

Now look at the open-loop phase/frequency and gain/frequency responses for the 741 op-amp in Fig. 5-12, (reproduced from Appendix 1-1). It is seen that because of the internal compensation, the phase shift for the 741 remains at $-90°$ for much of the frequency range during which the gain is falling by 20 dB per decade.

Substituting $-90°$ for θ in the equation for closed-loop gain gives

$$A_v \approx \frac{M\angle-90°}{1 + (M\angle-90°\,\beta)} = \frac{-jM}{1 - jM\beta}$$

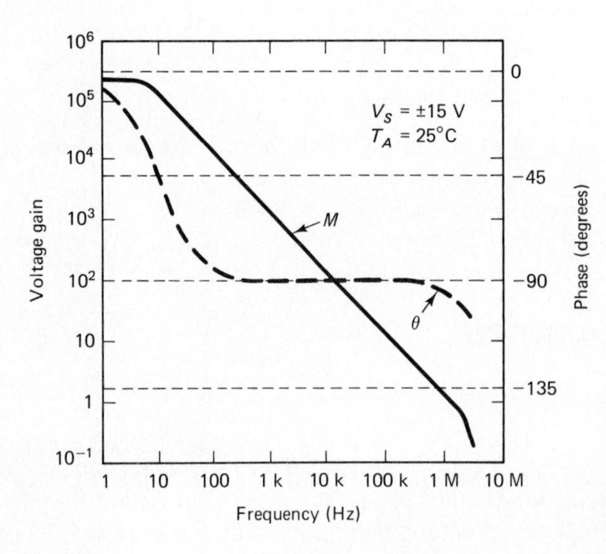

Figure 5-12 Open-loop gain/frequency and phase/frequency responses for a 741 operational amplifier. The phase shift remains approximately $-90°$ for much of the frequency range.

or, $$|A_v| \approx \frac{M}{\sqrt{1^2 + (M\beta)^2}}$$

When $M \approx A_v \approx 1/\beta$, the closed-loop gain becomes

$$|A_{v2}| \approx \frac{A_v}{\sqrt{2}} = A_v - 3 \text{ dB}$$

Therefore, f_2 occurs at $$M \approx A_v \tag{5-5}$$

If an amplifier has no external compensating components, f_2 is found by drawing a horizontal line at A_v on the open-loop gain/frequency response graph [see Fig. 5-13(a)]. When the amplifier is externally compensated, the compensated open-loop gain/frequency response must be employed [Fig. 5-13(b)].

The frequency response graphs published on manufacturers data sheets are typical for each particular type of operational amplifier. Like all typical devices characteristics and parameters, the precise frequency response differs from one op-amp to another. All frequencies derived from the response graphs should be taken as typical quantities.

Determining the upper cutoff frequency as the frequency at which M equals A_v implies that the open-loop phase shift is always 90°. This is clearly not true as illustrated by Figs. 5-4 and 5-12. However, as an approximation, the relationship is assumed to be correct at all points on the gain/frequency response graph.

Example 5-6

Determine the typical upper cutoff frequency for a noninverting amplifier with $A_v = 100$, (a) when a 741 op-amp is used and (b) when a 709 is employed.

Solution

$$A_v = 100 = 40 \text{ dB}$$

f_2 occurs at $$M = 40 \text{ dB}$$

(a) For the 741, refer to Fig. 5-13(a). Draw a horizontal line at $M \approx 40$ dB. From the intersection of the line and the open-loop frequency response,

$$f_2 \approx 8 \text{ kHz}$$

(b) For the 709, refer to Fig. 5-13(b). Draw a horizontal line at $M \approx 40$ dB. Now recall that a 709 op-amp was used in an inverting amplifier with $A_v = 100$ in Example 5-4. Also, that the compensating components were $C_1 = 100$ pF, $R_1 = 1.5$ kΩ and $C_2 = 3$ pF. The intersection of the $M \approx 40$ dB line and the frequency response curve using the above components occurs at

$$f_2 \approx 600 \text{ kHz}$$

Example 5-6 demonstrates the reason for using a compensated 709 operational amplifier instead of a 741 in some situations. The 709 clearly gives a much greater bandwidth than the 741. However, when the bandwidth of the 741 is suitable, the 741 should be used to avoid the additional compensating components.

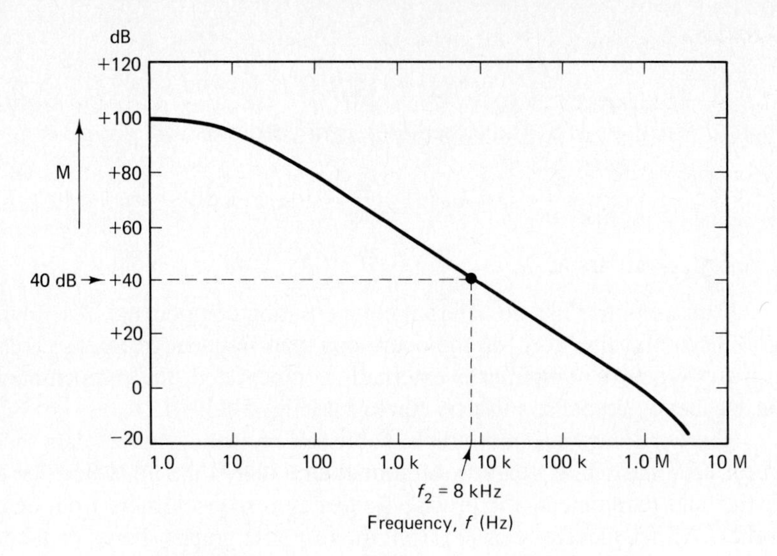

(a) Gain/frequency response for 741 operational amplifier

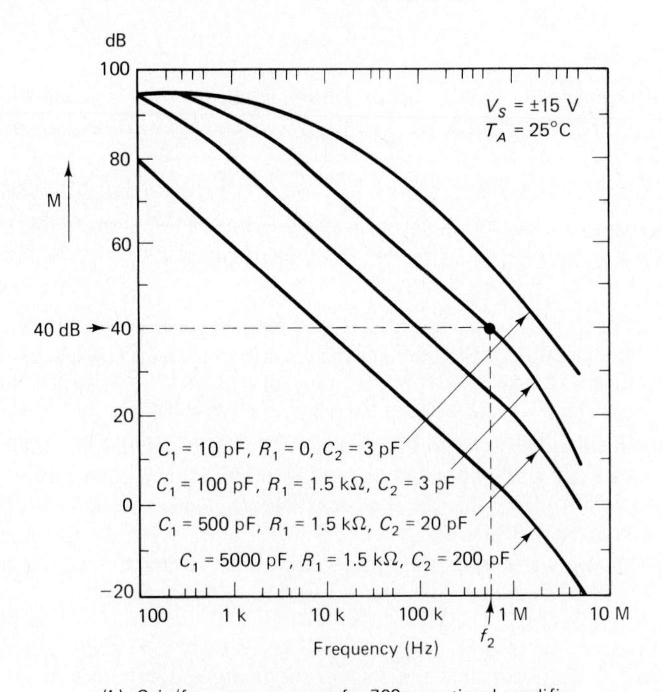

(b) Gain/frequency response for 709 operational amplifier using recommended compensation

Figure 5-13 Gain/frequency responses for 741 and 709 operational amplifiers. The upper cutoff frequency for any op-amp circuit is found by drawing a horizontal line at $M \approx A_v$.

Gain-Bandwidth Product

The *gain-bandwidth product* or *unity-gain bandwidth* of an op-amp is the closed-loop gain A_v multiplied by the cutoff frequency for that gain.

Refer to Fig. 5-13(a) again and note that an amplifier with A_v = 80 dB (or 10 000) has $f_2 \approx 80$ Hz. In this case the gain-bandwidth product is

$$A_v f_2 = 10\,000 \times 80$$

$$= 800\,000$$

Similarly, for A_v = 60 dB (or 1000), $f_2 \approx 800$ Hz; with A_v = 40 dB (or 100), $f_2 \approx 8$ kHz. Both of these give

$$A_v f_2 = 800\,000$$

Finally, for $A_v = 1$ (unity), the *unity-gain frequency* is $f_u \approx 800$ kHz,

or

$$A_v f_u = 1 \times 800\,000$$

$$= 800\,000$$

So, the gain-bandwidth product equals the unity-gain frequency.

$$A_v f_2 = f_u$$

or

$$f_2 = \frac{f_u}{A_v} \tag{5-6}$$

Thus, when the unity-gain frequency is known, the cutoff frequency of an amplifier can be determined simply by dividing the closed-loop gain into the unity-gain frequency.

It is important to note that Eq. 5-6 applies only to operational amplifiers which have a gain that falls off at 20 dB per decade to the unity-gain frequency. In other situations, f_2 must be determined by drawing a horizontal line on the frequency response graph at A_v as already discussed.

Consider the 709 op-amp frequency response illustrated in Fig. 5-13(b). The gain is seen to fall off at 20 dB per decade to unity only when maximum frequency compensation is employed. For a 709 using less than maximum compensation, the gain falls off at a rate greater than 20 dB per decade as it approaches unity. Also, for two of the 709 response curves, the unity-gain frequency cannot be determined. In these situations, the gain-bandwidth product cannot be applied.

Example 5-7

Using the gain-bandwidth product, estimate the upper cutoff frequencies for the amplifiers in Example 5-6.

Solution

(a) 741 op-amp,

Eq. 5-6,
$$f_2 \approx \frac{f_u}{A_v} \approx \frac{800 \text{ kHz}}{100}$$

$$\approx 8 \text{ kHz}$$

(b) 709 op-amp, with $C_1 = 100$ pF, $R_1 = 1.5$ kΩ, and $C_2 = 3$ pF.
Equation 5-6 cannot be used to calculate the cutoff frequency for this circuit, because, with the above compensating components, the voltage gain does not fall off at 20 db per decade throughout the frequency response. See Fig. 5-13(b).

Bandwidth of Inverting and Noninverting Amplifiers

The gain/frequency response graph given on manufacturer's data sheets are for noninverting amplifiers. However, as already discussed in Section 5-4, the graphs can be assumed to apply equally to high-gain inverting amplifiers. This is because, with a high-gain circuit, the inverting amplifier gain (R_2/R_1) does not differ significantly from that of a noninverting amplifier $[(R_1 + R_2)/R_1]$; see Figs. 5-1(a) and 5-2(a).

With low-gain inverting amplifiers, the noninverting gain must be calculated in order to determine the upper cutoff frequency.

Example 5-8

Determine the upper cutoff frequency for (a) a voltage follower circuit using a 741 op-amp and (b) a unity-gain inverting amplifier using a 741 op-amp.

Solution

(a) For the 741,

$$f_u \approx 800 \text{ kHz}$$

Eq. 5-6

$$f_2 \approx \frac{f_u}{A_v} \approx 800 \text{ kHz}$$

(b) For unity gain,

$$R_1 = R_2$$

Converting to the noninverting gain,

$$A_v = \frac{R_1 + R_2}{R_1}$$

$$= 2$$

Eq. 5-6,

$$f_2 \approx \frac{f_u}{A_v} \approx \frac{800 \text{ kHz}}{2}$$

$$\approx 400 \text{ kHz}$$

5-6 SLEW RATE EFFECTS

Slew Rate Effect on Bandwidth and Output Amplitude

The foregoing discussion of upper cutoff frequency refers only to small signal amplifiers. Manufacturer's specifications on this topic are normally stated for an amplifier with an output voltage less than ±1 V. When large output voltages are involved, the circuit upper cutoff frequency is likely to be substantially lower than for small signal circuits. This is due to the op-amp slew rate (see Section 2-5) limiting the cutoff frequency. For op-amps with large output voltage swings, the manufacturer sometimes specifies a *full-power bandwidth*.

Figure 5-14(a) shows a sinusoidal output voltage waveform from an op-amp and illustrates the fact that the fastest rate of change of the waveform occurs as the voltage crosses zero. At this point,

$$\text{rate of change} = 2\pi f V_p \text{ (volts/second)}$$

where f is the frequency, and V_p is the peak amplitude.

The kind of waveform distortion that occurs when the slew rate is too slow for the output amplitude and frequency is illustrated in Fig. 2-9. Figure 5-14(b) shows that the slew rate limits the circuit maximum operating frequency for a given distortion-free output amplitude. Alternatively, the slew rate can be said to limit the distortion-free output amplitude at the circuit cutoff frequency. For a distortion-free

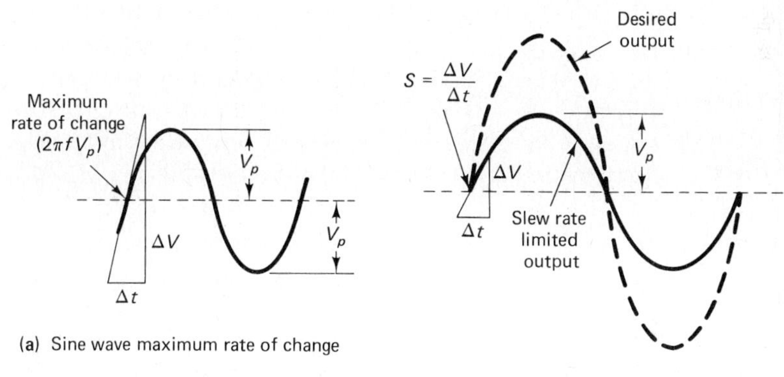

(a) Sine wave maximum rate of change

(b) Sine wave output amplitude can be limited by the op-amp slew rate

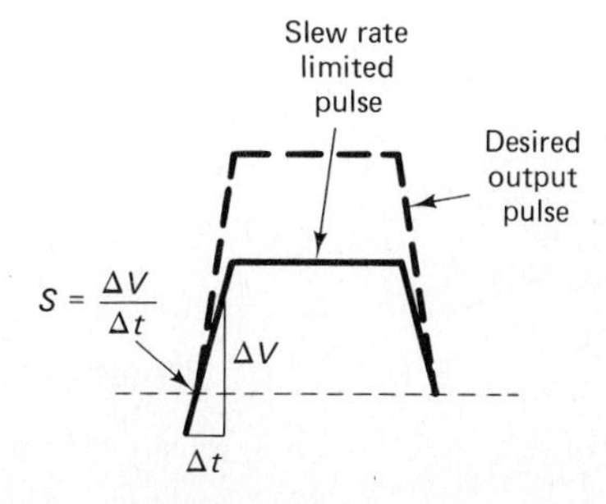

(c) A pulse output waveform can be limited by the op-amp slew rate

Figure 5-14 The op-amp slew rate can distort output waveforms. Thus, the slew rate limits the maximum distortion-free output amplitude with a given circuit cutoff frequency or a given output rise time.

output, the op-amp slew rate (S) must be equal to, or greater than, the maximum rate of change of the waveform. Therefore

$$S = 2\pi f V_p$$

or, the *slew rate-limited frequency* for a given output amplitude is,

$$f_s = \frac{S}{2\pi V_p} \tag{5-7}$$

Equation 5-7 can be rewritten to give a *slew rate-limited output amplitude* for a given cutoff frequency (f_2),

$$V_p = \frac{S}{2\pi f_2}$$

Example 5-9

(a) Calculate the slew rate-limited cutoff frequency for a voltage follower circuit using a 741 op-amp if the peak of sine wave output is to be 5 V.

(b) Determine the maximum peak value of the sinusoidal output voltage that will allow the 741 voltage follower circuit to operate at the 800 kHz unity-gain cutoff frequency.

(c) Calculate the maximum peak value of sine wave output voltage that can be produced by the amplifier in part (a) of Example 5-6.

Solution

(a) The typical slew rate for the 741 op-amp is (from Appendix 1-1),

$$S = 0.5\ \text{V}/\mu\text{s}$$

Eq. 5-7,

$$f_s = \frac{S}{2\pi V_p} = \frac{0.5\ \text{V}/1\ \mu\text{s}}{2\pi \times 5\ \text{V}}$$

$$= 15.9\ \text{kHz}$$

(b) The slew rate-limited frequency should equal the cutoff frequency.

Therefore $\quad f_s$ (due to slew rate) = unity-gain frequency f_2

$$= 800\ \text{kHz}$$

From Eq. 5-7,

$$V_p = \frac{S}{2\pi f_2} = \frac{0.5\ \text{V}/1\ \mu\text{s}}{2\pi \times 800\ \text{kHz}}$$

$$= 99\ \text{mV}$$

(c) From Example 5-6, the op-amp is a 741 and f_2 is 8 kHz.

From Eq. 5-7,

$$V_p = \frac{S}{2\pi f_2} = \frac{0.5\ \text{V}/1\ \mu\text{s}}{2\pi \times 8\ \text{kHz}}$$

$$= 9.9\ \text{V}$$

Slew Rate Effect on Output Pulse Rise Time and Amplitude

When a pulse-type signal is to be amplified, the output rise time is related to the circuit cutoff frequency. The equation for *cutoff frequency limited rise time* is

$$t_{r(f2)} = \frac{0.35}{f_2} \qquad (5\text{-}8)^*$$

However, as in the case of sinusoidal signals, a pulse waveform can be distorted by the op-amp slew rate [see Fig. 5-14(c)]. The output rise time and amplitude are directly related to the slew rate. The equation for *slew rate limited rise time* is

$$t_{r(S)} = \frac{V_p}{S} \qquad (5\text{-}9)$$

Similar to the case of the amplifier with a sinusoidal signal, the slew rate can be said to limit the minimum output rise time for a given output amplitude. Alternatively, it can be stated that the slew rate limits the output amplitude for a given rise time. Slew rate effects are further considered in Section 7-5.

Example 5-10

(a) Calculate the cutoff frequency-limited rise time for a voltage follower circuit using a 741 op-amp. Also determine the slew rate-limited rise time if the output amplitude is to be 5 V.

(b) Determine the maximum undistorted pulse output amplitude for the 741 voltage follower if the output rise time is not to exceed 1 μs.

(c) Calculate the minimum output rise time and the maximum pulse amplitude at that rise time for a 741 amplifier with an upper cutoff frequency of 100 kHz.

Solution

(a) Eq. 5-8,

$$t_{r(f2)} = \frac{0.35}{f_2} = \frac{0.35}{800\text{ kHz}}$$

$$\approx 0.4\ \mu\text{s}$$

Eq. 5-9,

$$t_{r(S)} = \frac{V_p}{S} = \frac{5\text{ V}}{0.5\text{ V}/\mu\text{s}}$$

$$= 10\ \mu\text{s}$$

(b) From Eq. 5-9,

$$V_p = t_{r(S)}S = 1\ \mu\text{s} \times 0.5\text{ V}/\mu\text{s}$$

$$= 0.5\text{ V}$$

* David A. Bell, *Solid State Pulse Circuits,* 3rd ed. (Englewood Cliffs, NJ: Prentice-Hall, Inc., 1988), p. 39.

(c) Eq. 5-8, $$t_{r(f2)} = \frac{0.35}{f_2} = \frac{0.35}{100 \text{ kHz}}$$
$$= 3.5\ \mu\text{s}$$
From Eq. 5-9, $$V_p = t_{r(S)}S = 3.5\ \mu\text{s} \times 0.5\ \text{V}/\mu\text{s}$$
$$= 1.75\ \text{V}$$

5-7 STRAY CAPACITANCE EFFECTS

Circuit Instability Due to Stray Capacitance

Stray capacitance can occur anywhere in a circuit. However, the worst place for stray capacitance in an op-amp circuit is at the inverting input terminal. An inverting amplifier with stray capacitance C_s between the inverting and noninverting input terminals is illustrated in Fig. 5-15(a). Stray capacitance can also exist from each input terminal to ground. For simplicity, only the type illustrated in Fig. 5-15(a) will be

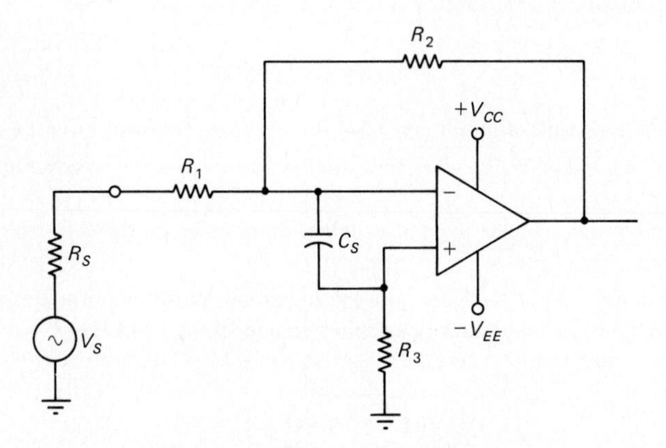

(a) Stray capacitance at the op-amp input terminals can cause circuit instability

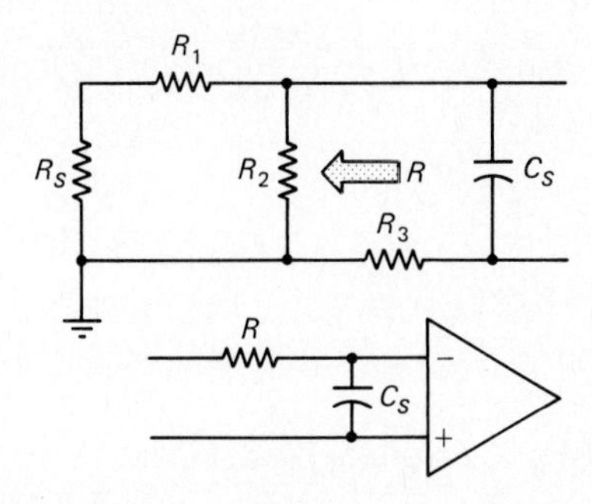

(b) The stray capacitance C_S combines with the network resistance R to constitute a lag network

Figure 5-15 Stray capacitance (C_S) at the input terminals of an op-amp combines with the resistive components to constitute a phase lag network which can produce circuit instability.

considered. However, the compensating methods described below are appropriate for all types of input stray capacitance.

Figure 5-15(b) shows that the resistance "seen" when "looking back" from the capacitance into the resistor network is

$$R = R_3 + R_2 \| (R_1 + R_s)$$

When R_3 is not present and R_s is very much smaller than R_1,

$$R \approx R_1 \| R_2$$

The stray capacitance C_s and the network resistance R constitute a phase lag network. This introduces additional phase lag in the feedback loop which might cause circuit oscillations.

The question arises: How much stray capacitance can be tolerated before the circuit becomes unstable? It is possible that the integrated circuit manufacturer could be working with as little as 30° of phase margin with the recommended compensating components. Therefore, the circuit may become unstable if the stray capacitance introduces a very small additional phase lag. An additional lag of 6° is usually taken as that which might make the circuit oscillate.

Analysis of a CR-phase shift network [Fig. 5-16(a)] shows that when $X_c = R$, the voltage across C lags the applied voltage by 45° [Fig. 5-16(b)]. Also, when $X_c = 10\,R$, the phase lag is approximately 6° [Figure 5-16(c)]. So, in the case of stray capacitance at the op-amp input, the circuit might become unstable when

$$X_{cs} = 10\text{ R} \tag{5-10}$$

or, the stray capacitance should be very much less than,

$$C_s = \frac{1}{2\pi f(10\text{ R})} \tag{5-11}$$

where f is the frequency at which $M\beta = 1$, (the frequency at which the circuit is likely to oscillate).

Note from Eq. 5-11 that when very large value resistors are used in the op-amp circuit, R is a large quantity. Consequently, the calculated value of C_s is quite small. When small value resistors are used, C_s comes out as a larger value capacitance. This means where the largest possible resistor values are used with an op-amp circuit, a very small quantity of input stray capacitance might make the circuit oscillate. When small resistor values are employed, relatively large amounts of stray ca-

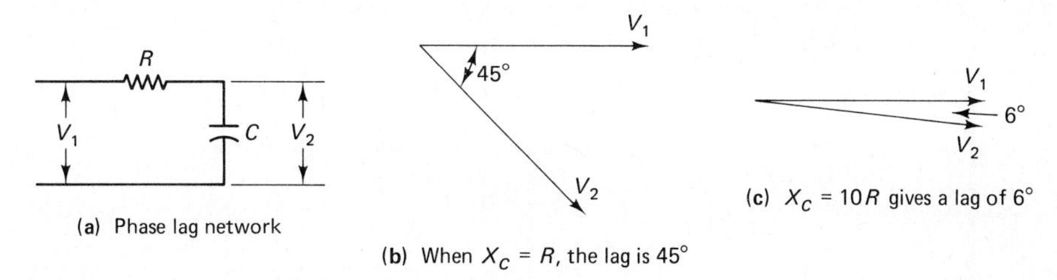

Figure 5-16 A phase lag network introduces a lag of 45° when $X_C = R$ and a lag of approximately 6° when $X_C = 10$ R.

pacitance must occur before the circuit becomes unstable. Therefore, *for greatest circuit stability the smallest possible resistor values should be selected.* This contradicts the usual circuit design approach, which is to select the largest possible resistor values to minimize current demand on the supply and to maximize circuit input resistance.

Circuit stability can be further improved if resistor R_3 in Fig. 5-15(a) is eliminated to reduce the total calculated value of R. Where this is not possible, R_3 might be bypassed with a capacitor which has an impedance much smaller than R_3 at the frequency at which $M\beta = 1$.

Note from Fig. 5-15 that, when the signal source is open-circuited, the network resistance becomes

$$R = R_2 + R_3$$

Thus, R is a large resistance. The stray capacitance C_s, which might make the circuit unstable (as calculated from Eq. 5-11), becomes an extremely small quantity. Therefore, *an op-amp circuit is very likely to oscillate when the signal source is open-circuited.*

Example 5-11

Determine the input stray capacitance that might make the circuit of Example 5-4 unstable. Assume that R_s is 220 Ω.

Solution

$$R = R_3 + R_2 \| (R_1 + R_s)$$

$$= 2.2 \text{ k}\Omega + 220 \text{ k}\Omega \| (2.2 \text{ k}\Omega + 220 \ \Omega)$$

$$\approx 4.6 \text{ k}\Omega$$

Refer to the 709 op-amp frequency response graph illustrated in Fig. 5-13(b). The frequency at which $M\beta = 1$ for an amplifier which has a gain of 40 dB and which uses the compensating components determined in Example 5-4 ($C_1 = 100$ pF, $R_1 = 1.5$ kΩ, $C_2 = 3$ pF) is approximately 600 kHz.

Eq. 5-11, $$C_s = \frac{1}{2\pi f(10 \text{ R})} = \frac{1}{2\pi \times 600 \text{ kHz} \times 10 \times 4.6 \text{ k}\Omega}$$

$$\approx 5.8 \text{ pF}$$

Example 5-12

For the circuit of Example 5-4, calculate the input stray capacitance that might make the circuit unstable (a) when the signal source is open-circuited, (b) when R_3 is short-circuited, and (c) when R_3 is present and all resistors are reduced by a factor of 10.

Solution

(a) $$R = R_2 + R_3 = 220 \text{ k}\Omega + 2.2 \text{ k}\Omega$$

$$= 222.2 \text{ k}\Omega$$

Eq. 5-11, $$C_s = \frac{1}{2\pi f(10 \text{ R})} = \frac{1}{2\pi \times 600 \text{ kHz} \times 10 \times 222.2 \text{ k}\Omega}$$

$$\approx 0.12 \text{ pF}$$

(b) $$R = R_2 \| (R_1 + R_s) = 220\ \text{k}\Omega \| (2.2\ \text{k}\Omega + 220\ \Omega)$$
$$\approx 2.4\ \text{k}\Omega$$

$$C_s = \frac{1}{2\pi f(10\ \text{R})} = \frac{1}{2\pi \times 600\ \text{kHz} \times 10 \times 2.4\ \text{k}\Omega}$$
$$= 11\ \text{pF}$$

(c) $$R = R_3 + R_2 \| (R_1 + R_s)$$
$$= 220\ \Omega + 22\ \text{k}\Omega \| (220\ \Omega + 22\ \Omega)$$
$$= 459\ \Omega$$

$$C_s = \frac{1}{2\pi f(10\ \text{R})} = \frac{1}{2\pi \times 600\ \text{kHz} \times 10 \times 459\ \Omega}$$
$$\approx 58\ \text{pF}$$

Part (a) of Example 5-12 demonstrates that, as already discussed in this section, op-amp circuits are likely to oscillate when the signal source is open-circuited because an extremely small input stray capacitance will render the circuit unstable. Parts (b) and (c) of Example 5-12, together with Example 5-11, show that the circuit stability can be improved by reducing the resistance values and thus increasing the input stray capacitance required for instability. An oscilloscope, with its typical 40 pF input capacitance (or any other electronics instrument) should never be connected directly to the input terminals of an op-amp. Such instrument capacitance is almost certain to cause circuit instability.

Compensating for Stray Capacitance

As already explained, stray capacitance effects can be minimized by using the lowest possible resistor values. The stray capacitance can also be kept low by keeping connecting leads short at the op-amp input terminals. The bodies of resistors connected to these terminals should be positioned close to the terminals so that the connecting leads can be kept as short as possible.

Compensation for stray capacitance at an op-amp input can be provided by means of a capacitor connected across the feedback resistor. C_2 in Fig. 5-17 is such a capacitor. To eliminate the phase shift introduced by stray capacitance C_s, the division of the output voltage produced by C_s and C_2 in series must be equal to that caused by resistors R_1 and R_2. Thus,

$$\frac{X_{CS}}{X_{C2}} = \frac{R_1}{R_2}$$

or

$$\frac{2\pi f C_2}{2\pi f C_S} = \frac{R_1}{R_2}$$

giving

$$C_2 = \frac{R_1}{R_2} C_S \tag{5-12}$$

Note the above equation does not allow for resistor R_3 and signal source resistance

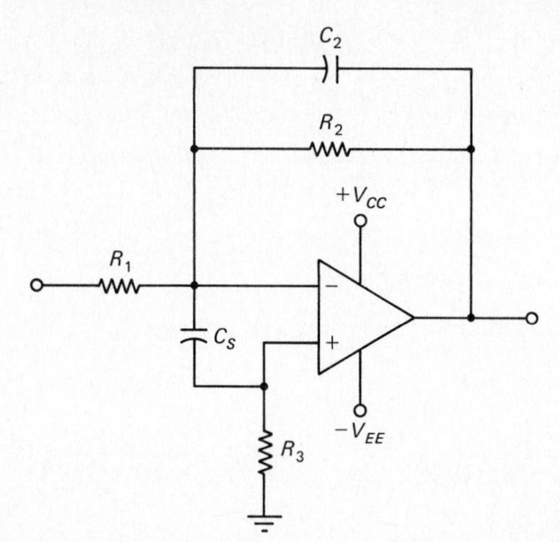

Figure 5-17 A feedback capacitor (C_2) can be employed to compensate for the effects of stray capacitance.

R_S. If R_3 must be in a circuit that requires this type of compensation, it should be bypassed as already explained. Where R_S is not much smaller than R_1, it should be added to R_1 in Eq. 5-12.

Another way of writing Eq. 5-12 is,

$$C_2 R_2 = C_S R_1$$

This can be stated as: *The time constant $C_2 R_2$ should be equal to, or greater than, the time constant $C_S R_1$.*

Example 5-13

Determine the feedback capacitor value required to compensate for the stray capacitance calculated in part (c) of Example 5-12.

Solution

Eq. 5-12,
$$C_2 = \frac{R_1}{R_2} C_S = \frac{220\ \Omega}{22\ \text{k}\Omega} \times 58\ \text{pF}$$
$$= 0.58\ \text{pF}$$

The calculated value of C_2 in Example 5-13 is so small that something larger might exist as the terminal capacitance of resistor R_2. Larger values of C_2 are required with amplifiers that have a lower gain (lower ratio of R_2/R_1) than the circuit considered. When faced with an oscillating amplifier that might be stabilized by a feedback capacitor, the simplest approach is to start with a 5 pF capacitor and then try increasing values of capacitance as necessary.

5-8 LOAD CAPACITANCE EFFECTS

Circuit Instability Due to Load Capacitance

Capacitance connected at the output terminal of an op-amp circuit *(load capacitance)* can introduce additional phase lag in the feedback network and thus create instability. Figure 5-18(a) shows that the op-amp output resistance R_o and the load capacitance C_L constitute a phase-lag network. Just how large the load capacitance can be without causing circuit instability depends upon the output resistance R_o.

As explained in Section 5-7, an additional 6° of phase lag in the feedback network is undesirable. Also, as in the case of input stray capacitance, an additional 6° of lag occurs when the impedance of the capacitance is ten times the resistance in series with it.

$$X_{CL} = 10R_o$$

or

$$C_L = \frac{1}{2\pi f(10R_o)} \tag{5-13}$$

C_L determined from Eq. 5-13 is the minimum load capacitance that might make the circuit unstable. Therefore, to maintain circuit stability, the actual C_L should be much smaller than this calculated value.

If R_o is reduced, the value of C_L calculated from Eq. 5-13 becomes larger. That is, *an op-amp with a low output resistance can tolerate more load capacitance than one with a higher output resistance.*

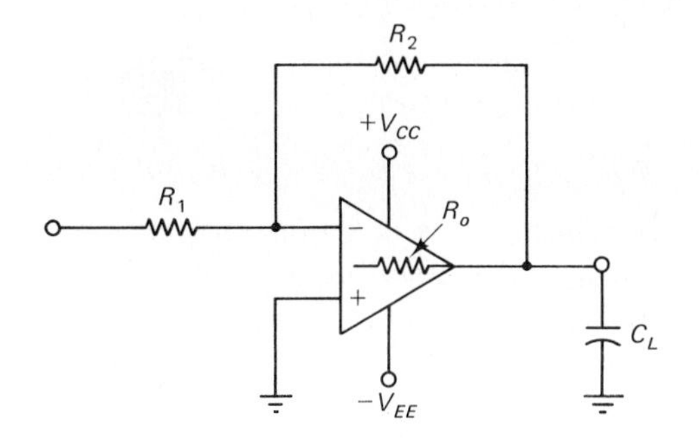

(a) Output resistance R_o and load capacitance C_L can create a lag network

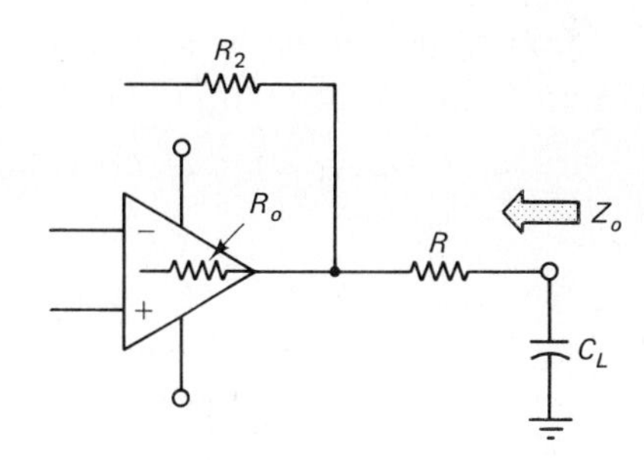

(b) Inserting a resistance R at the output can reduce the phase lag

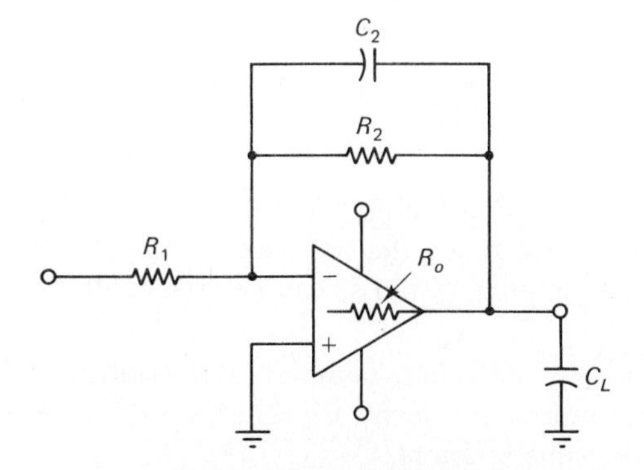

(c) A phase-lead capacitance C_2 can be used to compensate for the phase lag due to the load capacitance

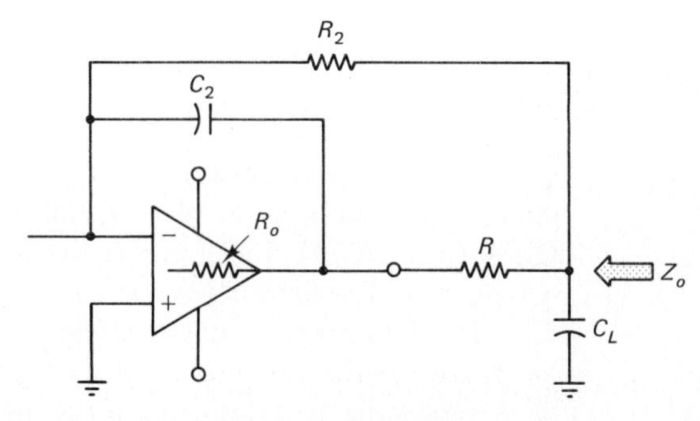

(d) C_2 and R should both be used to deal with the largest load capacitances

Figure 5-18 Load capacitance (C_L) at the output of an op-amp circuit combines with the output resistance (R_o) to constitute a phase lag network which can produce circuit instability.

Compensating for Load Capacitance Effects

One approach to improving the stability of circuits with capacitive loads is illustrated in Fig. 5-18(b). An additional resistor (R) is inserted in series with C_L and the feedback resistor R_2 is connected at the junction of R_o and R. In this position, R has the effect of reducing the phase lag generated by R_o and C_L. If a phase lag of 6° occurs because of R_o and C_L, then including $R = R_o$ as illustrated, reduces the phase lag to approximately 3°. Larger values of R produce greater reductions in the phase lag.

An undesirable effect of R is that the output impedance of the circuit is increased by the value of R. The output impedance of op-amp circuits is normally very low because of negative feedback (see Section 2-4). So, the effective output impedance of the circuit in Fig. 5-18(b) is equal to R. If $R = 47\ \Omega$, $Z_o \approx 47\ \Omega$. This may be unacceptable in some circumstances. However, in circuits where it is acceptable, this technique can be employed to improve circuit stability.

Figure 5-18(c) shows a method of using a feedback capacitor (C_2) to introduce a phase lead which will counter the phase lag generated by C_L and R_o. The equation for calculating the feedback capacitor is similar to the one used in the case of stray capacitance at the op-amp input.

$$C_2 = \frac{R_o}{R_2} C_L \tag{5-14}$$

Like Eq. 5-12, Eq. 5-14 can be stated in terms of time constants. *The time constant $C_2 R_2$ should be equal to (or greater than) the time constant $C_L R_o$.*

A modification which provides a larger feedback capacitor is shown in Fig. 5-18(d), where an additional resistor (R) is inserted in series with the load capacitance. Because R_2 is connected to the junction of R and C_L, R is within the feedback loop. As already discussed, the presence of R within the feedback loop should impair the circuit stability. But the feedback capacitor connected from the op-amp output to the inverting input terminal provides a phase lead which counters the lag generated by R_o, R, and C_L. In this case, the feedback capacitor value is calculated as

$$C_2 = \frac{(R_o + R)}{R_2} C_L \tag{5-15}$$

Because the feedback resistor R_2 is connected at the junction of R and C_L, the circuit output impedance is not significantly affected by the presence of resistor R. [$Z_o = (R_o + R)/(1 + M\beta)$.] Resistor values as high as 470 Ω can be employed for R without any serious change in Z_o.

It is important to note from Eqs. 5-14 and 5-15 that, as in the case of stray capacitance, smaller values of R_2 give larger, more convenient, capacitance values for C_2. Once again, it is desirable to use lower value resistors.

Example 5-14

Determine the load capacitance that might create instability in the inverting amplifier circuit designed in Example 5-4.

Solution

From the 709 data sheet in Appendix 1-5,

$$R_o = 150\ \Omega$$

From Example 5-11, the frequency at which $M\beta = 1$ for this circuit is approximately $f = 600$ kHz.

Eq. 5-13,
$$C_L = \frac{1}{2\pi f(10R_o)} = \frac{1}{2\pi \times 600\text{ kHz} \times 10 \times 150\ \Omega}$$
$$= 177\text{ pF}$$

Example 5-15

(a) For the circuit of Example 5-4, calculate the feedback capacitance required to compensate for a load capacitance of 0.1 μF, as in Fig. 5-18(c).

(b) If an additional resistor $R = 470\ \Omega$ is included in the circuit, as in Fig. 5-18(d), calculate the feedback capacitance required to compensate for $C_L = 0.1\ \mu$F.

Solution

(a) Eq. 5-14,
$$C_2 = \frac{R_o}{R_2}C_L = \frac{150\ \Omega}{220\text{ k}\Omega} \times 0.1\ \mu\text{F}$$
$$= 68\text{ pF (standard value)}$$

(b) Eq. 5-15,
$$C_2 = \frac{(R_o + R)}{R_2}C_L = \frac{(150\ \Omega + 470\ \Omega)}{220\text{ k}\Omega} \times 0.1\ \mu\text{F}$$
$$= 282\text{ pF (use 300 pF standard value—slightly larger than calculated, for over-compensation)}$$

5-9 Z_{in} MOD COMPENSATION

The *input impedance modification* (Z_{in} *Mod*) technique of frequency compensation is normally used only as a method of increasing the bandwidth of an op-amp circuit. The method, as illustrated in Fig. 5-19(a), involves the connection of a resistor (R_4) and capacitor (C_4) across the op-amp input terminals. The capacitor serves only to ac couple R_4 into the circuit. So, C_4 is selected to have an impedance much smaller than the resistance of R_4 at the frequency at which $M\beta = 1$. Thus, at the possible oscillation frequency of the circuit (and higher), R_4 appears in parallel with R_1 to give the feedback network illustrated in Fig. 5-19(b). For simplicity, R_3 is assumed to be a short-circuit and R_S is taken to be much smaller than R_1.

Without R_4 and C_4 in the circuit, the feedback factor is

$$\beta = \frac{R_1}{R_1 + R_2}$$

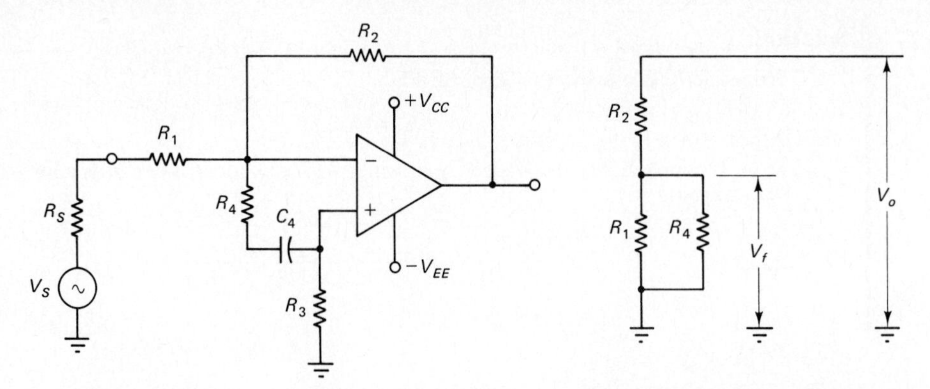

(a) Capacitor C_4 and resistor R_4 are used at the op-amp input to alter the circuit loop gain

(b) Neglecting R_S and R_3, the feedback network is changed by Z_{in} Mod as shown

Figure 5-19 The Z_{in} mod technique of frequency compensation uses resistor R_4 and capacitor C_4 to change the feedback factor from $\beta \approx R_1/(R_1 + R_2)$ to $\beta' \approx (R_1 \parallel R_4)/[(R_1 \parallel R_4) + R_2]$. This allows smaller compensating components to be used and thus increases the circuit bandwidth.

With R_4 in the circuit, the feedback factor becomes

$$\beta' = \frac{R_1 \parallel R_4}{(R_1 \parallel R_4) + R_2}$$

So, the voltage fed back from the output to the input is attenuated by the presence of R_4.

For circuit stability, the op-amp gain (that makes $M\beta'$ equal to 1) is now

$$\frac{1}{\beta'} = \frac{(R_1 \parallel R_4) + R_2}{R_1 \parallel R_4} \tag{5-16}$$

As explained below, the voltage gain from the signal source to the output remains,

$$A_v = \frac{R_2}{R_1}$$

The compensating components for an op-amp are really selected to suit the loop gain $1/\beta$. This usually corresponds with A_v but is not the case when Z_{in} Mod is used. Because denominator $R_1 \parallel R_4$ in Eq. 5-16 is less than R_1, $1/\beta'$ is greater than A_v, and consequently, the compensating components can be selected for the higher gain $(1/\beta')$.

For a better understanding of this situation, look once again at the circuit designed in Example 5-4. Compensating components were selected for $A_v = 100$ (or 40 dB). In part (b) of Example 5-6, the upper cutoff frequency for this circuit was found to be $f_2 \approx 600$ kHz (at a gain of 40 dB). Suppose now that by the use of Z_{in} Mod, $1/\beta'$ is made equal to 1000 so that compensating components are used for a gain of 60 dB [$C_1 = 10$ pF $R_1 = 0$, $C_2 = 3$ pF (see Fig. 5-10)]. The frequency response graph for these components in Fig. 5-13(b) shows that at $A_v = 40$ dB, $f_2 \approx 2$ MHz. The circuit bandwidth has been increased from 600 kHz to 2 MHz.

The impedance at the op-amp input terminals is obviously modified by the presence of R_4 and C_4. C_4 behaves as an ac short-circuit. But, as a dc open-circuit, it ensures that R_4 does not alter the dc conditions of the circuit. Recall that for an inverting amplifier, the inverting input terminal of the op-amp remains close to ground level, a virtual ground, because the noninverting terminal is grounded. Thus, the input impedance of the inverting amplifier in Fig. 5-19(a) remains equal to R_1 despite the presence of the Z_{in} Mod components. For a noninverting amplifier, R_4 appears in parallel with the op-amp Z_i, so the input impedance becomes

$$Z_{in} = (1 + M\beta)(Z_i \| R_4) \tag{5-17}$$

which is still usually a very high value.

In both inverting and noninverting amplifiers, some of the signal current would seem to be diverted through resistor R_4. However, only a very small differential voltage is developed across the op-amp input terminals ($V_d = V_o/M$). So, the current diversion is normally insignificant and the amplifier voltage gain is not noticably affected.

Example 5-16

Calculate suitable values of C_4 and R_4 in Fig. 5-19 to apply Z_{in} Mod to the circuit designed in Example 5-4. Assume that R_3 is a short-circuit, and that R_S is much smaller than R_1.

Solution

The circuit was originally designed to have,

$$A_v = \frac{1}{\beta} = 100 = 40 \text{ dB}$$

To reduce the compensating components to the next smaller recommended values, select R_4 to give

$$\frac{1}{\beta'} = 1000 = 60 \text{ dB}$$

Eq. 5-16,

$$\frac{1}{\beta'} = \frac{(R_1 \| R_4) + R_2}{R_1 \| R_4} = 1000$$

or

$$1 + \frac{R_2}{R_1 \| R_4} = 1000$$

giving

$$R_1 \| R_4 = \frac{R_2}{1000 - 1} = \frac{220 \text{ k}\Omega}{999}$$

$$\approx 220 \ \Omega$$

$$\frac{1}{R_4} = \frac{1}{220 \ \Omega} - \frac{1}{R_1} = \frac{1}{220 \ \Omega} - \frac{1}{2.2 \text{ k}\Omega}$$

$$R_4 = 244 \ \Omega \text{ (use 220 } \Omega \text{ standard value)}$$

Referring to the 709 op-amp data sheet portion in Fig. 5-10, the compensating components for $1/\beta' = 60$ dB are: $C_1 = 10$ pF, $R_1 = 0$, $C_2 = 3$ pF.

From Figure 5-13(b), the frequency at which $M\beta = 1$ (same as the cutoff frequency) is found from the intersection of the horizontal line at $1/\beta = 100 = 40$ dB and the gain/frequency response for $C_1 = 10$ pF, $R_1 = 0$, $C_2 = 3$ pF.

$$f_2 \approx 2 \text{ MHz}$$

at $f_2 \approx 2$ MHz,
$$X_{C4} \ll R_4$$

let
$$X_{C4} = \frac{R_4}{10} = \frac{220\ \Omega}{10}$$
$$= 22\ \Omega$$

$$C_4 = \frac{1}{2\pi f_2 X_{C4}} = \frac{1}{2\pi \times 2\text{ MHz} \times 22\ \Omega}$$
$$\approx 3600 \text{ pF (standard value)}$$

5-10 CIRCUIT STABILITY PRECAUTIONS

Feedback along supply lines is another source of op-amp circuit instability. This is minimized by connecting 0.01 μF high frequency *supply-decoupling* capacitors from each supply terminal to ground, as shown in Fig. 5-20. These must be connected right at the IC terminals. Sometimes one pair of decoupling capacitors can be used with several op-amps which use the same supply lines and which are located close together. In some cases, 0.1 μF capacitors might be necessary.

The following list of precautions should be observed for op-amp circuit stability.

1. Where only low frequency circuit performance is required, use an internally compensated operational amplifier like the 741.
2. When using an op-amp that must be compensated, use the manufacturer's recommended methods and components. Where possible use over-compensation.
3. Keep all component leads as short as possible and take care with component placement. Resistors connected to the op-amp input terminals should have the body of the resistor placed close to the input terminal.
4. Use 0.01 μF high frequency capacitors (or 0.1 μF if necessary) to bypass the supply terminals of op-amp (or groups of op-amp) to ground. Connect these capacitors close to the IC terminals.
5. Always have the signal source connected to a circuit to be tested. Alternatively, ground the circuit input.
6. Do not connect oscilloscopes or other instrument at the op-amp (inverting or noninverting) input terminals.
7. If a circuit is unstable after all of the above precautions have been observed, reduce the value of all circuit resistors (except compensating resistors). Also, reduce the signal source resistance if possible. Then, if necessary, try compensating for stray capacitance or load capacitance, as appropriate.

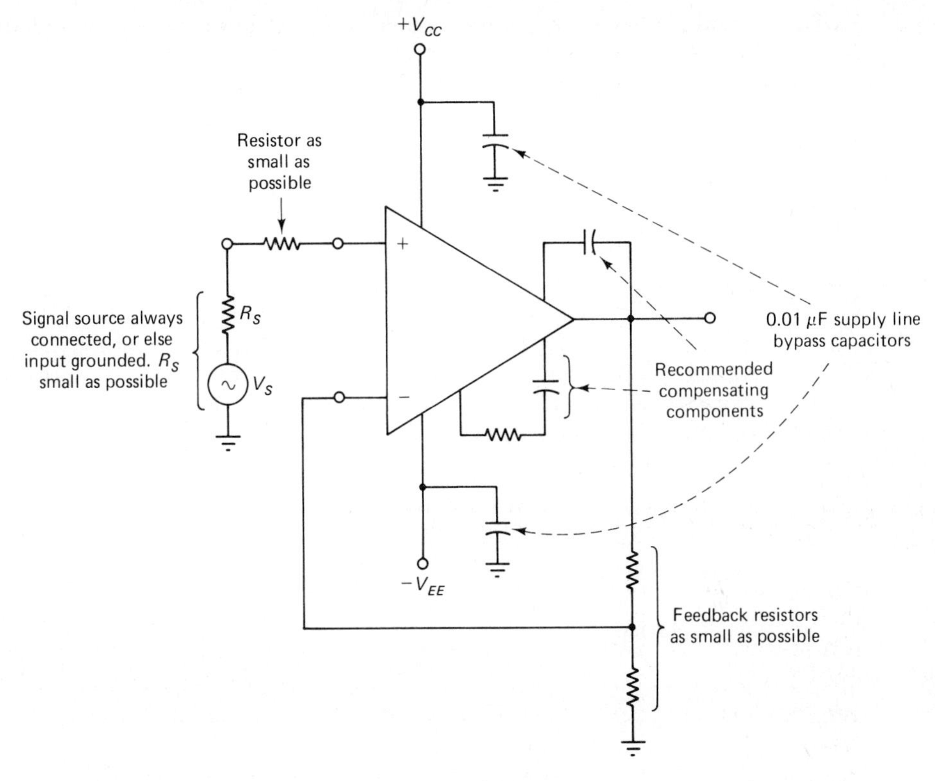

Figure 5-20 For operational amplifier circuit stability, use the manufacturer's recommended compensating components, keep all resistor values to a minimum, connect 0.01 μF capacitors to bypass the supply lines to ground, and always have the signal source connected.

8. If the circuit bandwidth is not large enough with the recommended compensating components, try Z_{in} Mod.

REVIEW QUESTIONS

5-1. Discuss operational amplifier circuit stability and show how feedback in an inverting amplifier can produce instability. Explain the conditions necessary for oscillations to occur in an op-amp circuit.

5-2. Show how feedback in a noninverting amplifier circuit can produce instability.

5-3. Define *loop phase shift, loop gain, open-loop gain, closed-loop gain.*

5-4. Sketch typical *gain/frequency response* and *phase/frequency response* graphs for a single-stage transistor amplifier at the high frequency end of the frequency band. Briefly explain.

5-5. Sketch typical *gain/frequency response* and *phase/frequency response* graphs for an operational amplifier at the high frequency end of the frequency band. Identify the pole frequencies and rates of fall of voltage gain. Also state the typical phase shift at each pole frequency. Briefly explain.

5-6. Explain why a low-gain amplifier using an operational amplifier is more likely to be unstable than a high-gain circuit.

5-7. Sketch the circuit of a *lag compensation* network. Explain its operation and show how it affects the frequency response of an operational amplifier.

5-8. Sketch the circuit of a *lead compensation* network. Explain its operation and show how it affects operational amplifier frequency response.

5-9. Explain *Miller effect*. Derive the equation relating the input capacitance of an inverting amplifier to the capacitance connected between input and output terminals.

5-10. Sketch circuits to show how Miller-effect capacitors can be employed for phase lag compensation and for phase lead compensation. Briefly explain.

5-11. Explain why manufacturer's recommended compensating components are normally suitable for both inverting and noninverting amplifiers. State the precautions that should be taken with low-gain inverting amplifier circuits.

5-12. Discuss *over-compensation* and explain its effects.

5-13. Show that the upper cutoff frequency for an op-amp circuit occurs when the open-loop gain equals the closed-loop gain.

5-14. Explain how the upper cutoff frequency of an op-amp circuit may be determined from the open-loop gain/frequency response graph.

5-15. Define *gain-bandwidth product* and discuss its application to inverting and noninverting amplifiers.

5-16. Show how the *slew rate* of an op-amp can produce distortion in a sinusoidal output waveform. Also, explain how the slew rate can limit the amplitude of the distortion-free sine wave output for a given op-amp cutoff frequency.

5-17. Show how the *slew rate* of an op-amp can produce distortion in a pulse-type output waveform. Also, explain how the slew rate can limit the amplitude of the distortion-free pulse output for a given op-amp cutoff frequency.

5-18. Discuss the effects of *stray capacitance* on op-amp circuit stability. Write the equation to determine the value of input stray capacitance that might produce circuit instability.

5-19. State the precautions that should be taken to minimize stray capacitance and its effects. Sketch a circuit to show the method of compensating for stray capacitance and derive the appropriate equation.

5-20. Discuss the effects of *load capacitance* on op-amp circuit stability. Write the equation to determine the value of load capacitance that might produce circuit instability.

5-21. Discuss methods of compensating for load capacitance effects. Sketch the appropriate circuits, and write the equations for calculating the value of compensating components.

5-22. Sketch a circuit to show the Z_{in} *mod* method of frequency compensation. State the application of the circuit, explain its operation, and write the equation for the feedback factor.

5-23. List the precautions that should be observed for operational amplifier circuit stability; briefly explain each one.

PROBLEMS

5-1. The noninverting amplifier circuit designed in Example 3-4 is to use a 709 operational amplifier. Select suitable compensating components. Determine the circuit upper cutoff frequency when a 709 op-amp is used and when a 741 is used.

5-2. The inverting amplifier circuit designed in Example 3-7 is to use a 715 operational amplifier. Select suitable compensating components. Determine the circuit upper cutoff frequency when a 715 op-amp is used and when a 741 is used.

5-3. Using a 715 operational amplifier, design an inverting amplifier to have a voltage gain of 100 with $V_i \simeq 100$ mV. Select suitable compensating components, determine the upper cutoff frequency for the circuit, and calculate the maximum distortion-free output amplitude.

5-4. Using a 715 operational amplifier, design a difference amplifier (as in Fig. 3-20). The input voltage levels are approximately 0.5 V and the voltage gain is to be 5. Select suitable compensating components.

5-5. Determine the upper cutoff frequency for the circuit designed for Problem 5-4. Also determine the upper cutoff frequencies when 741 and LF353 op-amps are used.

5-6. Determine the upper cutoff frequency and the maximum distortion-free output amplitude for a voltage follower (a) when a 741 op-amp is used and (b) when an LF353 is used.

5-7. Determine the upper cutoff frequency for an inverting amplifier circuit with $A_v = 10$, using a 715 op-amp as in Example 5-1. Also, calculate the maximum undistorted output voltage amplitude.

5-8. An inverting amplifier is to be designed to have a voltage gain of 15 with a minimum input signal of 0.1 V. The upper cutoff frequency is to be at least 200 kHz. Select a suitable operational amplifier, design the circuit, and determine the correct compensating components.

5-9. Using an LM108 op-amp, design an inverting amplifier to have a voltage gain of 40. Select a suitable compensating capacitor and estimate the upper cutoff frequency for the circuit.

5-10. Determine the upper cutoff frequency for the circuit in Problem 3-9.

5-11. Modify the circuit designed for Problem 3-9 to use a 715 op-amp. Select suitable compensating components and determine the upper cutoff frequency.

5-12. Modify the circuit designed for Problem 3-9 to use an LF353 op-amp. Select a suitable compensating capacitor, and estimate the upper cutoff frequency for the circuit.

5-13. A noninverting amplifier is to have a voltage gain of 120 with $V_o = 10$ V. Design the circuit to use a 741 op-amp and investigate the frequency response.

5-14. Repeat Problem 5-13 using a 715 op-amp.

5-15. Modify the circuit designed for Problem 3-13 to use a 715 op-amp. Select suitable compensating components, and determine the upper cutoff frequency.

5-16. Investigate the upper cutoff frequency for the circuit designed for Problem 3-16.

5-17. Modify the circuit designed for Problem 3-16 to use a 715 op-amp. Determine the correct compensating components and the upper cutoff frequency.

5-18. Investigate the upper cutoff frequency of the summing amplifier designed for Problem 3-17.

5-19. Investigate the upper cutoff frequency for the circuit designed for Problem 3-17 when a 715 op-amp is used.

5-20. Calculate the minimum rise time and the maximum undistorted output pulse amplitude at that rise time for an amplifier with a closed-loop gain of 50, using (a) a 741, (b) a 715, and (c) an LF353.

5-21. Determine the output rise time with each of the amplifiers in Problem 5-20 for an output pulse amplitude of 1.5 V.

5-22. A difference amplifier is to be designed to have a gain of 100 and an upper cutoff frequency of at least 1 MHz. The output voltage is to be 1 V maximum. Select a suitable op-amp, design the circuit, and determine the correct compensating components.

5-23. Determine the input stray capacitance that might produce instability in the circuit designed in Example 5-1. Assume that the signal source impedance is very much smaller than R_1.

5-24. Repeat Problem 5-23 for the case of (a) all resistors reduced by a factor of 10, and (b) all resistors as calculated for Problem 5-23, but with R_S open-circuted.

5-25. Calculate the feedback capacitance required to compensate for the stray capacitance determined for Problem 5-23.

5-26. Determine the input stray capacitance that might produce instability in the circuit designed for Problem 5-3. Calculate the feedback capacitance required to compensate for the input stray capacitance.

5-27. Determine the load capacitance that might produce instability in the circuit designed for Example 5-1. Calculate the feedback capacitance required to compensate for the load capacitance.

5-28. **(a)** For the circuit designed in Example 5-1, determine the feedback capacitance required to compensate for a load capacitance of 0.5 μF.
(b) Recalculate the feedback capacitance value when a resistance of $R = 390\ \Omega$ is included at circuit output as in Fig. 5-18(d).

5-29. The circuit designed in Example 5-1 is to have its bandwidth extended by the Z_{in} mod technique. Calculate suitable values of C_4 and R_4 (as in Fig. 5-19) and determine the new upper cutoff frequency.

5-30. An inverting amplifier with a gain of 1 is to be used as a substitute for a voltage follower. Using a 709 op-amp, design the circuit to have an output of 1 V. Select suitable compensating components and determine the circuit upper cutoff frequency.

5-31. Using Z_{in} mod, modify the circuit designed for Problem 5-30 to extend the upper cutoff frequency to approximately 3 MHz.

COMPUTER PROBLEMS

5-32. Write a computer program to determine the peak undistorted sine wave output, the pulse output rise time, and the maximum undistorted pulse amplitude from an op-amp circuit. Given: A_v, f_2, and S.

5-33. Write a computer program to calculate the input stray capacitance that might produce instability in an op-amp circuit. Given: Circuit component values, f_2 and A_v.

5-34. Write a computer program to solve the type of stray capacitance problem presented in Example 5-12.

5-35. Write a computer program to solve the type of Z_{in} *mod* problem presented in Example 5-16.

LABORATORY EXERCISES

5-1 Bandwidth of Amplifier Using an LM108

1. Construct the LM108 inverting amplifier circuit shown in Fig. 5-9(a). Use a supply of $\pm$9V to $\pm$15 V and component values of $R_1 = 470\ \text{k}\Omega$, $R_2 = 1.5\ \text{M}\Omega$, $R_3 = 470\ \text{k}\Omega$, $C_f = 10$ pF, as designed in Example 5-3.

2. Apply a 1 kHz sinusoidal signal input, adjusting it to give a peak output of 500 mV as displayed on an oscilloscope. Measure the input voltage, calculate the circuit voltage gain, and compare it to the designed gain.
3. Maintaining the input voltage (V_i) constant, increase the signal frequency until the output is approximately 0.707 of the v_o level at 1 kHZ. Note the upper cutoff frequency (f_2).
4. Estimate the approximate cutoff frequency from the gain/frequency response graph for the LM108 in Fig. 5-9(b). Compare to the measured f_2.

5-2 Bandwidth of Amplifier Using a 741

1. Construct the 741 inverting amplifier circuit shown in Fig. 3-13, changing R_2 to 100 kΩ to give a voltage gain of 100. Use a supply of ±9 V to ±15 V.

2–3. Same as Procedures 2 through 3 in Laboratory Exercise 5-1.

4. Compare the measured f_2 to the frequency determined in part (a) of Example 5-6.

5-3 Bandwidth of Amplifier Using a 709

1. Construct the 709 inverting amplifier circuit shown in Fig. 5-11, as designed in Example 5-4. Use a supply of ±9 V to ±15 V.

2–3. Same as Procedures 2 through 3 in Laboratory Exercise 5-1.

4. Compare the measured f_2 to the frequency determined in part (b) of Example 5-6.

5-4 Rise Time and Slew Rate Effects

1. Connect a 741 op-amp as a voltage follower circuit using 1 kΩ resistors in series with each input terminal. Use a supply of ±15 V.
2. Apply a 100 Hz sinusoidal signal input, adjusting it to give a peak output of 5 V.
3. Maintaining the input voltage constant, increase the signal frequency until the output just begins to show distortion. (Superimpose the output and input waveforms.)
4. Note the signal frequency and compare it to the frequency determined in part (a) of Example 5-9.
5. Reduce the signal amplitude to zero and increase the frequency to 800 kHz.
6. Monitor the output waveform while carefully increasing the signal level until the output just begins to show distortion.
7. Measure the peak output voltage and compare it to the voltage level determined in part (b) of Example 5-9.
8. Replace the sinusoidal signal with a square wave input with a peak-to-peak amplitude of 5 V and frequency of 10 kHz.
9. Measure the slew rate-limited rise time [$t_{r(s)}$] of the output waveform, and compare to the value determined in part (a) of Example 5-10.
10. Reduce the signal amplitude to 100 mV and adjust the frequency as necessary to obtain a measurement of the cutoff frequency-limited rise time [$t_{r(f2)}$] of the output.
11. Compare the measured value of $t_{r(f2)}$ to that determined in part (a) of Example 5-10.

CHAPTER 6

Miscellaneous Op-Amp Linear Applications

Objectives

- Sketch the following op-amp circuits and explain the operation of each: *voltage source, precision voltage source, current source, current sink, precision current source, precision current sink, current-to-voltage converter, current amplifier-attenuator, dc voltmeter, ac voltmeter, linear ohmmeter, differential input/output amplifier, instrumentation amplifier*.
- Design each of the above circuits using bipolar and BIFET op-amps, together with transistors, FETs, and MOSFETs, where appropriate.
- Analyze each of the above circuits to determine its performance.

INTRODUCTION

Operational amplifiers may be employed on a wide variety of applications, from the dc and ac amplifiers, which have already been studied, to filters, signal generators, and voltage regulators, which are treated in succeeding chapters. Miscellaneous applications covered in this chapter include voltage sources, current sources, current amplifiers, dc voltmeters, ac voltmeters, linear ohmmeters, and instrumentation amplifiers.

6-1 VOLTAGE SOURCES

Low Resistance Voltage Source

The voltage source circuit in Fig. 6-1(a) is simply a potential divider and voltage follower (as in Fig. 3-6(b)) with transistor Q_1 included at the output. Because the maximum output current that can be supplied by most operational amplifiers is approximately 25 mA, the transistor is necessary for higher load current levels. The maximum load current is now the maximum emitter current for the transistor, and the op-amp has only to supply the much smaller transistor base current. With the in-

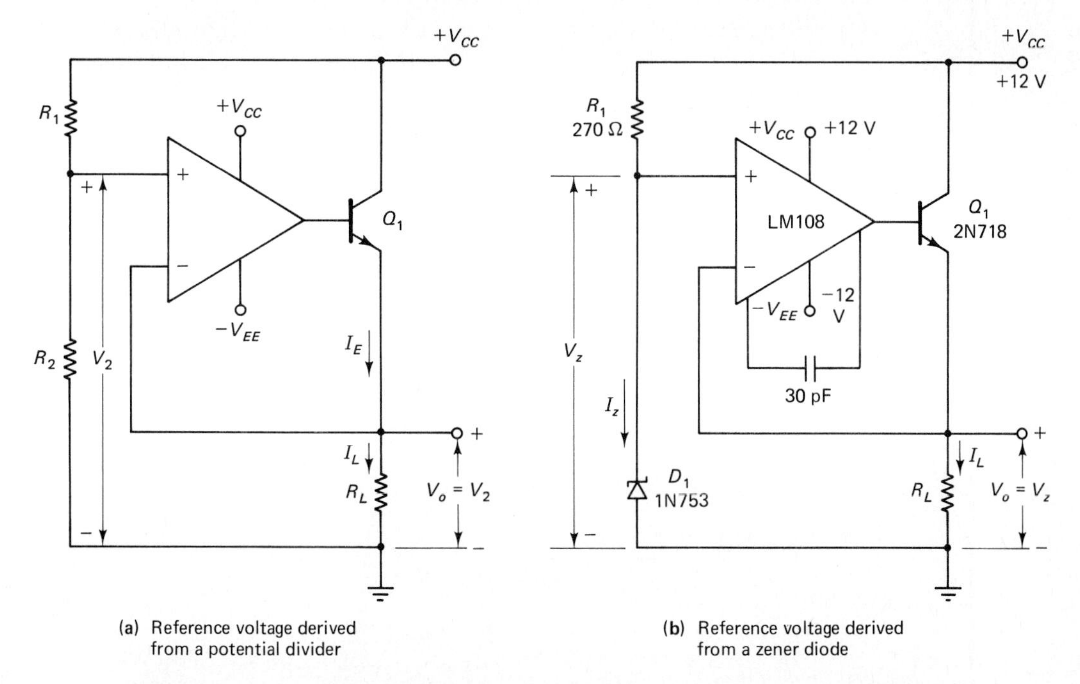

Figure 6-1 Low resistance voltage sources using bipolar transistors to increase the level of output current. The reference voltage can be derived from a potential divider or from a Zener diode.

verting input terminal of the op-amp connected to the transistor emitter, the emitter terminal is maintained at the same voltage level as the noninverting terminal (at voltage V_2). Instead of using a potential divider to provide the desired reference voltage, a Zener diode can be employed as in Fig. 6-1(b). The potential divider voltage is stable only if the supply voltage remains constant, while the Zener diode voltage is largely independant of any variations in supply voltage. The circuit in Fig. 6-1(b) is essentially a dc voltage regulator. These are discussed in Chapter 12.

Design of the voltage source circuit in Fig. 6-1(a) involves selecting a transistor to handle the desired load current, then calculating the potential divider resistors in the usual way. To use a Zener diode as the reference voltage, a diode with the desired breakdown voltage is first selected, then R_1 is calculated as

$$R_1 = \frac{V_{CC} - V_Z}{I_Z}$$

Example 6-1

A voltage source is to be designed to provide a constant output voltage of approximately 6 V. The load resistance has a minimum value of 150 Ω and the available supply voltage is ±12 V. Design a suitable circuit as in Fig. 6-1(b).

Solution

A 1N753 Zener diode has $V_Z = 6.3$ V. Use a 1N753.

The recommended current for best voltage stability for all low-current Zener diodes is $I_Z = 20$ mA.

$$R_1 = \frac{V_{CC} - V_Z}{I_Z} = \frac{12\text{ V} - 6.3\text{ V}}{20\text{ mA}}$$

$$= 285\ \Omega \qquad \text{(use 270 Ω standard value)}$$

$$I_{L(\max)} = \frac{V_Z}{R_{L(\min)}} = \frac{6.3\text{ V}}{150\ \Omega}$$

$$= 42\text{ mA}$$

Transistor specification is

$$npn \text{ device, } I_{E(\max)} > 42\text{ mA},\ V_{CE(\max)} > V_{CC} = 12\text{ V},$$

$$P_D = I_{E(\max)}(V_{CC} - V_{RL}) = 42\text{ mA}(12\text{ V} - 6\text{ V})$$

$$\approx 0.25\text{ W (a 2N718 is suitable)}$$

Op-amp specification is

$$V_{CC} = \pm 12\text{ V},$$

$$I_{O(\max)} \approx \frac{I_{L(\max)}}{h_{FE(\min)}} = \frac{42\text{ mA}}{20}$$

$$\approx 2\text{ mA}$$

Use a 741 or a LM108 with a compensating capacitor.

Precision Voltage Source

One problem with the simple voltage source discussed above is that the Zener diode voltage can vary slightly because its current will change if the supply voltage does not remain constant. The circuit shown in Fig. 6-2 uses the output voltage to maintain a constant current through the diode, thus providing a precise unchanging voltage, and consequently, a precise output level. If the required load current is too great for the op-amp to supply, a transistor can be included at the output as in Fig. 6-1.

In Fig. 6-2(a) both the Zener diode circuit (D_1 and R_1) and the potential divider (R_2 and R_3) are supplied from the op-amp output terminal. Suppose the Zener diode voltage is 4.5 V and the voltage drop across R_1 is 4.5 V. The total voltage drop across D_1 and R_1 is 9 V. This is also the output voltage V_0. Since the op-amp inverting and noninverting terminal voltages must be equal when the op-amp is operating as a linear circuit, $V_{R3} = V_Z$, and $V_{R2} = V_{R1}$. Now, if the output were to somehow increase by ΔV_o, the voltage change at the op-amp inverting input would be

$$\Delta V_{R3} = \frac{\Delta V_o \times R_3}{R_2 + R_3}$$

This would cause the output to move in a negative direction to correct the original increase in V_0. Similarly, any negative change in V_0 would be corrected by the resultant voltage drop at the inverting input terminal. So, there is negative feedback which holds the output voltage at the designed level.

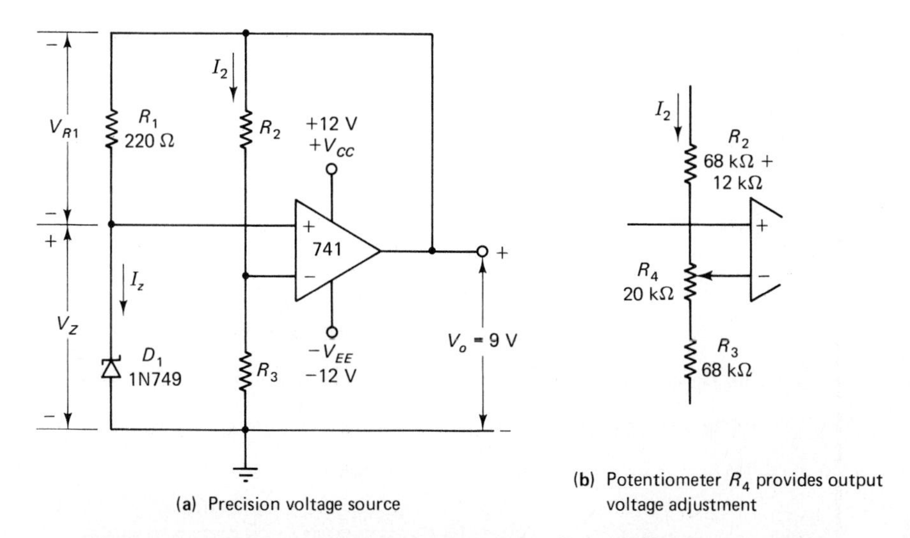

Figure 6-2 Precision voltage source using a Zener diode supplied from the output. The constant output voltage maintains the Zener current constant and thus maintains a precise level of Zener voltage and output voltage.

To find the relationship between V_o and V_Z, recall that $V_{R3} = V_Z$, and $V_o = V_{R2} + V_{R3}$.

So $$I_2 = \frac{V_Z}{R_3}$$

and $$V_o = I_2(R_2 + R_3)$$

giving $$V_o = \frac{V_Z(R_2 + R_3)}{R_3} \tag{6-1}$$

Equation 6-1 shows that the exact level of the Zener voltage is not important because the desired output level can be obtained by suitable selection of the resistor values. Because of the usual ±5% or ±10% tolerance on the Zener diode voltage, it is often necessary to include a potentiometer for output adjustment as shown in Fig. 6-2(b).

To design a precision voltage source, a Zener diode is first selected, usually with a voltage approximately equal to half the output voltage. Then the resistance in series with the Zener is calculated and the potential divider resistor values are deter mined in the usual way. The only concern in selecting a suitable op-amp is that its output voltage range must be greater than V_o.

Example 6-2

Design a precision voltage source as in Fig. 6-2, to provide an output of 9 V. The available supply is ±12 V. Allow for approximately ±10% tolerance on the Zener diode voltage.

Solution

$$V_Z \approx \frac{V_o}{2} = \frac{9\text{ V}}{2}$$

$$\approx 4.5\text{ V}$$

Use a 1N749 which has $V_Z = 4.3$ V

$$R_1 = \frac{V_o - V_Z}{I_Z} = \frac{9\text{ V} - 4.3\text{ V}}{20\text{ mA}}$$

$$= 235\ \Omega \qquad \text{(use 220 } \Omega \text{ standard value)}$$

For R_2, R_3, and R_4, $\qquad I_2 \gg I_{B(max)}$ of the op-amp

Using a 741 op-amp,

let $$I_2 \approx 100 \times 500\text{ nA} = 50\ \mu\text{A}$$

$$R_3 + R_4 = \frac{V_Z + 10\%}{I_2} = \frac{4.3\text{ V} + 10\%}{50\ \mu\text{A}}$$

$$= 94.6\text{ k}\Omega$$

and $$R_4 \approx 20\% \text{ of } (R_3 + R_4) = 20\% \text{ of } 94.6 \text{ k}\Omega$$

$$= 18.9 \text{ k}\Omega \qquad \text{(use a 20 k}\Omega\text{ potentiometer)}$$

$$R_3 = (R_3 + R_4) - R_4 = 94.6 \text{ k}\Omega - 20 \text{ k}\Omega$$

$$= 74.6 \text{ k}\Omega \qquad \text{(use 68 k}\Omega\text{ standard value, and recalculate } I_2\text{)}$$

$$I_2 = \frac{V_Z + 10\%}{R_3 + R_4} = \frac{4.3 \text{ V} + 10\%}{68 \text{ k}\Omega + 20 \text{ k}\Omega}$$

$$\approx 53.8 \ \mu\text{A}$$

$$R_2 = \frac{V_o - (V_{R3} + V_{R4})}{I_2} = \frac{9 \text{ V} - (4.3 \text{ V} + 10\%)}{53.8 \ \mu\text{A}}$$

$$= 79.4 \text{ k}\Omega \qquad \text{(use 68 k}\Omega\text{ and 12 k}\Omega\text{ in series)}$$

6-2 CURRENT SOURCES AND CURRENT SINKS

Current Sources

A *current source* is a circuit that supplies a constant current in the conventional direction from positive to negative. Thus, current flows out of the positive terminal of a current source. A *current sink* also has a constant current level, but the current direction is into the ungrounded terminal of the current sink.

A current source circuit in which the load resistor is floating (i.e., not grounded) is shown in Fig. 6-3(a). The circuit functions like a noninverting amplifier with V_1 appearing across R_1, so the load current is always $I_L = V_1/R_1$. This is independent of load resistance variations as long as the total voltage drop across R_1 and R_L does not exceed the output voltage range of the op-amp, and as long as the transistor has sufficient collector-emitter voltage to keep it operational. If the maximum load current is less than 25 mA, the transistor may not be required. Voltage V_1 can be derived from a potential divider or from a Zener diode, as in Fig. 6-1. The circuit could also be described as a *voltage-to-current converter,* because the load current is obviously proportional to the input voltage.

In Fig. 6-3(b), V_1 is the voltage between $+V_{CC}$ and the op-amp noninverting input terminal. The *pnp* transistor emitter terminal is connected to the op-amp inverting input, so that the circuit behaves as a voltage follower with V_1, once again, appearing across resistor R_1. The constant input voltage maintains a constant current through R_1, and thus maintains the transistor emitter and collector currents constant. Note that the voltage drop across R_L puts the transistor collector voltage above ground level. Allowing a minimum collector-emitter voltage of 3 V, the transistor emitter must be at least $(I_L R_L + 3 \text{ V})$ above ground level.

The load current is the transistor collector current in the circuit of Fig. 6-3(b). This is slightly different from I_1 because of the transistor base current. Substituting a

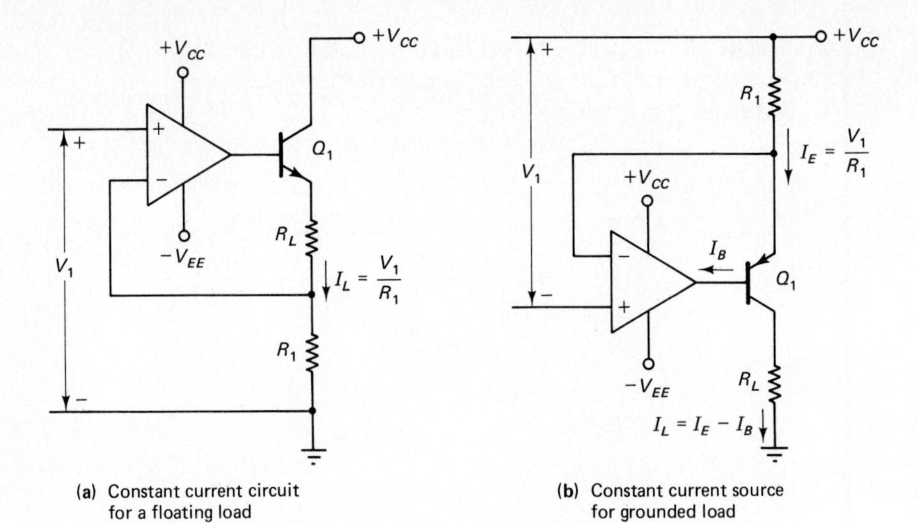

(a) Constant current circuit for a floating load

(b) Constant current source for grounded load

Figure 6-3 A constant current source with a floating load behaves like a noninverting amplifier circuit to keep $I_L = V_1/R_1$. For a grounded load, a *pnp* transistor is used and again V_1/R_1 keeps the load current constant.

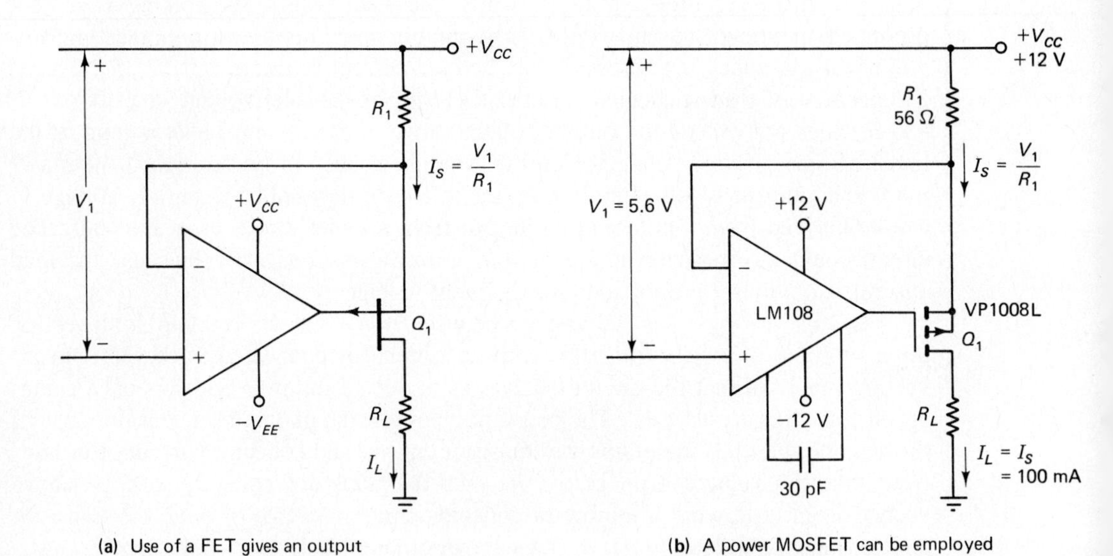

(a) Use of a FET gives an output current equal to I_S

(b) A power MOSFET can be employed for high output current levels

Figure 6-4 The use of a FET or a MOSFET in a constant current circuit produces an output current which precisely equals the current through resistor R_1.

p-channel FET in place of the bipolar transistor, as shown in Fig. 6-4(a), improves the circuit performance. Because the FET drain and source currents are equal, the load current is more precisely equal to the current through R_1. Figure 6-4(b) shows the use of a power MOSFET for high load current situations. As in the case of the bipolar transistor, the FET voltage drops must be considered when designing the circuit.

The drain-source voltage of the FET should be at least 1 V greater than the maximum *pinch-off voltage* ($V_{GS(\text{off})}$) when a junction FET is used.*

$$V_{DS(\text{min})} = V_{GS(\text{off})} + 1\text{ V} \tag{6-2}$$

For a MOSFET, V_{DS} should be at least 1 V greater than the drain current multiplied by the drain-source *on* resistance.*

$$V_{DS(\text{min})} = (I_D R_{D(\text{on})}) + 1\text{ V} \tag{6-3}$$

Example 6-3

Design a current source to produce an output of 100 mA to a load which has a maximum resistance of 40 Ω. Use an LM108 op-amp and a power MOSFET as in Fig. 6-4(b). The available supply voltage is ±12 V.

Solution

For the MOSFET, $V_{DS(\text{max})} = V_{CC} = 12\text{ V}$

and, $I_{D(\text{max})} = I_L = 100\text{ mA}$

The VP 1008L p-channel power MOSFET has $V_{DS(\text{max})} = 100\text{ V}$, $I_{D(\text{max})} = 210\text{ mA}$ and $R_{D(\text{on})} = 5\ \Omega$. So, the VP 1008L is a suitable device for Q_1.

$$V_{L(\text{max})} = I_L \times R_{L(\text{max})} = 100\text{ mA} \times 40\ \Omega$$

$$= 4\text{ V}$$

Eq. 6-3 $V_{DS(\text{min})} = (I_D R_{D(\text{on})}) + 1\text{ V} = (100\text{ mA} \times 5\ \Omega) + 1\text{ V}$

$$= 1.5\text{ V}$$

and $V_{R1(\text{max})} = V_{CC} - V_{L(\text{max})} - V_{DS(\text{min})}$

$$= 12\text{ V} - 4\text{ V} - 1.5\text{ V}$$

$$= 6.5\text{ V}$$

$$R_1 = \frac{V_{R1}}{I_L} = \frac{6.5\text{ V}}{100\text{ mA}}$$

$= 65\ \Omega$ (Use $R_1 = 56\ \Omega$, which is the next smaller standard value)

Then, $V_{R1} = I_L \times R_1 = 100\text{ mA} \times 56\ \Omega$

$$= 5.6\text{ V}$$

$$V_1 = V_{R1} = 5.6\text{ V}$$

* David A. Bell, *Electronic Devices and Circuits,* 3rd ed. (Englewood Cliffs, NJ: Prentice-Hall, Inc., 1986), p. 253 and p. 544.

Current Sinks

Two current sink circuits are illustrated in Figs. 6-5(a) and (b). These are similar to the current sources in Figs. 6-3(b) and 6-4(b) except that an *npn* bipolar transistor and an *n*-channel MOSFET are now used. The operation of the circuits and the design approach for each is the same as for the current source circuits.

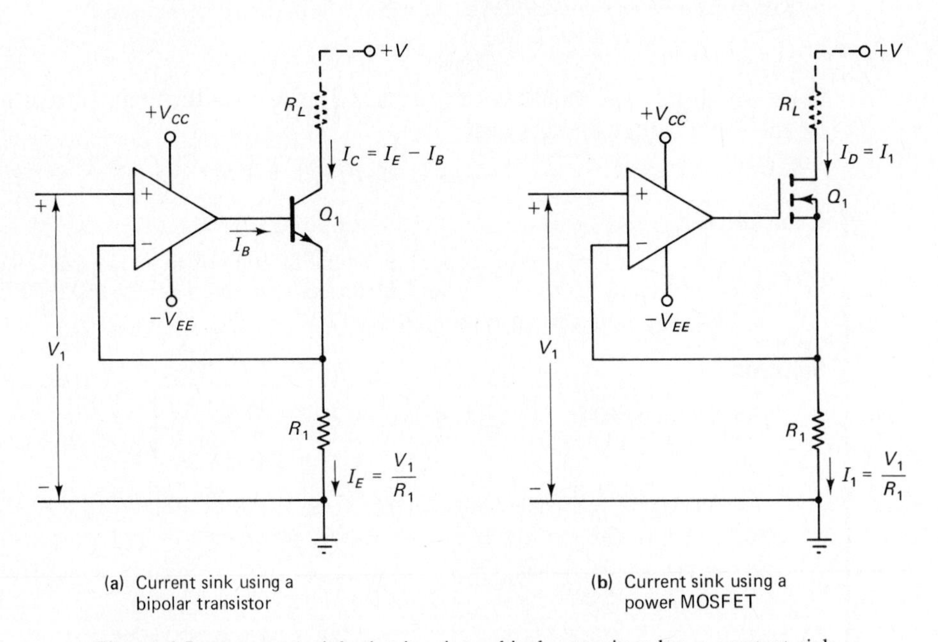

Figure 6-5 A current sink circuit using a bipolar transistor has a constant sink current because the emitter current is held at V_1/R_1. Use of a MOSFET (or a FET) gives a sink current precisely equal to the current through R_1.

6-3 PRECISION CURRENT SINK AND SOURCE CIRCUITS

The precision current sink circuit in Fig. 6-6 is the same as the precision voltage source in Fig. 6-2 except that an *n*-channel power MOSFET (Q_1) has been added at the op-amp output. A constant voltage is maintained at the output as explained in Section 6-1. In the case of Fig. 6-6, the output voltage at the source terminal of the FET is held constant. Thus, the current through R_5 is held constant and the MOSFET drain current is a constant quantity.

The current source circuit in Fig. 6-7 is a modified form of the circuit in Fig. 6-6. A *p*-channel MOSFET is used for Q_1 and components D_1 and R_1 have been interchanged in order to maintain a constant voltage across R_5 in its new position. Thus, once again, the drain current of the MOSFET is held constant.

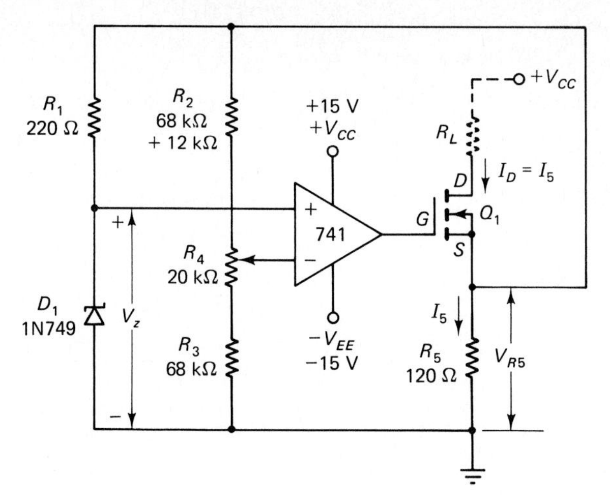

Figure 6-6 Precision current sink using a precision voltage source and a power MOSFET.

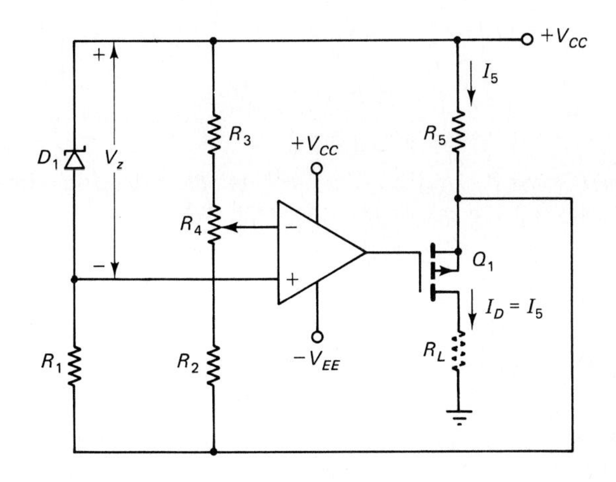

Figure 6-7 Precision current source using a precision voltage source and a power MOSFET.

To design a precision current source or sink circuit, the voltage drop across R_5 is first determined in the same way as V_{R1} is calculated in Example 6-3. Then the rest of the circuit is designed as for the precision voltage source in Example 6-2.

Example 6-4

Design a precision current sink, as in Fig. 6-6, to provide 75 mA through a load which has a maximum resistance of 50 Ω. The supply voltage is ±15 V and a 741 op-amp is to be used.

Solution

For the MOSFET, $$V_{DS(\max)} = V_{CC} = 15 \text{ V}$$

and $$I_{D(\max)} = I_L = 75 \text{ mA}$$

The VN2222L n-channel power MOSFET has $V_{DS(max)} = 60$ V, $I_{D(max)} = 150$ mA, and $R_{D(on)} = 7.5\ \Omega$. So, the VN2222L is a suitable device for Q_1.

$$V_{L(max)} = I_L \times R_{L(max)} = 75 \text{ mA} \times 50\ \Omega$$
$$= 3.75 \text{ V}$$

For satisfactory operation of Q_1,

let
$$V_{DS(min)} = (I_D R_{D(on)}) + 1 \text{ V} = (75 \text{ mA} \times 7.5\ \Omega) + 1 \text{ V}$$
$$\approx 1.6 \text{ V}$$

and
$$V_{R5(max)} = V_{CC} - V_{L(max)} - V_{DS(min)}$$
$$= 15 \text{ V} - 3.75 \text{ V} - 1.6 \text{ V}$$
$$\approx 9.6 \text{ V}$$
$$R_5 = \frac{V_{R5}}{I_L} = \frac{9.6 \text{ V}}{75 \text{ mA}}$$
$$= 128\ \Omega$$

Use $R_5 = 120\ \Omega$. Then,

$$V_{R5} = I_L \times R_5 = 75 \text{ mA} \times 120\ \Omega$$
$$= 9 \text{ V}$$

The circuit in Example 6-2 was designed for $V_o = 9$ V. So, the remaining component calculations for the current sink are exactly as for Example 6-2.

6-4 CURRENT AMPLIFIERS

Current-to-Voltage Converter

The current-to-voltage converter circuit in Fig. 6-8 can be thought of as an inverting amplifier without an input resistor. Because the op-amp noninverting input terminal is grounded, the inverting input remains at ground level. So, the input current source is effectively short-circuited to ground. With I_s much larger than the input bias current of the op-amp, virtually all of I_s flows through resistor R_1 to produce a

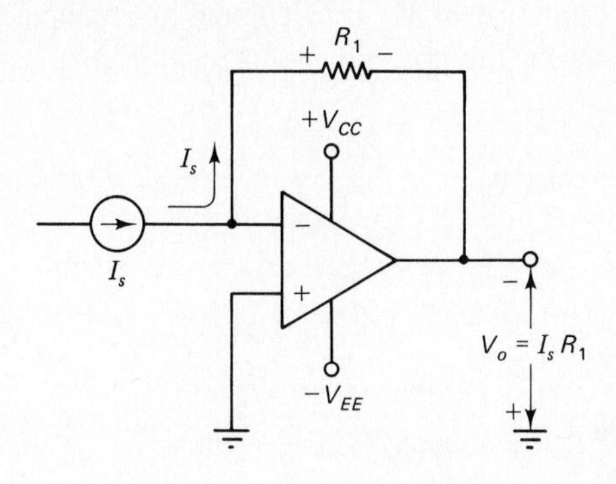

Figure 6-8 Current-to-voltage converter circuit which behaves like an inverting amplifier without an input resistor.

voltage drop $-I_sR_1$, which is also the negative voltage at op-amp output. If the current direction is reversed, the output voltage becomes positive.

Current Amplifier

In Fig. 6-9(a), resistor R_2 has been added to the circuit of Fig. 6-8. The current through R_2 is

$$I_2 = \frac{V_o}{R_2} = \frac{I_sR_1}{R_2}$$

So, the output current is R_1/R_2 times the input current and the circuit is a current amplifier if $R_1 > R_2$. Where $R_2 > R_1$, the circuit is a current attenuator.

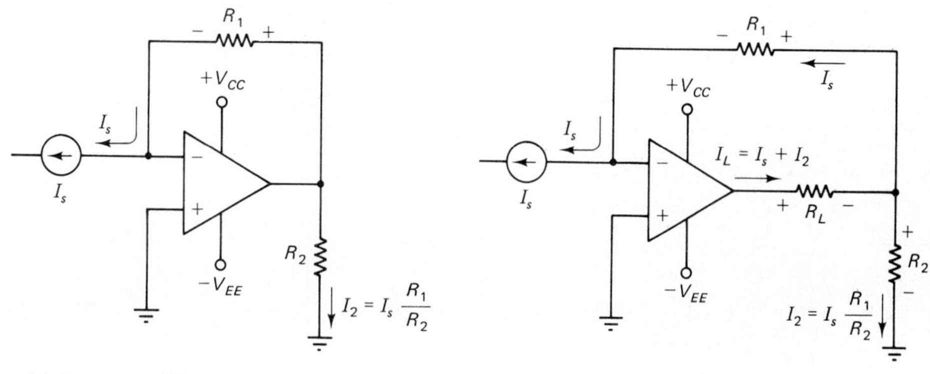

Figure 6-9 Current amplifier/attenuator circuits with grounded and floating loads. For the grounded load circuit, the current gain depends upon the resistance of load R_2. For the floating load circuit, the gain is independent of the load resistance.

One disadvantage of the circuit in Fig. 6-9(a) is that the current gain/attenuation is dependent on the resistance of R_2, which is the load resistance. If the load can be floating (ungrounded), then the circuit shown in Fig. 6-9(b) can be used. Here, the load current is

$$I_L = I_s + I_2$$

$$= I_s + I_s\frac{R_1}{R_2}$$

or

$$I_L = I_s\left(1 + \frac{R_1}{R_2}\right) \tag{6-4}$$

Note that the resistance of the load is not involved in Eq. 6-4. So, the circuit of Fig. 6-9(b) operates as a current amplifier with a gain which is independent of the load resistance. Of course the voltage drop across the load must not be so large that the op-amp output attempts to exceed its maximum output voltage range.

6-5 DC VOLTMETER

The voltage-to-current converter circuit in Fig. 6-10(a) has already been discussed in Section 6-2. However, this circuit has a moving-coil meter instead of a load resistor at the op-amp output terminal. The meter current is the current through resistor R_1 which is $I_m = V_i/R_1$. Therefore, the meter deflection is directly proportional to the applied voltage. A voltmeter must have a very high input resistance. This is also a characteristic of a noninverting amplifier, so the noninverting amplifier circuit makes a very satisfactory voltmeter.

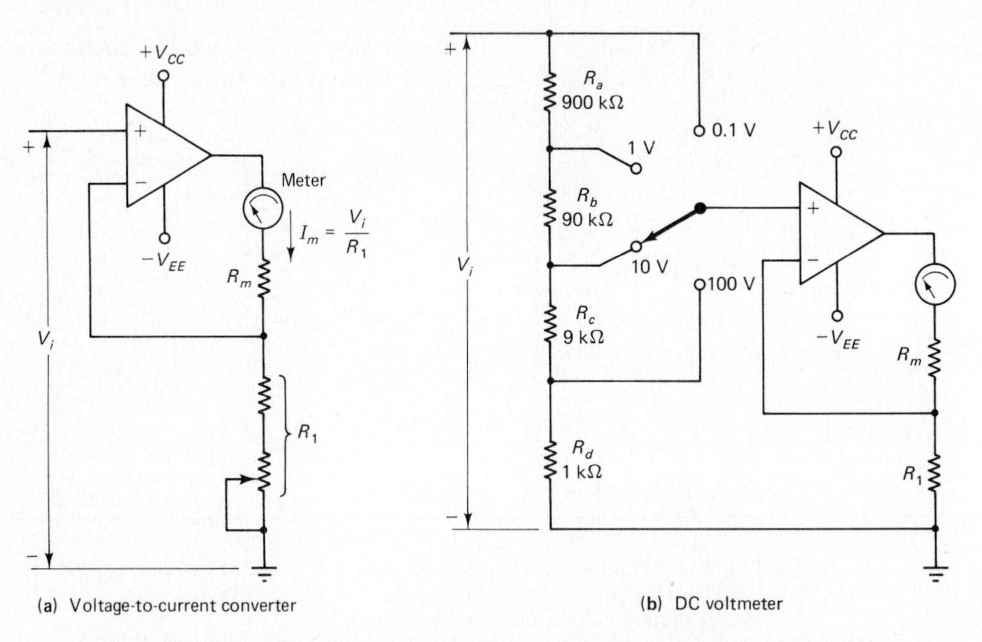

Figure 6-10 Op-amp dc voltmeters use a voltage-to-current converter circuit. Range changing is effected by means of a potential divider which offers a constant input resistance.

In addition to measuring low voltage levels, multirange voltmeters usually have to deal with voltages much larger than op-amps can handle. Fig. 6-10(b) shows the usual method of constructing a multirange voltmeter, using a switch and a potential divider. The current-to-voltage converter is designed to give full scale deflection on the meter when the input is 0.1 V. So, on the 0.1 V range, the input voltage is applied directly to the op-amp noninverting terminal. On the 1 V range, the input voltage is divided by a factor of 10, to give 0.1 V at the noninverting terminal when 1 V is applied to the voltmeter input terminal. Consequently, meter full-scale deflection can now be read as 1 V. Similarly, on the 10 V range the input is divided by 100, and on the 100 V range there is division by a factor of 1000. Note that the input resistance of the instrument on all ranges is the total resistance of the potential divider; 1 MΩ.

For accurate voltage measurement, precision resistors must be used in the potential divider of the voltmeter. Also, as illustrated, part of R_1 should be made adjustable to facilitate instrument calibration. The design of a such a dc voltmeter is only a matter of selecting a suitable moving coil meter and calculating the resistance of R_1.

6-6 AC VOLTMETER

Two op-amp ac voltmeter circuits are illustrated in Fig. 6-11. One uses half-wave rectification and the other employs a full-wave bridge rectifier. Apart from the use of the rectifiers, both circuits are similar to the dc voltmeter already discussed. Each circuit has a capacitor-coupled input to block dc voltages, since it is only ac quantities that are to be measured. Because of the capacitor, the noninverting terminal of the op-amp must be grounded via resistor R_1. (This is discussed in Section 4-3.)

With zero ac input voltage, there is no voltage across resistor R_2, and consequently, no current flow in the meter. When an ac input is applied to the half-wave rectifier voltmeter, the op-amp noninverting terminal goes positive during the positive half-cycles of the input voltage. This causes the op-amp output to go positive, thus forward-biasing diode D_1 and producing a current flow in the meter. When the input voltage enters its negative half-cycle, the op-amp output terminal goes negative, the diode is reverse biased, and no meter current flows. So, the meter current is a series of positive half-cycles with intervening spaces.

Note, that when the input voltage is at any instantaneous level v_i, the meter current is v_i/R_2. The diode voltage drop is not involved. Because of the presence of the op-amp, the voltage drop across the diode has no effect on the meter current. In fact, the input voltage can be smaller than the diode voltage drop.

The full-wave rectifier voltmeter in Fig. 6-11(b) operates in the same way as the half-wave circuit, except that during the positive half-cycle of the input voltage diodes D_1 and D_4 are forward biased, and during the negative half-cycle D_2 and D_3 are forward biased. In each case, current flows through the meter in a positive direction and the meter current is then a series of positive half-cycles with no intervening spaces.

Design of an op-amp rectifier voltmeter involves little more than calculation of resistor R_2. However, as demonstrated in Example 6-5, for the meter to indicate rms volts the calculation is more complicated than in the case of a dc voltmeter. The resistance of R_1 should be as high as possible for high input resistance. Depending upon the op-amp employed, R_1 might be determined by the use of Eq. 3-1 or might simply be selected as 1 MΩ when using a BIFET op-amp. Considerations in selection of the op-amp are; input resistance, supply voltage, input voltage range, output voltage range, and maximum frequency of the voltage to be measured. Capacitor C_1 should be calculated to have $X_{C1} \ll R_1$ at the minimum operating frequency of the voltmeter, so that the impedance of C_1 does not cause an error. The rectifiers should be low-current diodes, and the diode reverse recovery time might be important if high input frequencies are involved. Rectifier voltmeters are normally designed for measuring purely sinusoidal voltages; they are not suitable for use with nonsinusoidal waveforms.

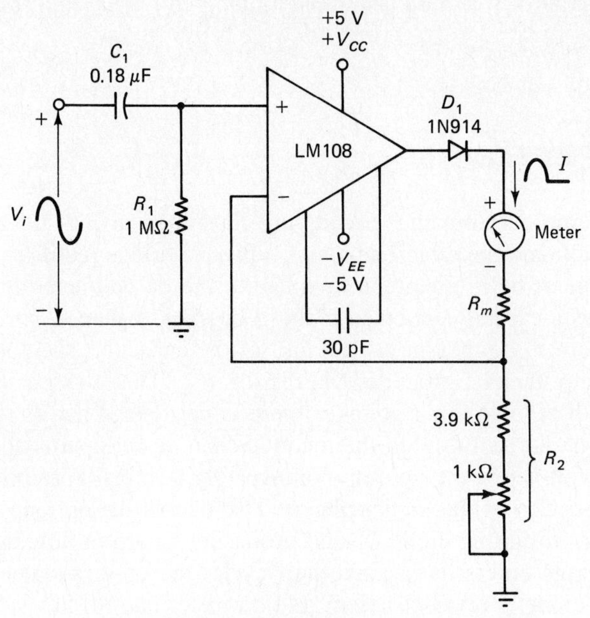

(a) Half-wave rectifier voltmeter

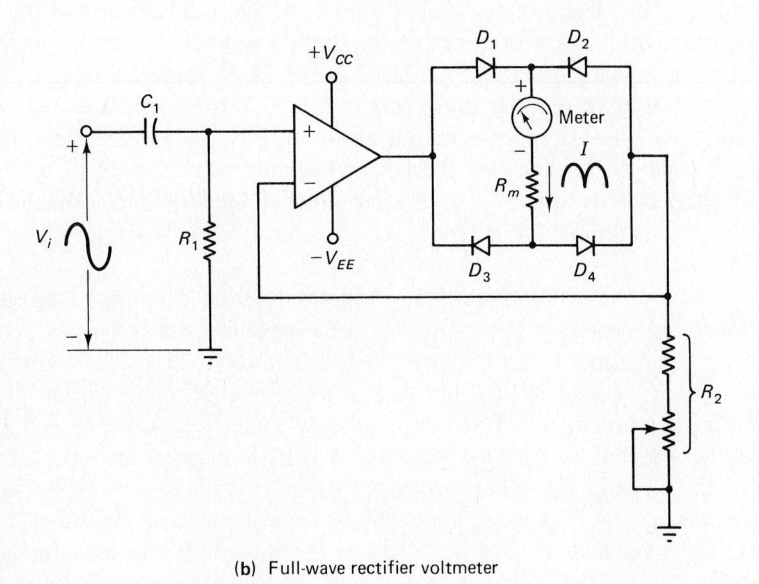

(b) Full-wave rectifier voltmeter

Figure 6-11 Half-wave and full-wave rectifier ac voltmeters. The action of the op-amp eliminates the effect of the rectifier voltage drops.

Example 6-5

Design the half-wave rectifier voltmeter in Fig. 6-11(a) to give full scale deflection for an rms input of 1 V. A 100 μA meter with a coil resistance of 2.5 kΩ is to be used, and the signal frequency range is to be 10 Hz to at least 1 kHz.

Solution

For full scale deflection, the average meter current is,

$$I_{av} = 100\ \mu\text{A}$$

and, with half-wave rectification,

$$I_{av} = 0.637\frac{I_p}{2}$$

or,

$$I_p = \frac{2I_{av}}{0.637} = \frac{2 \times 100\ \mu\text{A}}{0.637}$$

$$= 314\ \mu\text{A}$$

I_p occurs when the input voltage is at V_p,

and $\quad V_p = 1.414\ V_i = 1.414 \times 1\ \text{V} \quad$ (for a sine wave)

$$= 1.414\ \text{V}$$

$$R_2 = \frac{V_p}{I_p} = \frac{1.414\ \text{V}}{314\ \mu\text{A}}$$

$= 4.5\ \text{k}\Omega$ (use a 3.9 kΩ fixed value and a 1 kΩ variable in series)

For the op-amp.

output voltage range $\quad V_{o(\max)} = V_{D1} + I_p(R_m + R_2)$

$$= 0.7\ \text{V} + 314\ \mu\text{A}(2.5\ \text{k}\Omega + 4.5\ \text{k}\Omega)$$

$$\approx 2.9\ \text{V}$$

Input voltage range $\quad V_{i(\max)} = 1.414\ \text{V(peak)}$

upper cutoff frequency $\quad f_H > 1\ \text{kHz}$

Either the 741 or the LM108 would seem to be suitable. However, the LM108 can use much higher bias resistances than the 741. So, for high instrument input impedance, use an LM108.

For the LM108.

supply voltage can be $\quad V_{CC} = \pm 5\ \text{V to } \pm 20\ \text{V}$

select $\quad R_1 = 1\ \text{M}\Omega$

$$C_1 = \frac{1}{2\pi f_L R_1/10} = \frac{1}{2\pi \times 10\ \text{Hz} \times 1\ \text{M}\Omega/10}$$

$\approx 0.16\ \mu$F (use 0.18 μF standard value)

For the diodes.

$$I_F(\text{max}) = I_p = 314\ \mu\text{A, and } f_{(\text{max})} > 1\ \text{kHz}$$

Use a 1N914 diode.

6-7 LINEAR OHMMETER CIRCUIT

The block diagram and scales for an ohmmeter consisting of a constant current source and a dc voltmeter circuit are shown in Fig. 6-12(a) and the complete circuit is given in Fig. 6-12(b).

When a known constant current is passed through an unknown resistor, the voltage drop across the resistor can be used as a measure of the resistance. For example, a current of 100 μA gives full scale deflection on a 1 V meter when the resistance is 10 kΩ. So, the meter scale can be marked as 10 kΩ as illustrated. Half-scale deflection will be obtained if the resistor is 5 kΩ; quarter-scale with 2.5 kΩ. This demonstrates that the ohmmeter scale is linear, unlike nonelectronic ohmmeters. Resistance range changing is effected by changing the level of the constant current. This is achieved by switching resistor R_4 values as illustrated in Fig. 6-12(b). A 1 mA current gives full-scale on the 1 V meter when the measured resistance is 1 kΩ. For a current of 10 mA, meter full-scale represents 100 Ω.

Design of a linear ohmmeter involves design of a constant current source and a voltmeter, as already discussed. The constant current source should use precision resistors for R_4, and potentiometer R_2 should be included to provide some adjustment in voltage V_1 for instrument calibration. For minimum power dissipation in the measured resistors, the voltage applied to them should be kept to a minimum. This requires that the dc voltmeter be a low voltage instrument.

Example 6-6

Design a linear ohmmeter circuit, as in Fig. 6-12(b), to use a 100 μA moving coil meter with a 2.5 kΩ coil resistance. The required ohmmeter resistance ranges are: 100 Ω, 1 kΩ, and 10 kΩ.

Solution

Voltmeter

To keep the power dissipation to a minimum in the measured resistors, design the voltmeter to have a full-scale deflection of 1 V.

In Fig. 6-12(b),

$$R_5 = \frac{V_{RX}}{I_m} = \frac{1\ \text{V}}{100\ \mu\text{A}}$$

$$= 10\ \text{k}\Omega$$

For op-amp A_2,

$$V_{o(\text{max})} = V_{R5} + I_m R_m = 1\ \text{V} + (100\ \mu\text{A} \times 2.5\ \text{k}\Omega)$$

$$= 1.25\ \text{V}$$

and

$$V_{i(\text{max})} = V_{RX} = 1\ \text{V}$$

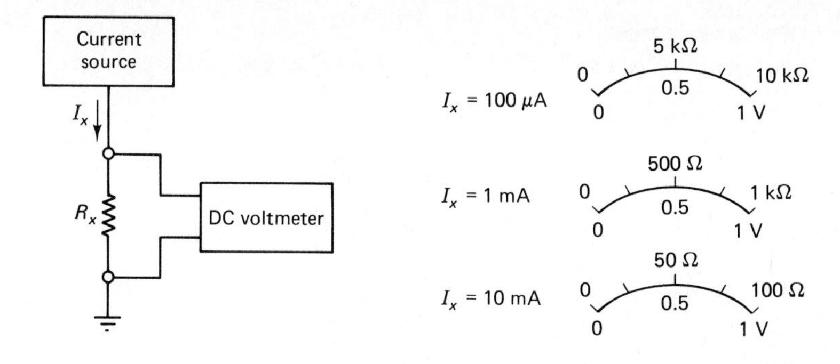

(a) Block diagram and scales for linear ohmmeter

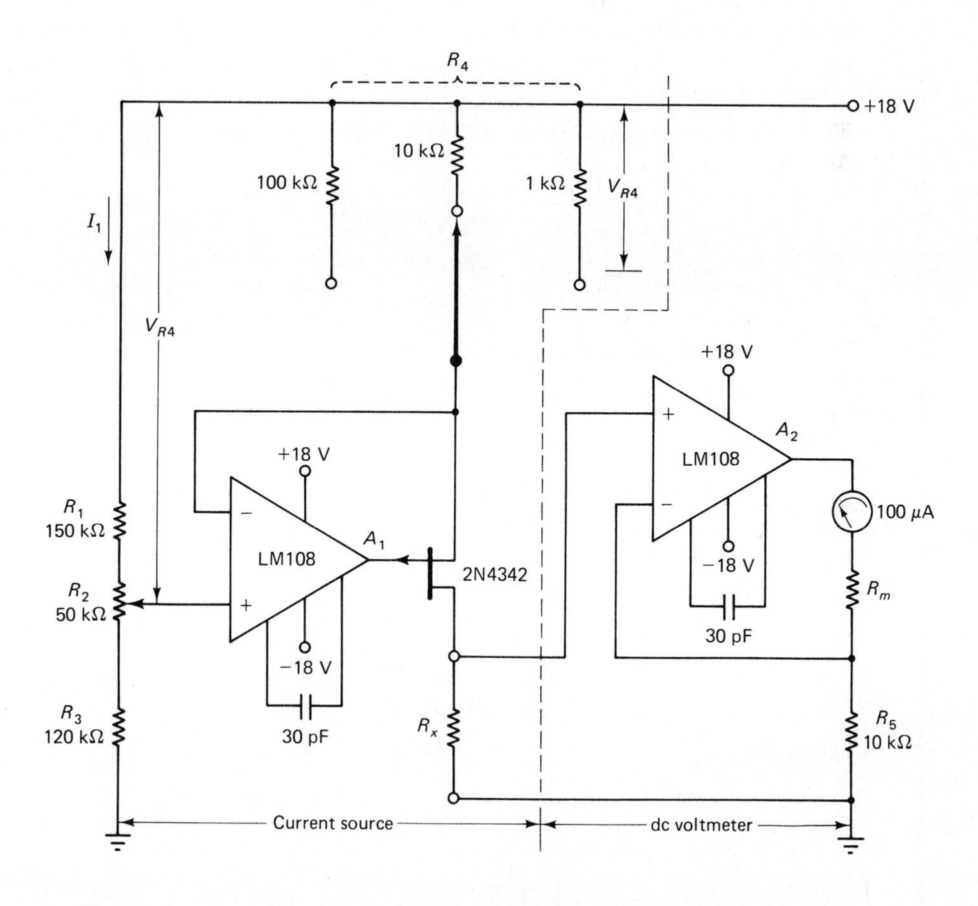

(b) Ohmmeter circuit diagram

Figure 6-12 Linear ohmmeter using a current source and a dc voltmeter. A known current is passed through the unknown resistance and the voltage drop across the resistance is measured using a meter with a scale calibrated in ohms.

Current Source

For the 10 kΩ, 1 kΩ, and 100 Ω resistance ranges, the three current ranges required are

$$I_x = \frac{1\ \text{V}}{10\ \text{k}\Omega},\ \frac{1\ \text{V}}{1\ \text{k}\Omega},\ \text{and}\ \frac{1\ \text{V}}{100\ \Omega}$$

$$= 100\ \mu\text{A},\ 1\ \text{mA},\ \text{and}\ 10\ \text{mA}$$

For the p-channel FET,

$$I_{D(\text{max})} = 10\ \text{mA},\ \text{and}\ V_{DS(\text{max})} = V_{CC}$$

The 2N4342 seems to be a suitable device. It has a specified maximum gate source cut-off voltage of

$$V_{GS(\text{off})} = 5.5\ \text{V}$$

This means the FET gate (the output of op-amp A_1) can be 5.5 V positive with respect to the source terminal. Allow the op-amp output to be at least 3 V below V_{CC} to keep it operating linearly.

$$V_{R4(\text{min})} = V_{GS(\text{off})} + 3\ \text{V} = 5.5\ \text{V} + 3\ \text{V}$$

$$= 8.5\ \text{V}$$

$$R_4 = \frac{V_{R4}}{I_x} = \frac{8.5\ \text{V}}{100\ \mu\text{A}},\ \frac{8.5\ \text{V}}{1\ \text{mA}},\ \text{and}\ \frac{8.5\ \text{V}}{10\ \text{mA}}$$

$$= 85\ \text{k}\Omega,\ 8.5\ \text{k}\Omega,\ \text{and}\ 850\ \Omega$$

Instead of 8.5 V, use $V_{R4} = 10$ V to get standard size resistors of

$$R_4 = 100\ \text{k}\Omega,\ 10\ \text{k}\Omega,\ \text{and}\ 1\ \text{k}\Omega$$

For satisfactory FET operation,

$$V_{DS(\text{min})} = V_{GS(\text{off})} + 1\ \text{V} = 5.5\ \text{V} + 1\ \text{V}$$

$$= 6.5\ \text{V}$$

Therefore,

$$V_{CC(\text{min})} = V_{RX} + V_{DS(\text{min})} + V_{R4}$$

$$= 1\ \text{V} + 6.5\ \text{V} + 10\ \text{V}$$

$$= 17.5\ \text{V}$$

use

$$V_{CC} = \pm 18\text{V}$$

For op-amp A_1,

$$V_{o(\text{max})} = V_{CC} - V_{R4} + V_{GS(\text{max})} = 18\ \text{V} - 10\ \text{V} + 5.5\ \text{V}$$

$$= 13.5\ \text{V}$$

and,

$$V_{1(\text{max})} = V_{CC} - V_{R4} = 18\ \text{V} - 10\ \text{V}$$

$$= 8\ \text{V}$$

Both the 741 and the LM108 can operate with a supply voltage of ±18 V and both have suitable input and output voltage ranges. Either one is suitable.

Potential Divider

$$I_1 \gg (I_{B(\max)} \text{ for } A_1)$$

let
$$I_1 = 50\ \mu\text{A}$$

$$V_{(R1+R2)} \approx V_{R4} + 10\% = 10\text{ V} + 10\%$$
$$= 11\text{ V}$$

$$R_1 + R_2 = \frac{11\text{ V}}{50\ \mu\text{A}} = 220\text{ k}\Omega$$

$$R_2 = 20\% \text{ of } (R_1 + R_2) = 44\text{ k}\Omega \qquad \text{(use 50 k}\Omega\text{ standard value potentiometer)}$$

$$R_1 = (R_1 + R_2) - R_2 = 220\text{ k}\Omega - 50\text{ k}\Omega$$
$$= 170\text{ k}\Omega \qquad \text{(use 150 k}\Omega\text{ standard value, and recalculate } I_1\text{)}$$

$$I_1 = \frac{V_{(R1+R2)}}{R_1 + R_2} = \frac{11\text{ V}}{(150\text{ k}\Omega + 50\text{ k}\Omega)}$$
$$= 55\ \mu\text{A}$$

$$R_3 = \frac{V_{CC} - V_{(R1+R2)}}{I_1} = \frac{18\text{ V} - 11\text{ V}}{55\ \mu\text{A}}$$
$$= 127\text{ k}\Omega \qquad \text{(use 120 k}\Omega\text{ standard value)}$$

6-8 INSTRUMENTATION AMPLIFIER

Differential Input/Output Amplifier

Figure 6-13 shows the circuit of an amplifier which accepts a differential input voltage and produces a differential output. As will be explained, this circuit has certain advantages over the difference amplifier introduced in Section 3-6. To understand the operation of the differential input/output amplifier, assume for the moment that the junction of resistors R_2 and R_3 in Fig. 6-13 is grounded. Op-amp A_1 together with resistors R_1 and R_2 would function as a noninverting amplifier. The input voltage to A_1 would be developed across R_2 and the voltage gain would be $(R_1 + R_2)/R_2$. Similarly, if the junction of resistors R_1 and R_2 was grounded, op-amp A_2 and resistors R_2 and R_3 would behave as a noninverting amplifier with a voltage gain of $(R_2 + R_3)/R_2$.

Now consider the operation of the complete circuit. Because of the noninverting amplifier configuration of A_1 and A_2 and the resistors, the voltage at the junction of R_1 and R_2 is equal to input voltage v_1. Also, the voltage at the junction of R_2 and R_3 equals input voltage v_2. Consequently, the voltage across resistor R_2 is

$$V_{R2} = v_1 - v_2 = V_i \text{ (the differential input)}$$

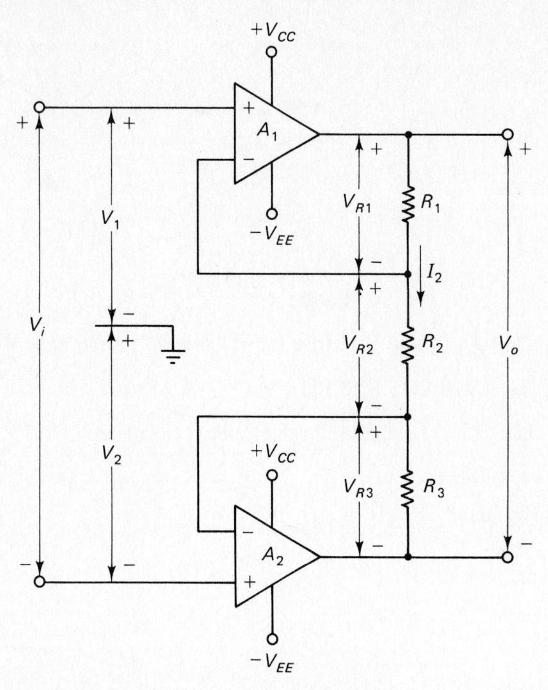

Figure 6-13 Differential input/output amplifier which accepts a differential input voltage and amplifies it to produce a differential output. With $R_3 = R_1$, the voltage gain is $A_v = (2R_1 + R_2)/R_2$.

This gives the current through R_2 as

$$I_2 = \frac{V_i}{R_2}$$

The differential output voltage is,

$$\begin{aligned} V_o &= V_{R1} + V_{R2} + V_{R3} \\ &= I_2(R_1 + R_2 + R_3) \\ &= \frac{V_i}{R_2}(R_1 + R_2 + R_3) \end{aligned}$$

The circuit differential voltage gain is,

$$A_{v(\text{dif})} = \frac{V_o}{V_i} = \frac{R_1 + R_2 + R_3}{R_2} \tag{6-5}$$

Normally

$$R_1 = R_3$$

so,

$$A_{v(\text{dif})} = \frac{2R_1 + R_2}{R_2} \tag{6-6}$$

Note that the voltage gain can be altered by adjusting a single resistor; R_2.

Like all amplifiers with a differential input, the common mode gain of the differential input/output amplifier is an important quantity. Suppose the two inputs are connected together and a common mode noise voltage (V_n) is applied to the two as shown in Fig. 6-14. The junction of R_1 and R_2 will be at the same voltage as the noninverting input terminal of A_1, and the junction of R_2 and R_3 will be at the same voltage as the noninverting input of A_2. That is, both resistor junctions will be at v_n volts with respect to ground. There will be no current flow through R_1, R_2, or R_3, and the output voltage of each amplifier will be v_n volts. This means the common mode gain is

$$A_{v(cm)} = 1 \tag{6-7}$$

So, common mode signals will be passed through but not amplified by the differential input/output amplifier.

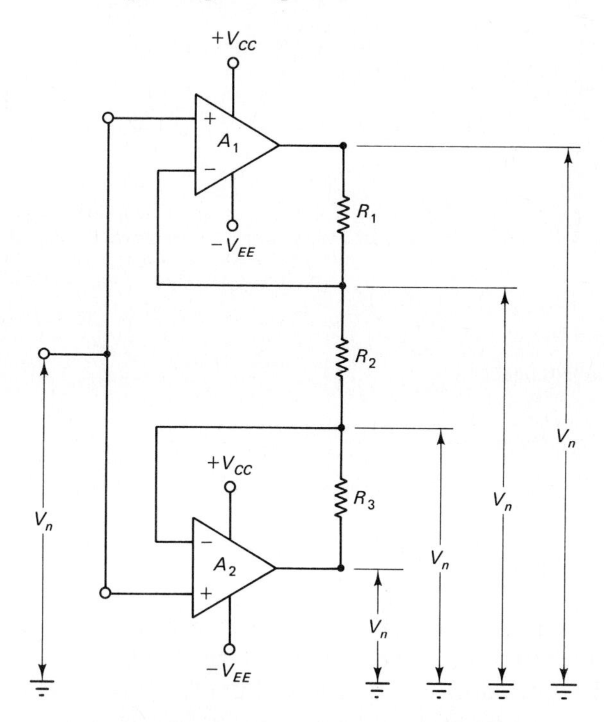

Figure 6-14 A common mode voltage applied to a differential input/output amplifier is reproduced at the output. $A_{v(cm)} = 1$.

The differential input/output amplifier is normally used in conjunction with the difference amplifier discussed in Section 3-6. The two circuits are reproduced in Fig. 6-15 for comparison. The resistance at each input terminal of the differential input/output amplifier is extremely high because of the noninverting amplifier configuration. The input resistance of the difference amplifier is $R_i = R_1$ at input terminal 1, and $R_i = (R_3 + R_4)$ at terminal 2. The voltage gain of the differential input/output amplifier can be changed by adjusting only one resistor: R_2. Changing the gain of the difference amplifier requires that R_2 and R_4 be adjusted together to maintain equal amplification of both inputs. The common mode gain of the differential input/output amplifier is 1, compared to a common mode gain of zero for the difference amplifier. As illustrated in Fig. 6-15, the differential input/output amplifier operates with a floating load, while the difference amplifier uses a grounded load.

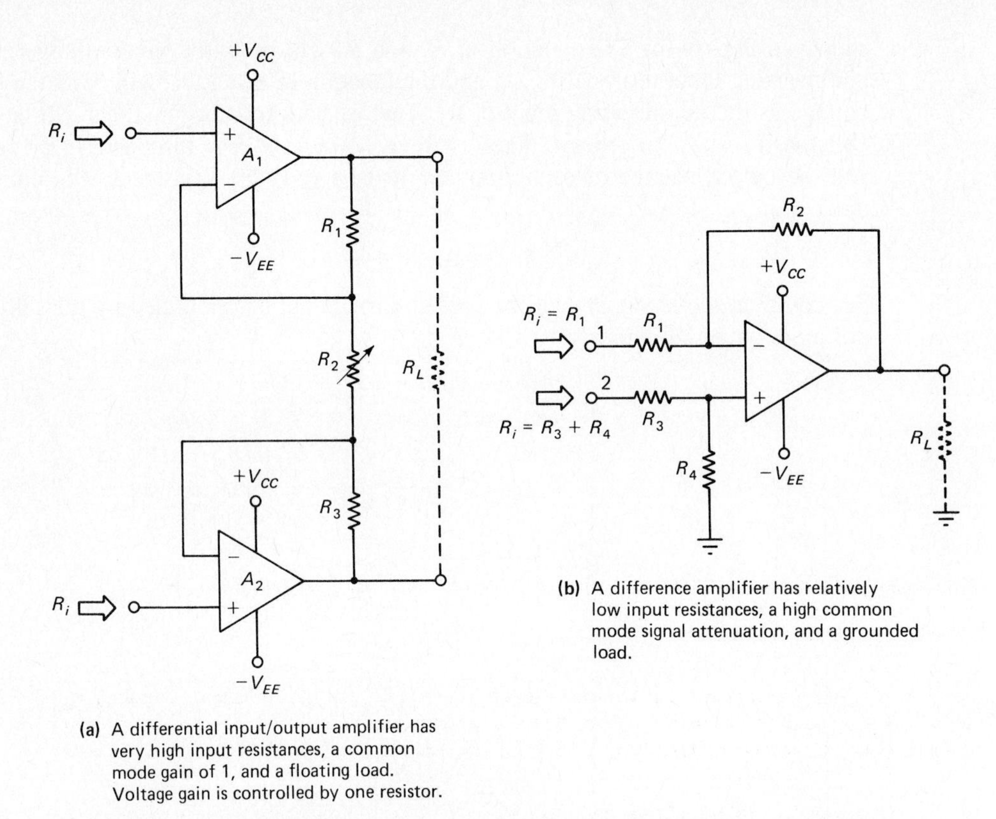

Figure 6-15 Comparison of a differential input/output amplifier and a difference amplifier.

Instrumentation Amplifier Circuit

The instrumentation amplifier circuit shown in Fig. 6-16 is a combination of the differential input/output amplifier (stage 1) and the difference amplifier (stage 2). The difference amplifier uses the differential output voltages from the differential input/output amplifier to drive a grounded load, as illustrated. For instrumentation purposes, most loads have one grounded terminal, otherwise ground loops and static electricity could cause problems. So, the ability to drive a grounded load is necessary. The differential input/output stage offers a very high input resistance at each input terminal. The voltage gain of the complete circuit is,

$$A_v = A_{v1} \times A_{v2}$$

where A_{v1} is the voltage gain of stage 1 and A_{v2} is the stage 2 gain.

$$A_v = \frac{2R_1 + R_2}{R_2} \times \frac{R_5}{R_4} \tag{6-8}$$

The overall voltage gain can be controlled by adjustment of R_2. The common mode signal attenuation for the instrumentation amplifier is that provided by the difference amplifier. As discussed in Section 3-6, this can be maximized by making resistor R_7

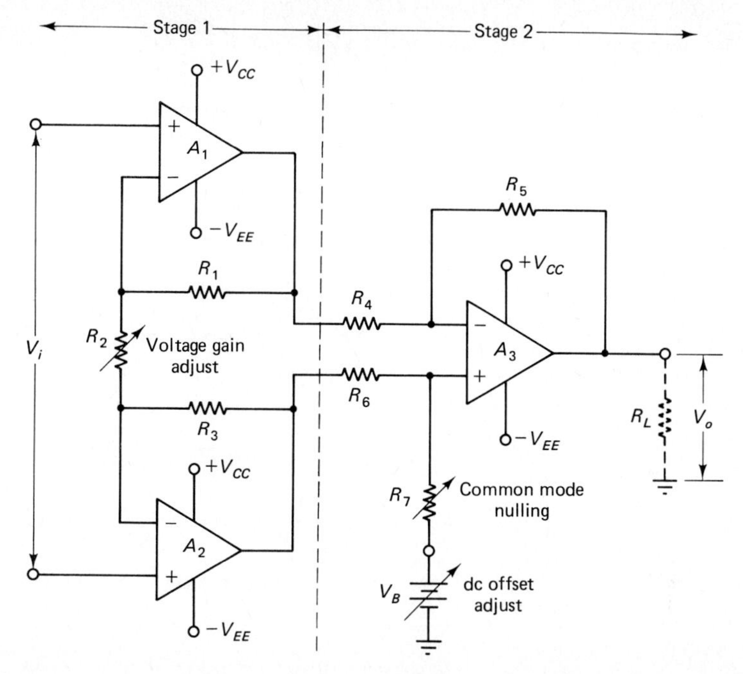

Figure 6-16 Instrumentation amplifier consisting of a differential input/output amplifier input stage and a difference amplifier output stage. The circuit has adjustable voltage gain, common mode output nulling, and dc output voltage level shifting.

adjustable. Also, as discussed in Section 3-6, the dc output voltage level can be controlled if R_7 is connected to an adjustable bias voltage instead of being directly grounded.

Example 6-7

Design an instrumentation amplifier to have an overall voltage gain of 900. The input signal amplitude is 15 mV, 741 op-amps are to be used, and the supply is ±15 V.

Solution

Stage 1

$$\text{let } A_{v1} \approx A_{v2} = \sqrt{A_v} = \sqrt{900} = 30$$

$$I_2 \gg I_{B(\max)}$$

let

$$I_2 = 100 I_{B(\max)} = 50\ \mu\text{A}$$

$$R_2 = \frac{V_i}{I_2} = \frac{15\text{ mV}}{50\ \mu\text{A}} = 300\ \Omega \qquad \text{(use 500 } \Omega \text{ variable)}$$

Eq. 6-6,

$$A_{v(\text{dif})} = \frac{2R_1 + R_2}{R_2}$$

which gives, $R_1 = \frac{R_2}{2}[A_{v(\text{dif})} - 1] = \frac{270\ \Omega}{2}[30 - 1]$

$$\approx 3.9\ \text{k}\Omega \qquad \text{(standard value)}$$

$$R_3 = R_1 = 3.9\ \text{k}\Omega$$

Stage 2 $V_o = A_v V_i = 900 \times 15\ \text{mV}$

$$= 13.5\ \text{V}$$

$$I_{5(\min)} \gg I_{B(\max)}$$

let $I_5 = 100 I_{B(\max)} = 50\ \mu\text{A}$

$$R_5 = \frac{V_o}{I_5} = \frac{13.5\ \text{V}}{50\ \mu\text{A}}$$

$$= 270\ \text{k}\Omega \qquad \text{(standard value)}$$

$$R_4 = \frac{R_5}{A_{v2}} = \frac{270\ \text{k}\Omega}{30}$$

$$= 9\ \text{k}\Omega \qquad \text{(use } 8.2\ \text{k}\Omega + 820\ \Omega \text{ or } 9.1\ \text{k}\Omega \pm 5\%)$$

$$R_6 = R_4 = 9\ \text{k}\Omega$$

$$R_7 = R_5 \pm 20\% \approx 270\ \text{k}\Omega \pm 20\%$$

$$\approx 216\ \text{k}\Omega \text{ to } 324\ \text{k}\Omega \qquad \text{(use a } 220\ \text{k}\Omega \text{ fixed resistor and a } 100\ \text{k}\Omega \text{ variable)}$$

Integrated Circuit Instrumentation Amplifier

Instrumentation amplifiers are available in a single integrated circuit package. Figure 6-17 shows the simplified circuit of the LH0036 IC instrumentation amplifier. In this circuit, difference amplifier resistors R_3, R_4, R_5, and R_6 are all equal value components; so that the voltage gain of that stage is 1. The overall voltage gain for the circuit is set by the externally connected resistor R_G. The equation for voltage gain is similar to Eq. 6-6,

$$A_v = \frac{2R_1 + R_G}{R_G}$$

With $R_1 = R_2 = 25\ \text{k}\Omega$, the equation for R_G becomes,

$$R_G = \frac{50\ \text{k}\Omega}{A_v - 1} \tag{6-9}$$

If R_G is left as infinity, the voltage gain is 1, while $R_G = 50\ \Omega$ give a voltage gain of approximately 1000. Resistor R_6 is trimmed by the manufacturer to yield a common mode rejection ratio greater than 80 dB. R_6 is also terminated at pin 9, which is usually grounded. Instead of grounding pin 9, it can be connected to a bias voltage source for dc output voltage level control. The LH0036 has an input impedance of

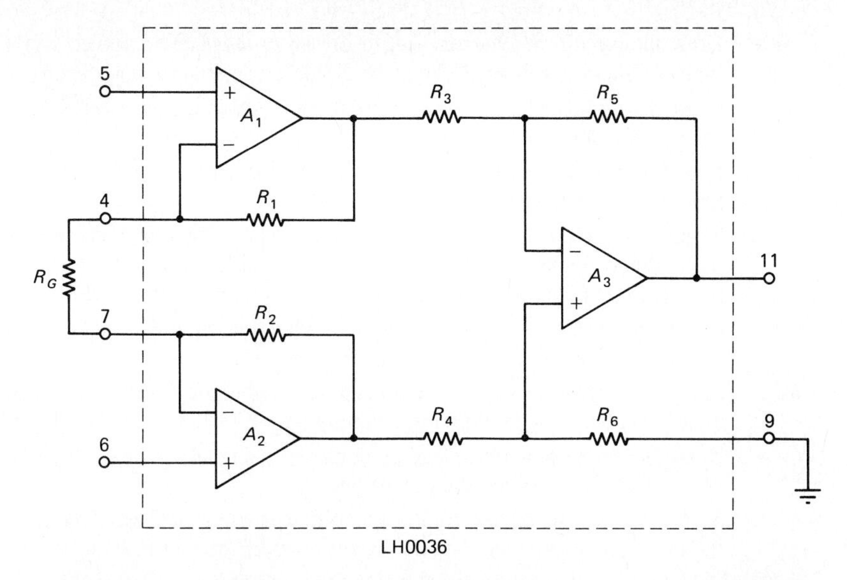

Figure 6-17 Simplified circuit of a LH0036 IC instrumentation amplifier. The voltage gain is programmed by selection of resistor R_G.

300 MΩ, a voltage gain adjustable from 1 to 1000, and it can operate with supply voltages ranging from ±1 V to ±18 V.

REVIEW QUESTIONS

6-1. Sketch the circuit of a low-resistance voltage source using an op-amp and a bipolar transistor. Show how a potential divider or a Zener diode may be used to determine the output voltage. Explain.

6-2. Draw the circuit of a percision voltage source using an op-amp and a Zener diode. Explain the circuit operation, and derive the equation relating V_o and V_z.

6-3. Draw the circuit of a current source for a floating load and the circuit of a current source for a grounded load, each using an op-amp and a bipolar transistor. Indicate typical voltage levels throughout each circuit and explain the operation of each circuit.

6-4. Draw the circuit of a current source using an op-amp and a low power FET. Indicate typical voltage levels throughout the circuit and explain its operation. Compare the performance of a current source using a FET to that of one using a bipolar transistor.

6-5. Draw the circuit of a current source using an op-amp and a power MOSFET. Indicate typical voltage levels throughout the circuit and explain its operation.

6-6. Sketch a current sink circuit that uses an op-amp and a bipolar transistor. Show typical voltage levels throughout the circuit and explain its operation.

6-7. Sketch the circuit of a current sink using an op-amp and a low power FET. Show typical voltage levels throughout the circuit.

6-8. Draw the circuit of a current sink using an op-amp and a power MOSFET. Indicate typical voltage levels throughout the circuit and explain its operation.

6-9. Draw the circuit of a precision current source with a power MOSFET output stage. Explain the circuit operation.

6-10. Sketch a precision current sink circuit with a power MOSFET output stage. Explain the circuit operation.

6-11. Sketch the circuit of a simple current-to-voltage converter, then show how it should be modified to function as a current amplifier/attenuator with a grounded load. Explain the operation of the amplifier/attenuator and derive an equation for current gain.

6-12. Sketch the circuit of a current amplifier with a floating load. Explain the circuit operation and derive an equation for current gain.

6-13. Sketch the circuit of an op-amp voltage-to-current converter used as a simple dc voltmeter. Explain its operation, and explain why it makes a satisfactory voltmeter.

6-14. Draw the circuit of a multirange dc voltmeter using an op-amp voltage-to-current converter. Carefully explain the circuit operation.

6-15. Sketch the circuit of an op-amp ac voltmeter using half-wave rectification. Explain the circuit operation and discuss the accuracy of the rectification process.

6-16. Sketch the complete circuit of an op-amp ac voltmeter using full-wave rectification. Carefully explain the circuit operation, and discuss the function of all components.

6-17. Draw the block diagram and scales for a linear ohmmeter. Explain the circuit operation and the scale markings.

6-18. Draw a complete circuit for a linear ohmmeter and explain its operation. Also, show how range changing and instrument calibration is effected.

6-19. Sketch the circuit of a differential input/output amplifier. Carefully explain the circuit operation and derive the equation for differential voltage gain. Also, show that the common mode gain is 1.

6-20. Compare the performance of a differential input/output amplifier with that of a difference amplifier.

6-21. Sketch the complete circuit of an instrumentation amplifier. Discuss the characteristics of the circuit, and show how the voltage gain can be varied. Also, show the method of nulling common mode outputs and how the dc output voltage can be level shifted.

PROBLEMS

6-1. Design a low-resistance voltage source as in Fig. 6-1(a), to provide an output of 8 V. A 741 op-amp with a ± 15 V supply is to be used, and the maximum output current is to be 60 mA.

6-2. Using a 741 op-amp and a $+20$ V supply (not $\pm$), design a voltage source to provide a constant output of 9 V to a load which can vary from a maximum of 450 Ω to a minimum of 225 Ω. Design the circuit to (a) use a potential divider, (b) use a Zener diode.

6-3. A precision voltage source as in Fig. 6-2 is to produce a 5 V output adjustable by $\pm 20\%$. Design the circuit using a BIFET op-amp with a ± 15 V supply.

6-4. A precision voltage source is to produce an output of −8 V. Design a suitable circuit using a 741 op-amp with a ±18 V supply. Provide adjustment for ±10% tolerance on the Zener diode.

6-5. Design a current source using a 741 op-amp and a bipolar transistor, as in Fig. 6-3(b). The supply voltage is ±18 V, the load resistance is 47 Ω, and the output current is to be 80 mA.

6-6. Redesign the circuit of Problem 6-5 to use a power MOSFET, as in Fig. 6-4(b).

6-7. Design a constant current circuit as in Fig. 6-3(b) to drive 45 mA through a load that can vary from 100 Ω to 150 Ω. A 741 op-amp with a ±15 V supply is to be used.

6-8. Design the current sink circuit in Fig. 6-5(a) to provide a current of 30 mA to a load with a maximum resistance of 200 Ω. Use a 741 op-amp with a ±18 V supply.

6-9. A current sink using a BIFET op-amp and a low-power junction FET is required to produce an output of 35 mA to a 140 Ω load. Design the circuit and specify a suitable FET. The available supply voltage is ±20 V.

6-10. Using a LF353 BIFET op-amp with a supply of ±22 V, design a precison current sink as in Fig. 6-6 to supply 50 mA to a load with a maximum resistance of 120 Ω.

6-11. A precision current source as in Fig. 6-7 is to produce an output of 120 mA to a 33 Ω load resistor. Design the circuit using an LF353 op-amp and a supply of ±18 V.

6-12. The circuit in Fig. 6-9(a) has an input current of approximately 1 mA and a 100 Ω load resistor (R_2). Resistor R_1 is to be made adjustable to give the amplifier a current gain ranging from 5 to 10. Calculate maximum and minimum resistances for R_1 and select a suitable op-amp.

6-13. A 500 μA current is to be amplified to a level of 10 mA by means of a current amplifier circuit as in Fig. 6-9(b). The load resistance is 1 kΩ, and a 741 op-amp is to be used with a ±20 V supply. Design a suitable circuit.

6-14. A 100 μA current (I_s) is to be monitored by means of a current amplifier as in Fig. 6-9(b). A 500 μA moving-coil meter with a resistance of 1.5 kΩ is to be substituted in place of the load resistor. Design the circuit using a 741 op-amp with a ±12 V supply. Also, calculate the typical voltage drop introduced in series with I_s, (a) when the current amplifier circuit is used, (b) when the 500 μA meter is connected directly in series with I_s.

6-15. A dc voltmeter as in Fig. 6-10(b) is to be designed to have ranges of 0.3 V, 3 V, 30 V, and 300 V. Design a suitable circuit to use an LM108 op-amp and a 1 mA moving-coil meter with a resistance of 1.3 kΩ.

6-16. The dc voltmeter in Problem 6-15 is converted into an ac voltmeter by simply including a rectifier [as in Fig. 6-11(a)] without making any other changes. Calculate the actual rms ac voltage that gives full-scale meter deflection on the 0.3 V range.

6-17. An ac voltmeter using a full-wave bridge rectifier [as in Fig. 6-11(b)] is to give full scale deflection for an input of 1 V rms. A 100 μA moving-coil meter with a coil resistance of 2.5 kΩ is to be used, and the signal frequency range is to be 10 Hz to at least 1 kHz. Design the circuit and select a suitable op-amp and supply voltage.

6-18. A linear ohmmeter circuit as in Fig. 6-12(b) is to use a 1 mA moving coil meter with a resistance of 1.3 kΩ. The resistance ranges are to be 30 Ω, 300 Ω, and 3 kΩ. Design the voltmeter portion of the circuit.

6-19. Design the current source section (without the potential divider) for the linear ohmmeter in Problem 6-18.

6-20. Design the potential divider for the linear ohmmeter in Problems 6-18 and 6-19.

6-21. An instrumentation amplifier with a differential input of 20 mV is to have a voltage gain adjustable from 500 to 600. A ±15 V supply and 741 op-amps are to be used. Design the circuit.

6-22. Determine the range of resistance of R_G for a LH0036 IC instrumentation amplifier to give a voltage gain adjustable from 30 to 300.

COMPUTER PROBLEMS

6-23. Write a computer program to design a precision voltage source as in Fig. 6-2. Given: V_o and V_{CC}.

6-24. Write a computer program to design a precision current sink as in Fig. 6-6. Given: I_D, R_L, and V_{CC}.

6-25. Write a computer program to design a full-wave rectifier voltmeter using a BIFET op-amp as in Fig. 6-11(b). Given: v_i for full scale, meter full-scale current, and meter resistance.

6-26. Write a computer program to design an instrumentation amplifier using bipolar op-amps as in Fig. 6-16. Given: $V_{i(\text{dif})}$, and the A_v range.

LABORATORY EXERCISES

6-1 Voltage and Current Sources

1. Construct the voltage source in Fig. 6-1(b) as designed in Example 6-1.
2. Carefully measure V_z and V_o, and compare to the design values.
3. Double the load resistance by connecting another 150 Ω resistor in series with R_L. Repeat Procedure 2.
4. Switch the supply *off,* then connect a milliammeter in series with the collector terminal of Q_1 and the positive supply.
5. Switch the supply *on* again. The circuit is now functioning as a current sink as in Fig. 6-5(a).
6. Measure the transistor collector current. Change the resistance in series with the emitter terminal from 300 Ω to 150 Ω and measure the collector current once more.
7. Construct the precision voltage source in Fig. 6-2 as designed in Example 6-2.
8. Measure the output voltage at the extremes of its range of adjustment and compare to the designed voltage range.

6-2 DC Voltmeter

1. Construct the voltage-to-current converter circuit in Fig. 6-10(a). Use an LM108 op-amp with a 30 pF compensating capacitor and a supply of ±15 V, a 100 μA permanent magnet moving coil (PMMC) meter with a coil resistance less than 2.5 kΩ, and R_1 = 8.2 kΩ in series with a 5 kΩ variable resistor.

2. Using an adjustable dc voltage source, set the circuit input to exactly 1 V, then adjust the variable resistor to produce full scale deflection on the PMMC instrument.
3. Adjust the input to 750 mV, 500 mV, and 250 mV, in turn. Note the scale deflection on the PMMC instrument for each case.

6-3 Instrumentation Amplifier

1. Construct the instrumentation amplifier in Fig. 6-16 using the component values determined in Example 6-7.
2. Ground the input terminals, set R_2 to maximum, set V_B to zero, and adjust R_7 for zero output.
3. Connect the two input terminals together and apply +10 V to them from a dc source. Measure the output voltage and calculate the common mode voltage gain.
4. Disconnect the input terminals and apply +9 mV to one terminal and −9 mV to the other. Monitoring the output on a voltmeter, adjust R_2 until V_o = 8.1 V. Then, adjust V_B to give V_o = 10 V.
5. Reverse the polarity at the two inputs and note the new output level.
6. Switch *off* the supply voltage. Disconnect R_2 without altering its setting and measure its resistance on an ohmmeter.
7. Analyze the circuit to determine its voltage gain and its common mode gain. Compare to the measured results and to the designed gain in Example 6-7.

CHAPTER 7

Signal Processing Circuits

Objectives

- Sketch the following op-amp circuits and explain the operation of each: *saturating half-wave precision rectifier, nonsaturating half-wave precision rectifier, two-output half-wave precision rectifier, full-wave precision rectifier, high input impedance precision rectifier, Zener diode peak clipper, dead zone circuit, precision limiters, precision clamper, precision peak detector, sample-and-hold circuit.*
- Design each of the above circuits, using bipolar and BIFET op-amps together with transistors, FETs, and MOSFETs where appropriate.
- Analyze each of the above circuits to determine input impedances, output impedances, and the responses to various input waveforms.

INTRODUCTION

The use of operational amplifiers can improve the performance of a wide variety of signal processing circuits. In rectifier circuits, the voltage drop that occurs with an ordinary semiconductor rectifier can be eliminated to give precision rectification. Waveforms can be limited and clamped at precise levels when op-amps are employed in clipping and clamping circuits. The errors that occur with peak detectors and sample-and-hold circuits can also be minimized by the use of op-amps.

7-1 PRECISION HALF-WAVE RECTIFIERS

Saturating Precision Rectifier

The circuit of an op-amp *precision half-wave rectifier,* illustrated in Fig. 7-1(a), is simply a voltage follower with a diode inserted between the op-amp output terminal and the circuit output point. When the input signal is positive, the diode is forward

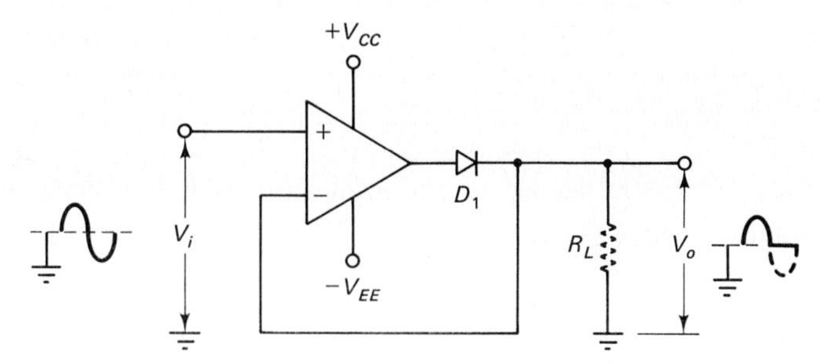

(a) Saturating precision half-wave rectifier circuit

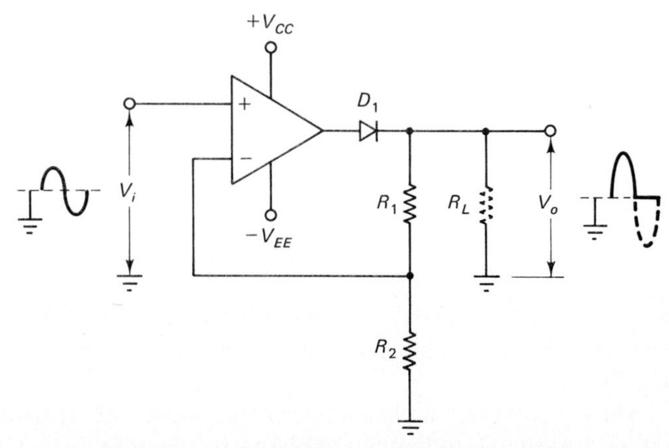

(b) Saturating precision half-wave rectifier with voltage gain

Figure 7-1 Including a diode at the output of a voltage follower or a noninverting amplifier converts the circuit into a precision half-wave rectifier. Because the op-amp output saturates when the diode is reversed, there is a relatively long recovery time.

biased and the output voltage follows the input. Recall that negative feedback causes the voltage at the inverting input terminal to follow that at the noninverting terminal. Since the output of the circuit and the inverting terminal are common, the output follows the input within microvolts. The diode forward voltage drop is not involved.

While the input voltage is in its negative half-cycle the op-amp output is negative and the diode is reverse biased. Consequently, the feedback path is interrupted, the inverting input terminal remains at ground level (because R_L is grounded), and the op-amp output is saturated in a negative direction. The negative half-cycle of the input does not pass to the output; it is clipped off. If the diode polarity is reversed in Fig. 7-1(a), the negative half-cycle of the input waveform will be passed to the output and the positive half-cycle will be clipped off.

The advantages of the op-amp precision rectifier circuit over a simple diode rectifier are (1) no diode voltage drop between input and output, (2) the ability to rectify very small voltages (less than the typical 0.7 V diode forward voltage drop), (3) amplification, if required, and (4) low output impedance. Items (1) and (2) indicate that the precision rectifier is a close approximation to an ideal diode.

While the input waveform is in its negative half-cycle, the output of the op-amp in Fig. 7-1(a) is saturated in a negative direction. Some time is required to get the op-amp out of saturation and this will limit the frequency response of the circuit. For high frequency performance, a nonsaturating precision rectifier circuit must be used.

Figure 7-1(b) shows the circuit of a precision rectifier with voltage gain. This is a noninverting amplifier with the diode included. The circuit is designed exactly as a noninverting amplifier, except that the current through R_1 and R_2 should be a minimum of about 100 μA to ensure that the diode is operating correctly in the absence of a load current. A minimum diode current of 500 μA is a good design objective.

Nonsaturating Precision Rectifier

The precision rectifier circuit in Fig. 7-2 uses an inverting amplifier configuration. Diode D_1 is reverse biased and D_2 is forward biased when the op-amp output terminal is positive. This occurs when the input signal is negative. While D_2 is forward biased, the circuit output is

$$V_o = -V_i\frac{R_2}{R_1}$$

If R_2 equals R_1, V_o equals $-V_i$ during the negative half-cycle of the input. If R_2 is greater than R_1, the output is an amplified (inverted and half-wave rectified) version of the input.

During the positive half-cycle of the input, the op-amp output terminal goes negative, causing D_2 to be reverse biased. Without D_1 in the circuit, the op-amp output would be saturated in a negative direction. However, the negative voltage at the op-amp output forward biases D_1. This tends to pull the op-amp inverting input ter-

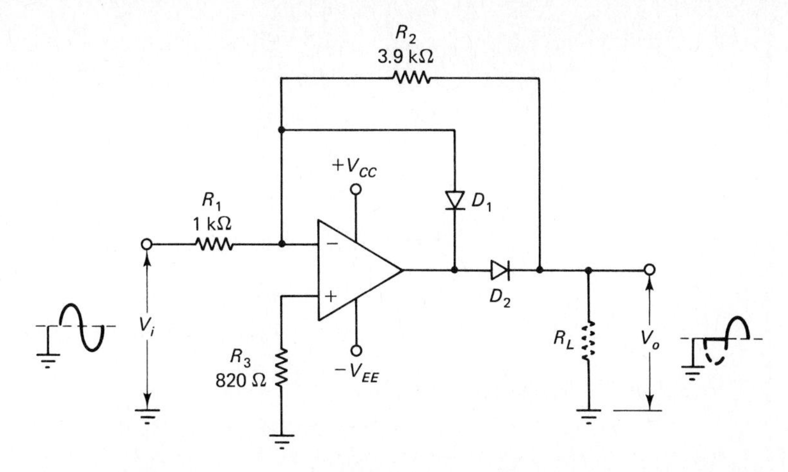

Figure 7-2 An inverting amplifier circuit is converted into a nonsaturating precision half-wave rectifier circuit by the inclusion of two diodes. When D_2 is reversed, D_1 provides negative feedback to keep the output from saturating.

minal in a negative direction. But, such a move would cause the op-amp output to go positive. So, the output settles at the voltage that keeps the inverting input terminal close to ground level. In this case, that voltage is the diode forward voltage drop below ground, approximately -0.7 V. The sole purpose of D_1 is to keep the op-amp output from going into saturation and thus maintain the best possible circuit frequency response.

In Fig. 7-2, the negative half-cycle of the input is inverted and passed to the output while the positive half-cycle is clipped off. If the polarity of the two diodes is reversed, the negative half-cycle will be clipped, and the positive half-cycle will be inverted and reproduced at the output.

Design of a nonsaturating precision rectifier involves design of an inverting amplifier. The diodes should have a maximum reverse voltage greater than the supply voltage. They should also have a switching time that will not limit the circuit frequency response. This requires that the diode reverse recovery time (t_{rr}) be much smaller than the time period of the highest signal frequency to be processed. Because there is 100% feedback when D_1 is forward-biased (in Fig. 7-2), the op-amp should be compensated as a voltage follower.

Example 7-1

Design a nonsaturating precision half-wave rectifier as in Fig. 7-2, to produce a 2 V peak output from a sine wave input with a peak value of 0.5 V and frequency of 1 MHz. Use a bipolar op-amp with a supply voltage of ± 15 V.

Solution

$$I_1 \gg I_{B(\max)}$$

let $\quad I_1 = 500\ \mu\text{A} \quad$ (for adequate diode current)

$$R_1 = \frac{V_i}{I_1} = \frac{0.5\text{ V}}{500\ \mu\text{A}}$$

$$= 1\text{ k}\Omega \qquad \text{(standard value)}$$

$$R_2 = \frac{V_o}{I_1} = \frac{2\text{ V}}{500\ \mu\text{A}}$$

$$= 4\text{ k}\Omega \qquad \text{(use 3.9 k}\Omega\text{ standard value)}$$

$$R_3 = R_1 \parallel R_2 = 1\text{ k}\Omega \parallel 3.9\text{ k}\Omega$$

$$= 796\ \Omega \qquad \text{(use 820 }\Omega\text{ standard value)}$$

For diodes D_1 and D_2,

$$V_R > [V_{CC} - (-V_{EE})] > [15\text{ V} - (-15\text{ V})]$$

$$> 30\text{ V}$$

$$t_{rr} \ll T$$

let

$$t_{rr(\max)} = \frac{T}{10} = \frac{1}{10 \times f}$$

$$= \frac{1}{10 \times 1\text{ MHz}} = 0.1\ \mu\text{s}$$

Compensate the op-amp as a voltage follower.

Two-Output Precision Rectifier

Although the precision rectifier circuit illustrated in Fig. 7-3 has two output terminals, it is just a modification of the circuit in Fig. 7-2. Diode D_2 is connected in series with the op-amp output terminal and resistor R_2 exactly as in Fig. 7-2. Diode D_1 is also connected as in Fig. 7-2, except that resistor R_4 has been added in series with D_1.

During the positive half-cycle of the input, the op-amp output terminal goes negative, causing D_1 to be forward-biased, and D_2 to be reversed. In this situation, the op-amp together with resistors R_1 and R_4 functions as an inverting amplifier to give an output at terminal B

$$V_B = -(+V_i)\frac{R_4}{R_1}$$

No current flows in resistor R_2 at this time because D_2 is reverse-biased. Also, R_2 is connected to the virtual ground at the op-amp inverting input terminal. So, the output at point A remains at ground level for the duration of the positive half-cycle of the input.

When the input is in its negative half-cycle, the op-amp output goes positive, D_1 is reverse-biased, and D_2 is forward-biased. Now, resistors R_1 and R_2 combine

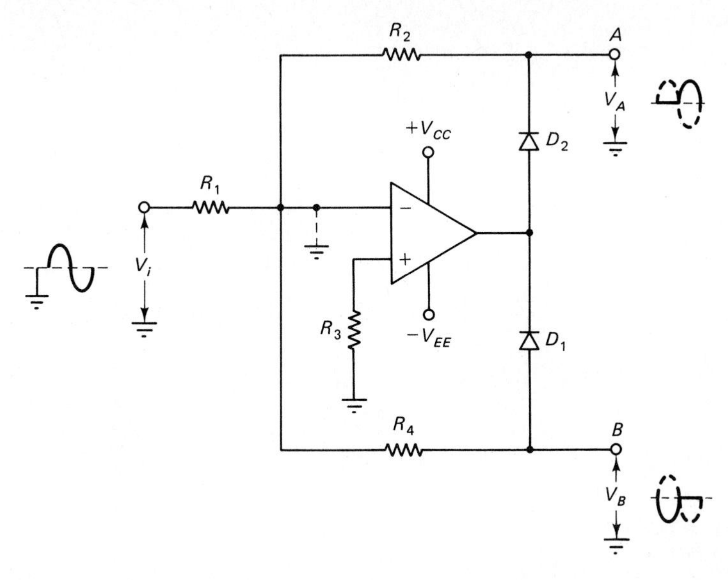

Figure 7-3 Two-output precision half-wave rectifier. D_1 is forward biased during the positive half-cycle, D_2 during the negative half-cycle. In each case, the circuit behaves as an inverting amplifier.

with the op-amp to perform as an inverting amplifier. The output at terminal A is

$$V_A = -(-V_i)\frac{R_2}{R_1}$$

With D_1 reversed, no current flows in resistor R_4; and with R_4 connected to the virtual ground at the op-amp inverting input, the output at terminal B is zero.

The circuit in Fig. 7-3 is seen to be a precision rectifier with positive half-cycles of output at terminal A, and negative half-cycles at terminal B.

7-2 PRECISION FULL-WAVE RECTIFIERS

Half-Wave Rectifier and Summing Circuit

The left side of the circuit in Fig. 7-4 is a precision half wave rectifier as in Fig. 7-2 but with the diodes reversed. The right side is an inverting summing amplifier circuit, (explained in Section 3-5). The input voltage is applied to terminal A of the summing amplifier and to the input of the precision rectifier. Note that resistor R_2 in the precision half-wave rectifier circuit has twice the resistance of R_1, so the rectified voltage applied to terminal B of the summing amplifier is -2 V_i, as illustrated.

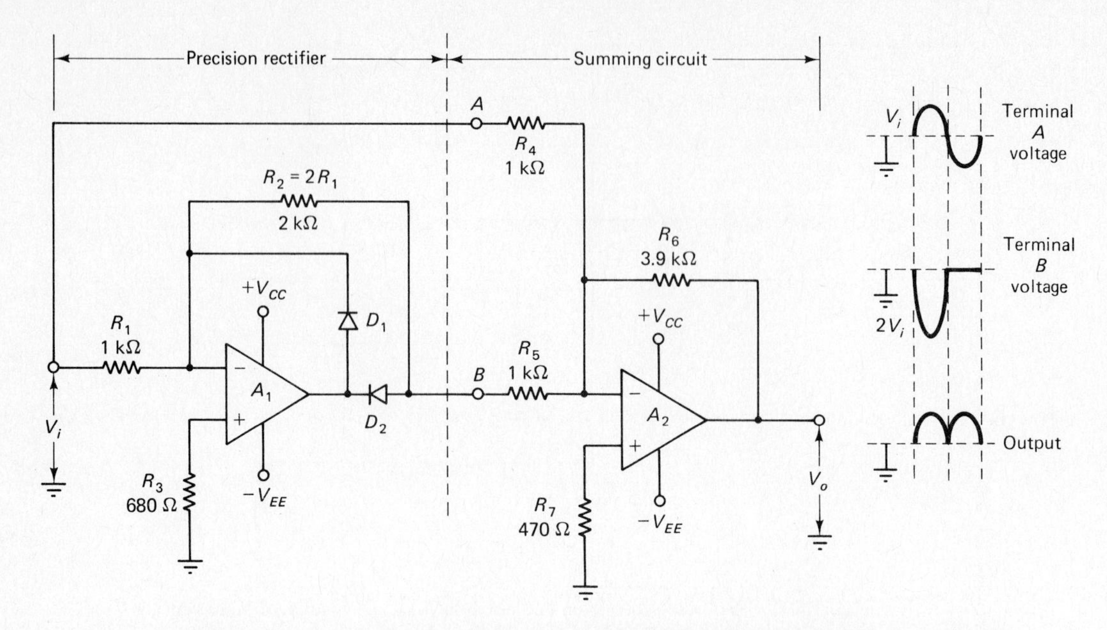

Figure 7-4 Full-wave precision rectifier consisting of a summing circuit and a precision half-wave rectifier which has a voltage gain of 2. During the input positive half-cycle, $-2V_i$ is summed with V_i and the result is inverted. During the negative half-cycle, V_i and zero are summed and inverted.

During the positive half-cycle of the input, the voltage at terminal A is $+V_i$, while that at terminal B is $-2\ V_i$. The output from the summing circuit with $R_5 = R_4$ is

$$V_o = -\frac{R_6}{R_4}(V_A + V_B)$$

$$= -\frac{R_6}{R_4}(V_i - 2\ V_i)$$

$$= \frac{R_6}{R_4}V_i$$

During the negative half-cycle of the input, $V_A = -V_i$, and $V_B = 0$. Consequently, the output is

$$V_o = -\frac{R_6}{R_4}(-V_i + o)$$

$$= \frac{R_6}{R_4}V_i$$

It is seen that the output is a full-wave rectified version of the input voltage. If resistor R_6 equals R_4 and R_5, the circuit has an overall voltage gain of 1. When R_6 is greater than R_4 and R_5, amplification and rectification both occur. A precision full-wave rectifier circuit is also known as an *absolute value circuit*. This means the circuit output is the absolute value of the input peak voltage regardless of the input polarity.

Example 7-2

Design a precision full-wave rectifier circuit as in Fig. 7-4, to produce a 2 V peak output from a sine wave input with a peak value of 0.5 V and a frequency of 1 MHz. Use bipolar op-amps with a supply voltage of ±15 V.

Solution

$$I_1 \gg I_{B(max)}$$

let $I_1 = 500\ \mu A$ (for adequate diode current)

$$R_1 = \frac{V_i}{I_1} = \frac{0.5\ V}{500\ \mu A}$$

$$= 1\ k\Omega \quad \text{(standard value)}$$

$$R_2 = 2\,R_1 = 2\ k\Omega \quad \text{(use two 1 k}\Omega\text{ resistors in series)}$$

$$R_3 = R_1 \| R_2 = 1\ k\Omega \| 2\ k\Omega$$

$$= 670\ \Omega \quad \text{(use 680 }\Omega\text{ standard value)}$$

$$R_4 = R_5 = R_1 = 1\ k\Omega \quad \text{(standard value)}$$

For the output to be 2 V when the input is 0.5 V,

$$R_6 = \frac{V_o}{V_i} \times R_5 = \frac{2\ V}{0.5\ V} \times 1\ k\Omega$$

$$= 4\ k\Omega \quad \text{(use 3.9 k}\Omega\text{ standard value)}$$

$$R_7 = R_4 \| R_5 \| R_6 = 1\ k\Omega \| 1\ k\Omega \| 3.9\ k\Omega$$

$$= 443\ \Omega \quad \text{(use 470 }\Omega\text{ standard value)}$$

For diodes D_1 and D_2, $V_R > 30$ V, and $t_{rr(max)} = 0.1\ \mu s$ as in Example 7-1.
Compensate A_1 as a voltage follower, and A_2 for a gain of

$$\frac{R_6 + R_4 \| R_5}{R_4 \| R_5} \simeq 9$$

High Input Impedance Full-Wave Precision Rectifier

A precision rectifier that uses a noninverting amplifier configuration to present a high input impedance to the signal is shown in Fig. 7-5. Op-amp A_1 together with resistors R_3 and R_4 constitutes a noninverting amplifier, as does A_2, R_5 and R_6. However, diodes D_1 and D_2 also affect the operation of the circuit.

Consider what occurs during the positive half-cycle of the input waveform. The output terminal of A_2 is positive, D_2 is forward-biased, and D_1 is reversed. The

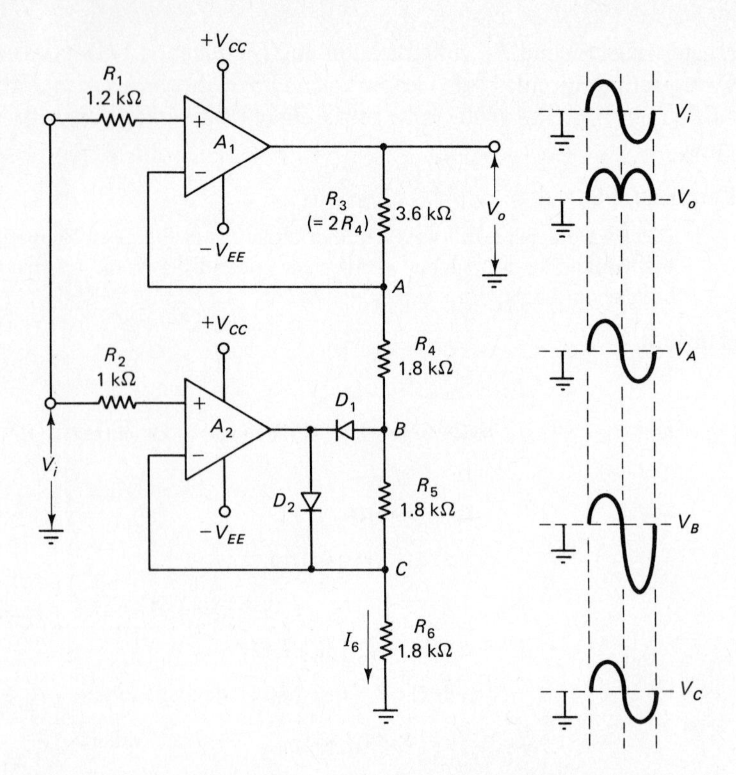

Figure 7-5 High input impedance full-wave rectifier circuit. During the positive half-cycle, V_A, V_B, V_C, and V_o all follow the input. During the negative half cycle, $V_A = V_i$, $V_B = 2V_i$, and $V_{R3} = 2V_i$, giving $V_o = V_i$ inverted.

voltage at the junction of resistors R_5 and R_6 (terminal C) follows V_i at the noninverting input terminal of A_2. The voltage at the junction of R_3 and R_4 (terminal A) also follows the input voltage applied to the noninverting input of A_1. With V_i appearing at terminal A and at terminal C, there is no voltage drop across resistors R_4 and R_5. Consequently, there is no current flowing through R_3 and no voltage drop across R_3. So, the circuit output voltage is the same as that at terminal A; that is, V_i.

When the input voltage goes negative, D_2 is reverse-biased and D_1 is forward-biased giving an output at the junction of R_4 and R_5 (terminal B). Because R_5 equals R_6, the voltage at terminal B is $2(-V_i)$. The circuit output voltage can be determined by first calculating the current flowing in each resistor. Another approach is to note that A_1, R_3, and R_4 constitute an inverting amplifier to voltages applied at the bottom of R_4, and a noninverting amplifier for voltages applied to resistor R_1. If the voltage at terminal B was zero when $-V_i$ is applied to R_1, the circuit output would be

$$V_x = (-V_i)\frac{R_3 + R_4}{R_4}$$

With $R_3 = 2R_4$,

$$V_x = -3V_i$$

If the voltage at the noninverting input terminal of A_1 was zero when $-2V_i$ occurs at terminal B, the output voltage would be

$$V_y = (-2V_i) \times \left(-\frac{R_3}{R_4}\right)$$

With $R_3 = 2R_4$,

$$V_y = 4V_i$$

Applying the superposition theorem, the output voltage is the sum of V_x and V_y.

$$\begin{aligned} V_o &= V_x + V_y \\ &= -3V_i + 4V_1 \\ &= V_i \end{aligned}$$

It is seen that the output voltage equals the input and is positive when the input is negative. As already explained, the output is also positive when the input is positive. So, the circuit functions as a full-wave rectifier.

To design this type of precision full-wave rectifier circuit, R_6 is first calculated, then $R_4 = R_5 = R_6$, and then $R_3 = 2R_4$.

Example 7-3

Using bipolar op-amps with $V_{CC} = \pm 15$ V, design the high input impedance precision full-wave rectifier circuit in Fig. 7-5. The input peak voltage is to be 1 V and no amplification is to occur.

Solution

let $I_6 = 500\ \mu\text{A}$ (for adequate diode current)

$$R_6 = \frac{V_i}{I_6} = \frac{1\ \text{V}}{500\ \mu\text{A}}$$

$$= 2\ \text{k}\Omega \qquad \text{(use 1.8 k}\Omega\text{ standard value)}$$

$$R_4 = R_5 = R_6 = 1.8\ \text{k}\Omega \qquad \text{(standard value)}$$

$$R_3 = 2R_4 = 3.6\ \text{k}\Omega \qquad \text{(use two 1.8 k}\Omega\text{ resistors in series)}$$

$$R_1 = R_3 \parallel R_4 = 3.6\ \text{k}\Omega \parallel 1.8\ \text{k}\Omega$$

$$= 1.2\ \text{k}\Omega \qquad \text{(standard value)}$$

$$R_2 = R_6 \parallel R_5 = 1.8\ \text{k}\Omega \parallel 1.8\ \text{k}\Omega$$

$$= 900\ \Omega \qquad \text{(use 1 k}\Omega\text{ standard value)}$$

Compensate the op-amps for $A_{v1} = 2$, and A_2 as a voltage follower.

7-3 LIMITING CIRCUITS

Peak Clipper

Back-to-back connected Zener diodes are used in the circuit of Fig. 7-6(a) to clip off the peaks of the output voltage waveform. One diode is forward-biased and the other is biased into reverse breakdown when the output voltage is greater than $(V_F + V_Z)$; that is, the forward voltage drop of one diode plus the Zener breakdown voltage of

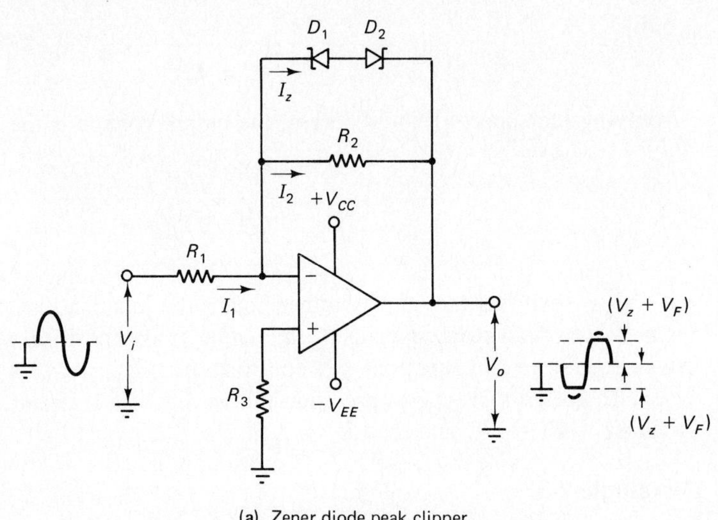

(a) Zener diode peak clipper

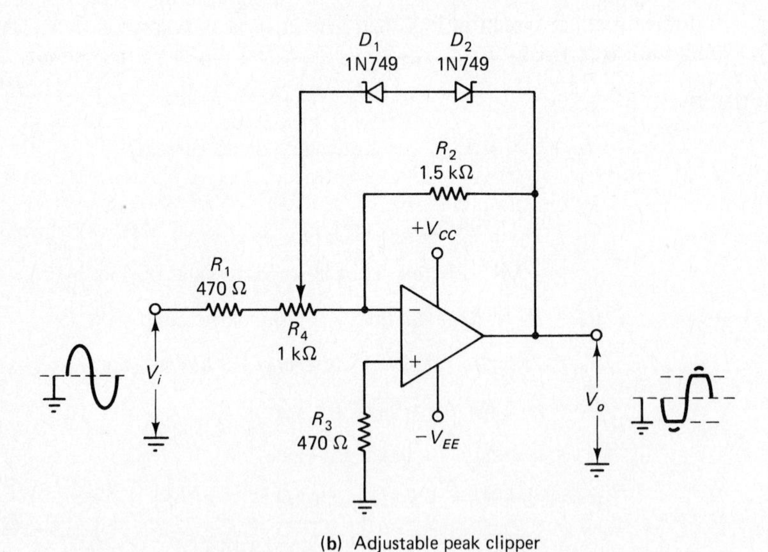

(b) Adjustable peak clipper

Figure 7-6 Two back-to-back, series-connected Zener diodes can be used to clip, or limit, the output amplitude of an inverting amplifier. The inclusion of potentiometer R_4 facilitates adjustment of the output voltage limits.

the other. Negative feedback causes the op-amp output to remain at the level that keeps the inverting input terminal close to the voltage at the (grounded) noninverting input. Thus, as illustrated, the output cannot exceed $\pm(V_F + V_Z)$. As long as the output voltage is less than this limit, the circuit behaves as an inverting amplifier unaffected by the diodes. This kind of circuit is typically used to protect a device that might be damaged by excessive input voltage.

Figure 7-6(b) shows a circuit modification in which variable resistor R_4 is connected in series with R_1. With the Zener diodes connected to the moving contact of R_4, as illustrated, the output limiting voltage can be adjusted. Suppose that $R_1 = R_4 = R_2$, and that $(V_Z + V_F) = 4$ V. With the moving contact at the right side of R_4,

$$V_{o(\max)} = V_Z + V_F = \pm 4\text{ V}$$

With the moving contact at the left side of R_4,

$$V_{R2} + V_{R4} = V_Z + V_F = \pm 4\text{ V}$$

and, with $R_2 = R_4$,

$$V_{R2} = V_{R4} = \pm 2\text{ V}$$

giving

$$V_{o(\max)} = V_{R2} = \pm 2\text{ V}$$

By means of the moving contact, the maximum output voltage can be adjusted between ± 2 V and ± 4 V in this case, or between $\pm(V_F + V_Z)$ and $\pm(V_F + V_Z)/2$ when $R_4 = R_2$. The circuit voltage gain remains $-R_2/(R_1 + R_4)$ until the output limit is reached.

To design a peak clipper, the Zener diodes are first selected to limit the output voltage at the desired level, bearing in mind that V_F is typically 0.7 V. Then, the inverting amplifier is designed to have the required voltage gain. Note that when the output reaches the limiting level, part of the current through resistor R_1 flows through R_2 and part flows through the diodes. The resistor current must be greater than the minimum level required for Zener breakdown, typically around 0.5 mA.

Example 7-4

Design an adjustable peak clipping circuit as in Fig. 7-6(b), to clip at approximately $\pm$(3 V to 5 V). The circuit is to have unity voltage gain before clipping.

Solution

$$V_{o(\max)} = V_Z + V_F \approx 5\text{ V}$$

$$V_Z = 5\text{ V} - V_F \approx 5\text{ V} - 0.7\text{ V}$$

$$\approx 4.3\text{ V} \qquad \text{(use a 1N 749 Zener diode)}$$

$$I_1 > (I_{Z(\min)} = 500\ \mu\text{A})$$

let

$$I_{1(\min)} \approx 2\text{ mA}$$

For $V_o \approx 3$ V,

$$R_2 = \frac{V_{o(\min)}}{I_{1(\min)}} = \frac{3\text{ V}}{2\text{ mA}}$$

$$= 1.5\text{ k}\Omega \qquad \text{(standard value)}$$

$$V_{R4} = V_{o(\max)} - V_{o(\min)} = 5\text{ V} - 3\text{ V}$$
$$= 2\text{ V}$$
$$R_4 = \frac{V_{R4}}{I_1} = \frac{2\text{ V}}{2\text{ mA}}$$
$$= 1\text{ k}\Omega \qquad \text{(standard potentiometer value)}$$

For $A_v = 1$, $\quad R_1 + R_4 = R_2$

or,
$$R_1 = R_2 - R_4 = 1.5\text{ k}\Omega - 1\text{ k}\Omega$$
$$= 500\ \Omega \qquad \text{(use 470 }\Omega\text{ standard value)}$$
$$R_3 = (R_1 + R_4) \parallel R_2 = (470\ \Omega + 1\text{ k}\Omega) \parallel 1.5\text{ k}\Omega$$
$$= 742\ \Omega \qquad \text{(use 680 }\Omega\text{ standard value)}$$

Dead Zone Circuit

The circuit in Fig. 7-7 is a precision half-wave rectifier (discussed in Section 7-1) with the addition of resistor R_1 and dc reference voltage V_{ref}. If R_1 and V_{ref} were absent, the circuit would simply produce an inverted half-wave rectified version of the input. Now note that if the diodes were absent (D_1 open-circuited and D_2 short-circuited), the arrangement would constitute an inverting summing circuit (Section 3-5). In fact, when the op-amp output goes positive, forward-biasing D_2 and reversing D_1, the circuit functions exactly as a summing circuit, giving an output of

$$V_o = -(V_{ref} + V_i)$$

V_{ref} is a positive quantity. If V_i equals zero, V_{ref} tends to drive the op-amp output in a negative direction. Of course, in this situation D_1 becomes forward-biased

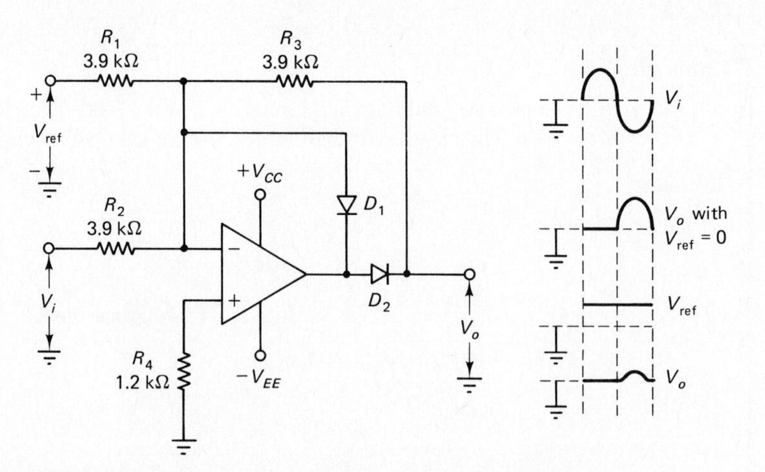

Figure 7-7 Dead zone circuit consisting of a precision half-wave rectifier and one additional resistor. During the positive half-cycle of the input, negative feedback via D_1 holds the output at zero. During the negative half-cycle, V_i and V_{ref} are summed and inverted, to give $V_o = -(V_{ref} + V_i)$.

and V_o remains at ground level. To drive the op-amp output terminal in a positive direction, V_i must be negative and have a level greater than the positive level of V_{ref}. For example, if V_{ref} equals +1 V, V_i must become slightly more negative (microvolts) than −1 V in order to drive the op-amp inverting input terminal below the level of the noninverting input and thus cause the output to go positive. So, the output will not move from ground level until

$$|-V_i| = |+V_{ref}|$$

When this occurs, D_2 is forward biased, D_1 is reversed, and as already stated, the output becomes the sum of V_{ref} and V_i. Suppose V_{ref} is +1 V and V_i is −4 V,

$$V_o = -(V_{ref} + V_i) = -[1\text{ V} + (-4\text{ V})]$$
$$= 3\text{ V}$$

Thus, as illustrated, the circuit output remains at ground level until $|-V_i|$ exceeds $|+V_{ref}|$. Then the output is the peak portion of the (negative) input that exceeds the reference voltage level. Only this part of the input wave is passed (and inverted). All other portions of the input wave are ineffective. The ineffective portions of the input are said to occupy a *dead zone.*

The reference voltage can be set to any convenient level, positive or negative, and can be made adjustable if desired. If the polarity of the diodes and the polarity of V_{ref} are reversed, the circuit produces an inverted version of the positive input peak. As an alternative to making V_{ref} adjustable, resistor R_1 could be adjustable. For example, $V_{ref} = 1$ V, and $R_1 = 2R_2$, has the same effect as setting $V_{ref} = 0.5$ V with $R_1 = R_2$.

Example 7-5

Using a BIFET op-amp, design a dead zone circuit to pass only the upper 1 V portion of the positive half-cycle of a sine wave input with a peak value of 3 V.

Solution

$$V_{ref} = V_p - 1\text{ V} = 3\text{ V} - 1\text{ V}$$
$$= 2\text{ V}$$

$$I_{R1(min)} = I_{D(min)} \approx 500\ \mu\text{A}$$

$$R_1 = \frac{V_{ref}}{I_{R1}} = \frac{2\text{ V}}{500\ \mu\text{A}}$$
$$= 4\text{ k}\Omega \qquad \text{(use 3.9 k}\Omega\text{ standard value)}$$

$$R_2 = R_3 = R_1 = 3.9\text{ k}\Omega$$

$$R_4 = R_1 \parallel R_2 \parallel R_3 = 3.9\text{ k}\Omega \parallel 3.9\text{ k}\Omega \parallel 3.9\text{ k}\Omega$$
$$= 1.3\text{ k}\Omega \qquad \text{(use 1.2 k}\Omega\text{ standard value)}$$

Select the diodes as in Example 7-1 and compensate the op-amp as a voltage follower.

Precision Clipper

The dead zone circuit discussed above is redrawn in Fig. 7-8 and its output is applied to an inverting summing circuit (terminal B). Input voltage V_i is applied to the other input (terminal A) of the summing circuit. The waveforms at terminals A and B are shown in the illustration and the resultant output is the inverted sum of the two. The output is seen to be the inverted input waveform with its positive peak precisely clipped off above the level of V_{ref}. The circuit is a precision clipper.

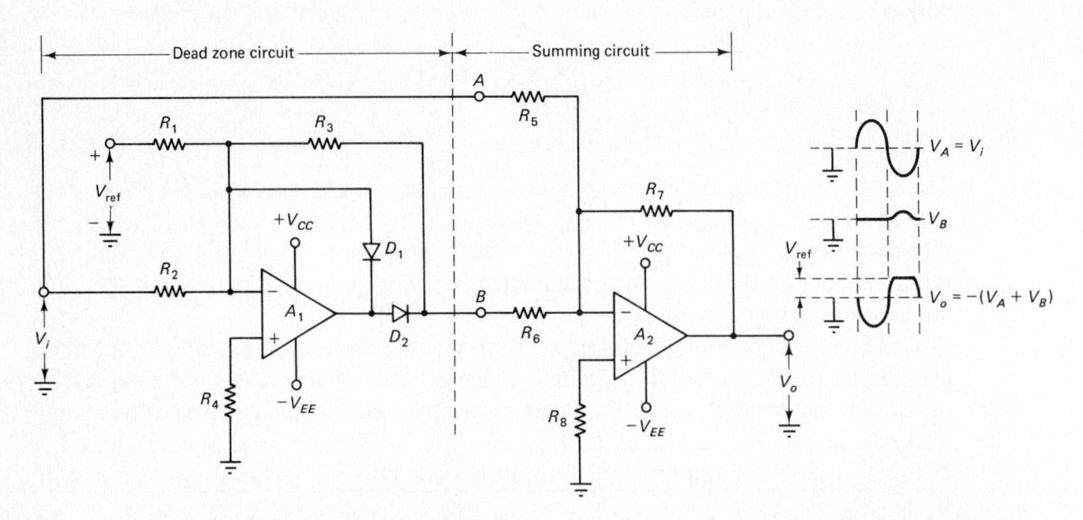

Figure 7-8 Precision clipping circuit consisting of a dead zone circuit and a summing circuit. The dead zone circuit output is summed with V_i to produce an output waveform with its positive half-cycle precisely clipped at V_{ref}.

Figure 7-9 shows the same circuit as Fig. 7-8, but with the addition of a third input to the summing circuit and with another dead zone circuit with the polarity of its diodes and reference voltage reversed. The output of the additional dead zone circuit, applied to terminal C of the summing circuit, is the negative peak of the inverted input. When the waveforms at terminals A, B, and C are summed and inverted, the output is an inverted version of V_i with both peaks precisely clipped at the levels of the reference voltages.

Design of a precision clipper circuit involves design of each dead zone circuit and design of the summing circuit. As already discussed, design of the dead zone circuit requires design of the inverting amplifier constituted by R_2, R_3, and the op-amp; selection of the diodes; selection of V_{ref}. Usually $R_1 = R_2$; and $R_4 = R_1 \parallel R_2 \parallel R_3$. The summing circuit design again involves design of an inverting amplifier (R_7, R_8, and A_3); selection of $R_5 = R_6 = R_7$; and $R_9 = R_5 \parallel R_6 \parallel R_7 \parallel R_8$. Suitable op-amps and diodes should be selected as discussed in Section 7-1.

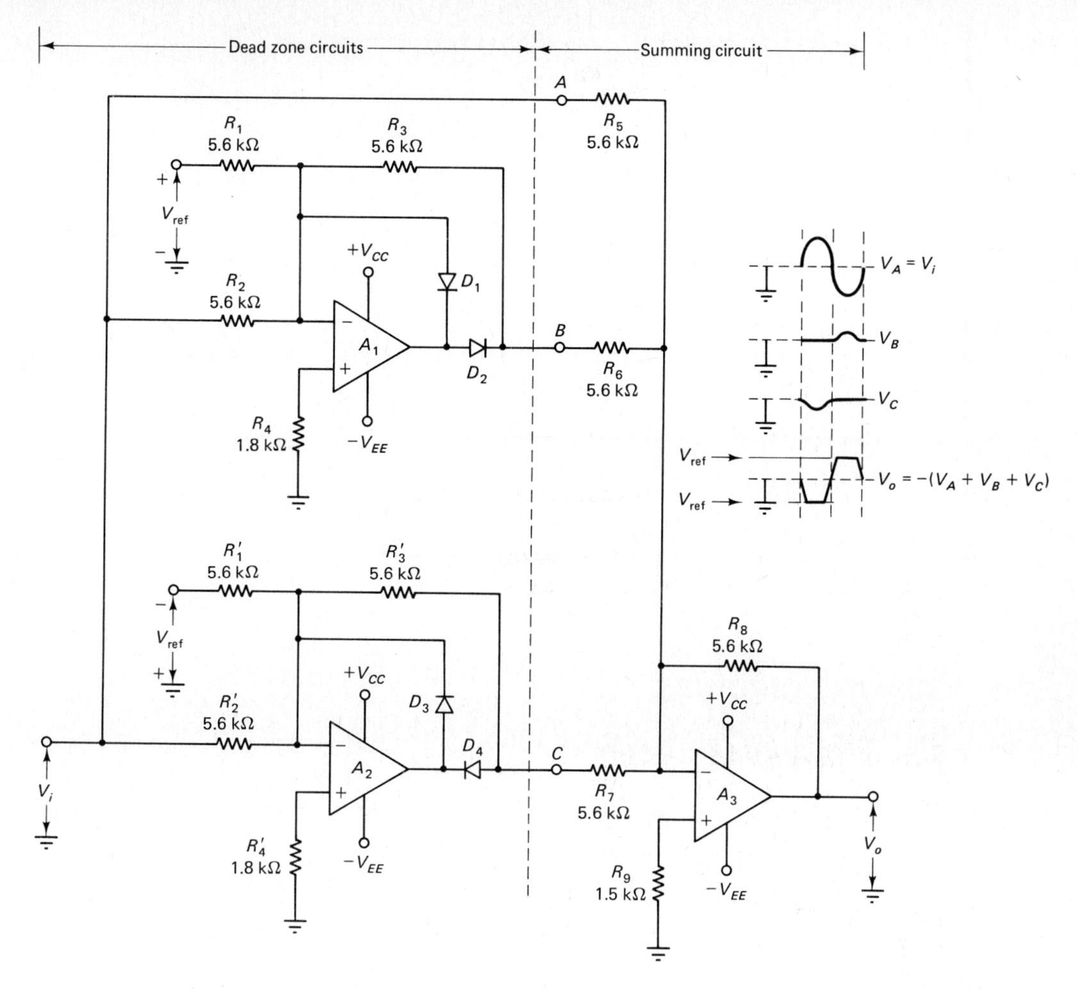

Figure 7-9 Symmetrical precision clipping circuit consisting of two dead zone circuit and a summing circuit. Summing the dead zone circuit outputs with V_i gives an output with both half-cycles clipped at V_{ref}.

Example 7-6

Using bipolar op-amps, design a precision clipping circuit as in Fig. 7-9 to clip a 100 kHz sine wave at the ± 3 V level.

Solution

$$V_{ref} \text{ for } A_1 = 3\text{ V}$$

$$V_{ref} \text{ for } A_2 = -3\text{ V}$$

$$I_{R1(min)} \gg I_{B(min)} \text{ for the op-amps}$$

let $\quad I_{R1(min)} = 500\ \mu\text{A} \quad$ (adequate diode current)

$$R_1 = \frac{V_{ref}}{I_{R1}} = \frac{3\text{ V}}{500\ \mu\text{A}}$$

$$= 6\ \text{k}\Omega \qquad \text{(use 5.6 k}\Omega\text{ standard value)}$$

$$R_2 = R_3 = R_1 = 5.6\ \text{k}\Omega$$

$$R_4 = R_1 \parallel R_2 \parallel R_3 = 5.6\ \text{k}\Omega \parallel 5.6\ \text{k}\Omega \parallel 5.6\ \text{k}\Omega$$

$$\approx 1.9\ \text{k}\Omega \qquad \text{(use 1.8 k}\Omega\text{ standard value)}$$

$$R_1' = R_2' = R_3' = R_1 = 5.6\ \text{k}\Omega$$

$$R_4' = R_4 = 1.8\ \text{k}\Omega$$

$$R_5 = R_6 = R_7 = R_8 = R_1 = 5.6\ \text{k}\Omega$$

$$R_9 = R_5 \parallel R_6 \parallel R_7 \parallel R_8 = 5.6\ \text{k}\Omega \parallel 5.6\ \text{k}\Omega \parallel 5.6\ \text{k}\Omega \parallel 5.6\ \text{k}\Omega$$

$$\approx 1.4\ \text{k}\Omega \qquad \text{(use 1.5 k}\Omega\text{ standard value)}$$

Select the diodes as in Example 7-1 and compensate A_1 and A_2 as voltage followers. Compensate A_3 for

$$A_v = \frac{R_8 + (R_5 \| R_6 \| R_7)}{R_5 \| R_6 \| R_7} = 4$$

7-4 CLAMPING CIRCUITS

Diode Clamping Circuit

A clamping circuit reproduces an input waveform without any clipping or distortion, but limits the upper or lower peak of the waveform to a predetermined level. Consider the diode clamping circuit in Fig. 7-10(a). When the input voltage is positive, diode D_1 is forward-biased and capacitor C_1 charges with the polarity shown. The peak input voltage (V_p) appears across C_1 and D_1 so the capacitor charges to

$$V_{C1} = V_p - V_F$$

At this time, the output voltage cannot exceed the voltage drop across the forward-biased diode

$$V_o = V_F$$

When the input goes to its negative peak, D_1 is reverse baised and the input and capacitor voltages combine to produce an output of

$$\begin{aligned} V_o &= -V_p - V_{C1} \\ &= -V_p - (V_p - V_F) \\ &= -2V_p + V_F \end{aligned}$$

As illustrated, the output waveform has the same peak-to-peak amplitude as the input, but its upper level is clamped at $+V_F$. If the diode is reversed, the lower

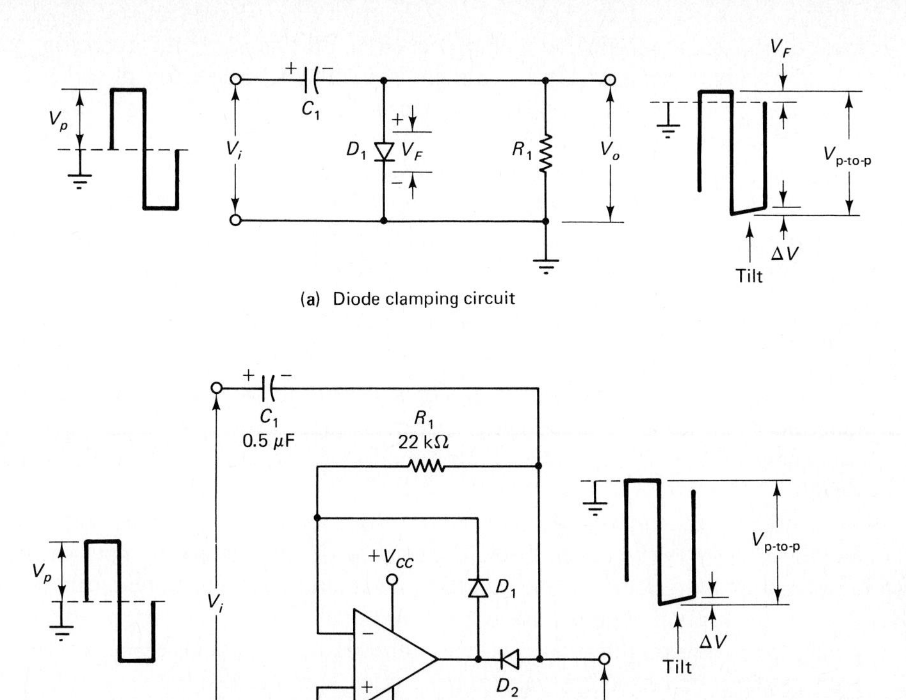

(a) Diode clamping circuit

(b) Precision clamping circuit

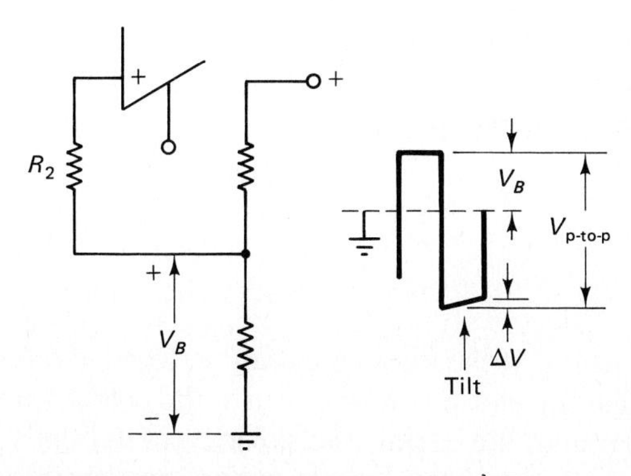

(c) A bias voltage can be included to clamp the output at V_B

Figure 7-10 Diode clamping circuit which clamps the positive peak of the input at $+V_F$, and precision clamping circuit which clamps the peak voltage at zero. The op-amp circuit behaves like a diode with $V_F = 0$.

level of the output would be clamped at $-V_F$. In this case, the capacitor polarity should also be reversed. A bias voltage (V_B) could be connected in series with the diode and ground to give a clamped output level of $+(V_B + V_F)$ or $(-V_B + V_F)$.

The function of resistor R_1 in Fig. 7-10(a) is to ensure the capacitor discharges when the peak input voltage drops to a lower level. R_1 produces *tilt* (or *slope*) on the unclamped peak of the output, as illustrated. This can be minimized by keeping R_1 as large as possible. Any load resistor at the circuit output will be in parallel with R_1 and will increase the tilt on the output waveform.

Precision Clamping Circuit

In the precision clamping circuit in Fig. 7-10(b) the op-amp circuit is required to function as an ideal diode. So, the circuit in Fig. 7-10(b) can be regarded as the equivalent of Fig. 7-10(a) if D_1 is assumed to be an ideal diode with V_F equal to zero.

When the input voltage in Fig. 7-10(b) is at $+V_p$, the circuit output tends to move in a positive direction. Thus, because the op-amp inverting input is connected to the output by resistor R_1, the inverting input terminal tends to be positive with respect to the grounded noninverting input terminal. This causes the op-amp output to go negative, resulting in diode D_2 being forward-biased and D_1 being reversed. Negative feedback via R_1 keeps the inverting input terminal and the anode of D_2 within microvolts of ground level. So, the circuit output is held close to ground level and C_1 charges via D_2 to $+V_p$. The output voltage at this time is

$$V_o = 0$$

When the input goes to its negative peak, the circuit output and the op-amp inverting input terminal tend to go negative. The negative voltage at the inverting input causes the op-amp output terminal to move in a positive direction, reversing D_2 and forward biasing D_1. Now, negative feedback via D_1 keeps the op-amp inverting input terminal close to the level of the grounded noninverting input terminal. With D_2 reverse-biased, the circuit output is completely free to move in a negative direction, giving an output of

$$\begin{aligned} V_o &= V_i + V_{C1} \\ &= -V_p + (-V_p) \\ &= -2V_p \end{aligned}$$

It is seen that the op-amp circuit performs the function of an ideal diode. The input voltage waveform is reproduced at the output with its upper level clamped precisely at ground. Reversing the polarity of D_1 and D_2 in Fig. 7-10(b) produces clamping of the lower level of the output waveform. Including a bias voltage (V_B) at the op-amp noninverting input terminal, as illustrated in Fig. 7-10(c), allows the output to be clamped at the level of V_B.

As in the case of the simple diode clamping circuit, resistor R_1 in Fig. 7-10(b) constitutes a load which tends to discharge the capacitor and thus produce tilt on the unclamped peak of the output waveform. The value of R_1 can be calculated to produce an acceptable amount of tilt, but the resistance must not be too large for the op-amp used (see Section 3-1). Minimum diode current might also be considered.

In any clamping circuit, the signal source resistance R_S is essentially the only resistance in series with the capacitor when it is charging. Allowing that the capacitor should be completely charged from zero to V_p in five cycles of the input waveform,*

$$5C_1R_S = 5 \times \frac{T}{2}$$

where T is the time period of the waveform.

So
$$C_1 = \frac{T}{2R_S} = \frac{1}{2R_Sf} \tag{7-1}$$

When C_1 is discharging via R_1, the voltage across R_1 is

$$V_o = 2V_p$$

giving a discharge current of

$$I = \frac{2V_p}{R_1}$$

and
$$C_1 = \frac{Q}{\Delta V} = \frac{It}{\Delta V}$$

Where ΔV is the discharge voltage, or tilt, on the unclamped peak, and t is the discharge time $T/2$.

so
$$C_1 = \frac{2V_p}{R_1} \times \frac{T}{2} \times \frac{1}{\Delta V}$$

or
$$R_1 = \frac{V_p}{C_1\,\Delta V f} \tag{7-2}$$

Although a square wave has been used to simplify the circuit explanation, the clamping circuit also functions with sinusoidal and other waveforms. It should be noted that in the precision clamping circuit (as in all other circuits in this chapter), the op-amp functions as a linear amplifier. Its output is not saturated. So, frequency compensation is required when an uncompensated op-amp is employed.

Example 7-7

A ±5 V, 10 kHz square wave from a signal source with a resistance of 100 Ω is to have its positive peak clamped precisely at ground level. Tilt on the output is not to exceed 1% of the peak amplitude of the wave. Design a suitable op-amp circuit as in Fig. 7-10(b), using a supply of ±12 V.

*David A. Bell, Solid State Pulse Circuits, 3rd. ed. (Englewood Cliffs, NJ: Prentice-Hall, Inc., 1988), p. 36.

Solution

$$\text{Eq. 7-1,} \qquad C_1 = \frac{1}{2R_s f} = \frac{1}{2 \times 100\ \Omega \times 10\ \text{kHz}}$$

$$= 0.5\ \mu\text{F} \qquad \text{(standard value)}$$

$$\Delta V = 1\%\ \text{of}\ 5\ \text{V}$$

$$= 0.05\ \text{V}$$

$$\text{Eq. 7-2,} \qquad R_1 = \frac{V_p}{C_1\ \Delta V f} = \frac{5\ \text{V}}{0.5\ \mu\text{F} \times 0.05\ \text{V} \times 10\ \text{kHz}}$$

$$= 20\ \text{k}\Omega \qquad \text{(use 22 k}\Omega\ \text{standard value)}$$

$$R_2 = R_1 = 220\ \text{k}\Omega$$

$$\text{For diodes } D_1 \text{ and } D_2, \qquad V_R > [V_{CC} - (-V_{EE})] > [12\ \text{V} - (-12\ \text{V})]$$

$$> 24\ \text{V}$$

$$t_{rr} \ll T$$

$$\text{let} \qquad t_{rr(\text{max})} = \frac{1}{10 \times f} = \frac{1}{10 \times 10\ \text{kHz}}$$

$$= 10\ \mu\text{s}$$

Compensate the op-amp as for a voltage follower.

7-5 PEAK DETECTORS

Precision Rectifier Peak detector

A *peak detector* monitors an input signal and holds its output voltage at the peak level of the input. This is achieved by charging a capacitor to the peak voltage level. A very simple peak detector can be constructed by the use of a diode and capacitor as illustrated in Fig. 7-11(a). When the input voltage increases to V_p, the capacitor is charged to $(V_p - V_F)$. When V_i falls below V_p, D_1 is reverse biased and C_1 retains its charge.

One obvious disadvantage of the simple peak detector circuit is that the diode voltage drop introduces considerable error. This can be eliminated by using a precision rectifier circuit, as in Fig. 7-11(b). Holding-capacitor C_1 is charged via the low output resistance of op-amp A_1. By making resistor R_2 greater than R_1, the signal may be amplified as well as peak detected. Op-amp A_2, connected as a voltage follower, isolates the capacitor from the discharging effect of any load resistance.

The holding-capacitor in a peak detector circuit normally should be kept as small as possible, so that it can be rapidly charged to allow its voltage to easily follow changes in the peak input. Another important requirement is that the capacitor should accurately hold its charge for a given time, usually until it is discharged by means of a FET switch. Consequently, all capacitor discharge currents must be kept

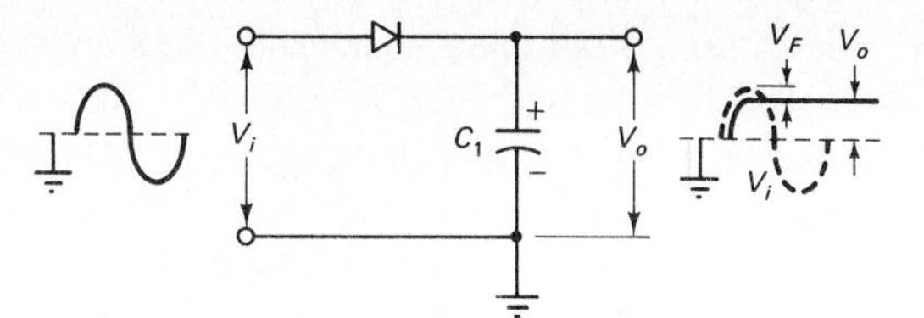

(a) Simple diode-capacitor peak detector

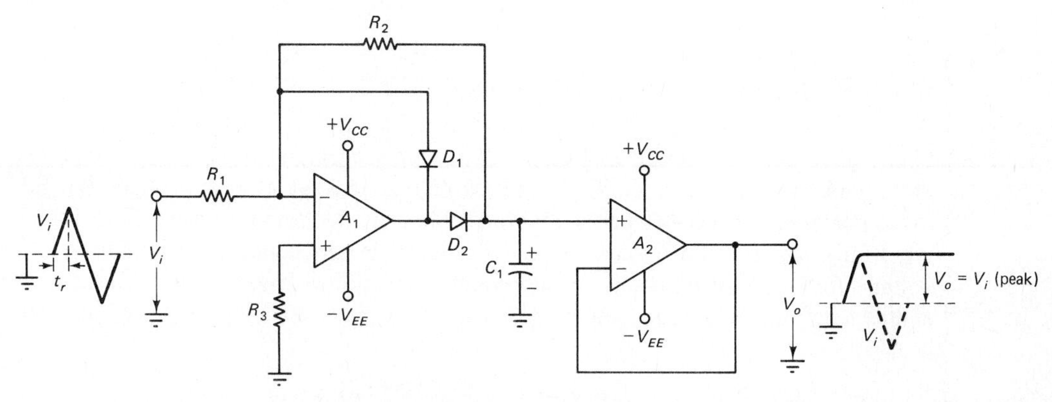

(b) Peak detector using a precision rectifier

Figure 7-11 Diode peak detector circuit which gives an output of $V_o = V_{i(\text{peak})} - V_F$, and precision peak detector which gives an output of $V_o = V_{i(\text{peak})}$. A_2 is a voltage follower, C_1 is the holding capacitor, and the rest of the circuit is a precision rectifier.

to a minimum while the charge is being held. The discharge currents can be minimized by using a BIFET op-amp for A_2 and by keeping resistor R_2 as large as possible. Diode D_2 should also be a low-leakage device and the capacitor itself should have a low-leakage dielectric. However, there is always some discharge current which introduces a voltage error. So, the minimum acceptable error is used to calculate the capacitor size.

If the acceptable holding voltage error is 1%, then the maximum capacitor discharge voltage is calculated as 1% of the peak voltage. For a capacitor discharge voltage ΔV, discharge current I_d, and holding time t_h, the capacitor is calculated as

$$C_1 = \frac{I_d t_h}{\Delta V} \tag{7-3}$$

Once the capacitor value has been determined, its maximum charging current should be calculated to arrive at the required maximum output current $I_{o(\max)}$ from op-amp A_1. For a peak voltage V_p and minimum signal rise time t_r [see Fig. 7-11(b)]

$$C_1 = \frac{I_{o(\max)} t_r}{V_p}$$

or

$$I_{o(\max)} = \frac{C_1 V_p}{t_r} \tag{7-4}$$

The frequency response of the operational amplifiers must be high enough to handle the maximum signal frequency (see Section 5-5). When a pulse-type input waveform is applied, the op-amp slew rate must be faster than the maximum rate of change of input signals (see Section 5-6). A slew rate which is three time faster than the rate of change of the signal will add approximately 5% to the signal rise time. This is usually acceptable. So, for peak signal voltage V_p and rise time t_r,

$$\text{op-amp minimum slew rate} \approx 3\frac{V_p}{t_r} \tag{7-5}$$

Voltage Follower Peak Detector

In the peak detector circuit in Fig. 7-12, the only capacitor discharge currents are the input bias current to op-amp A_2 and the reverse leakage current of diode D_2. Because op-amp A_1 is connected as a voltage follower, the circuit presents a very high input impedance to the signal source. Op-amp A_2 is also connected to function as a voltage follower, so output V_o always equals capacitor voltage V_C. With resistor R_2 connected from the circuit output to the inverting input terminal of A_1, V_C also appears at that terminal.

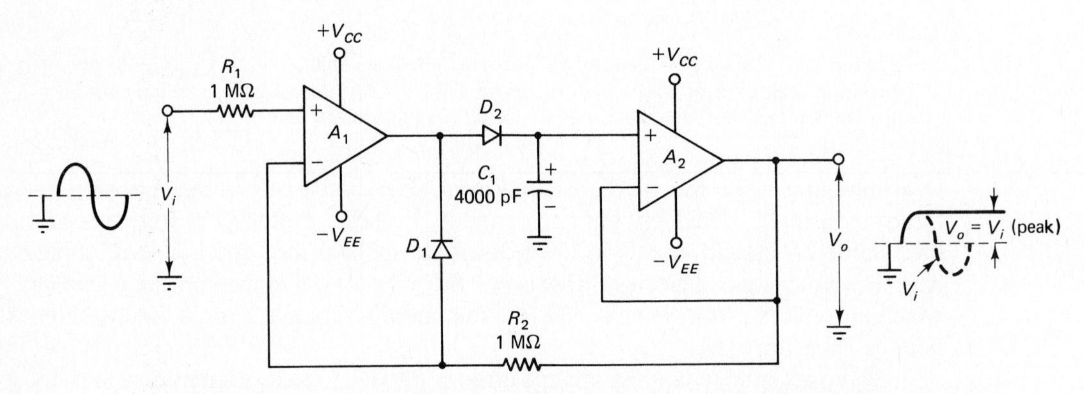

Figure 7-12 Voltage follower-type peak detector. Feedback from the output to the inverting input terminal of A_1 via R_2, causes the whole circuit to behave as a voltage follower during the positive half-cycle of the input. So, $V_o = V_C = V_{i(\text{peak})}$.

When V_i is greater than V_C, the output of A_1 is positive, D_2 is forward-biased, and A_1 behaves as a voltage follower, charging C_1 to V_p. When V_i falls below V_p, V_C remains at V_p, and consequently, the inverting input terminal of A_1 also remains at V_p. Therefore, the output of op-amp A_1 goes negative, reversing D_2 and forward-biasing D_1. Negative feedback via D_1 keeps A_1 from going into saturation.

In the circuit of Fig. 7-12, A_2 should be a BIFET op-amp for low input bias current, but A_1 can be any op-amp type that is otherwise suitable. Once again, diode D_2 should have a very low reverse leakage current. However, with the use of a BIFET op-amp for A_2, the reverse leakage of D_2 (perhaps a maximum of 1 μA) is likely to be much larger than the input bias current of A_2. Therefore, the diode reverse leakage current $I_{R(D2)}$ is more effective in discharging C_1 than $I_{B(MAX)}$ of A_2. Capacitor size and op-amp slew rate are calculated as already discussed, and C_1 should have a very high resistance dielectric.

Example 7-8

A peak detector circuit as in Fig. 7-12 is to be designed. The pulse-type signal voltage has a peak value of approximately 2.5 V with a rise time of 5 μs, and the output voltage is to be held at 2.5 V for a time of 100 μs. The maximum output error is to be approximately 1%. Calculate the required component values and specify the output current and slew rate of the operational amplifiers.

Solution

Use BIFET op-amps for minimum capacitor leakage current.

let $$R_1 = R_2 = 1\ \text{M}\Omega$$

C_1 discharge current, $$I_d \approx I_{R(D2)} \approx 1\ \mu\text{A}$$

$$\Delta V = 1\%\ \text{of}\ V_p = 1\%\ \text{of}\ 2.5\ \text{V}$$

$$= 25\ \text{mV}$$

Eq. 7-3, $$C_1 = \frac{I_d t_h}{\Delta V} = \frac{1\ \mu\text{A} \times 100\ \mu\text{s}}{25\ \text{mV}}$$

$$= 4000\ \text{pF} \qquad \text{(standard value)}$$

For op-amp A_1,

Eq. 7-4, $$I_{o(\max)} = \frac{C_1 V_p}{t_r} = \frac{4000\ \text{pF} \times 2.5\ \text{V}}{5\ \mu\text{s}}$$

$$= 2\ \text{mA}$$

Eq. 7-5, $$\text{min slew rate} \approx 3\frac{V_p}{t_r} = \frac{3 \times 2.5\ \text{V}}{5\ \mu\text{s}}$$

$$\approx 1.5\ \text{V}/\mu\text{s}$$

7-6 SAMPLE-AND-HOLD CIRCUITS

Op-amp Sample-and-Hold

A *sample-and-hold circuit* samples instantaneous amplitudes of a signal voltage at any point in its waveform and holds the voltage level constant until the next sample is acquired. Figure 7-13(a) shows a sample-and-hold circuit which is a modified version of the peak detector circuit of Fig. 7-12. The modifications are that FET switch Q_1 is included to alternately connect and disconnect the capacitor at the output of op-amp A_1. Diodes D_1 and D_2 are inverse-parallel connected to prevent A_1 from going into saturation when Q_1 is open-circuited. With Q_1 switched *on*, A_1 and A_2 act as a single-voltage follower, exactly as in the circuit of Fig. 7-12. The capacitor voltage then precisely follows the signal input.

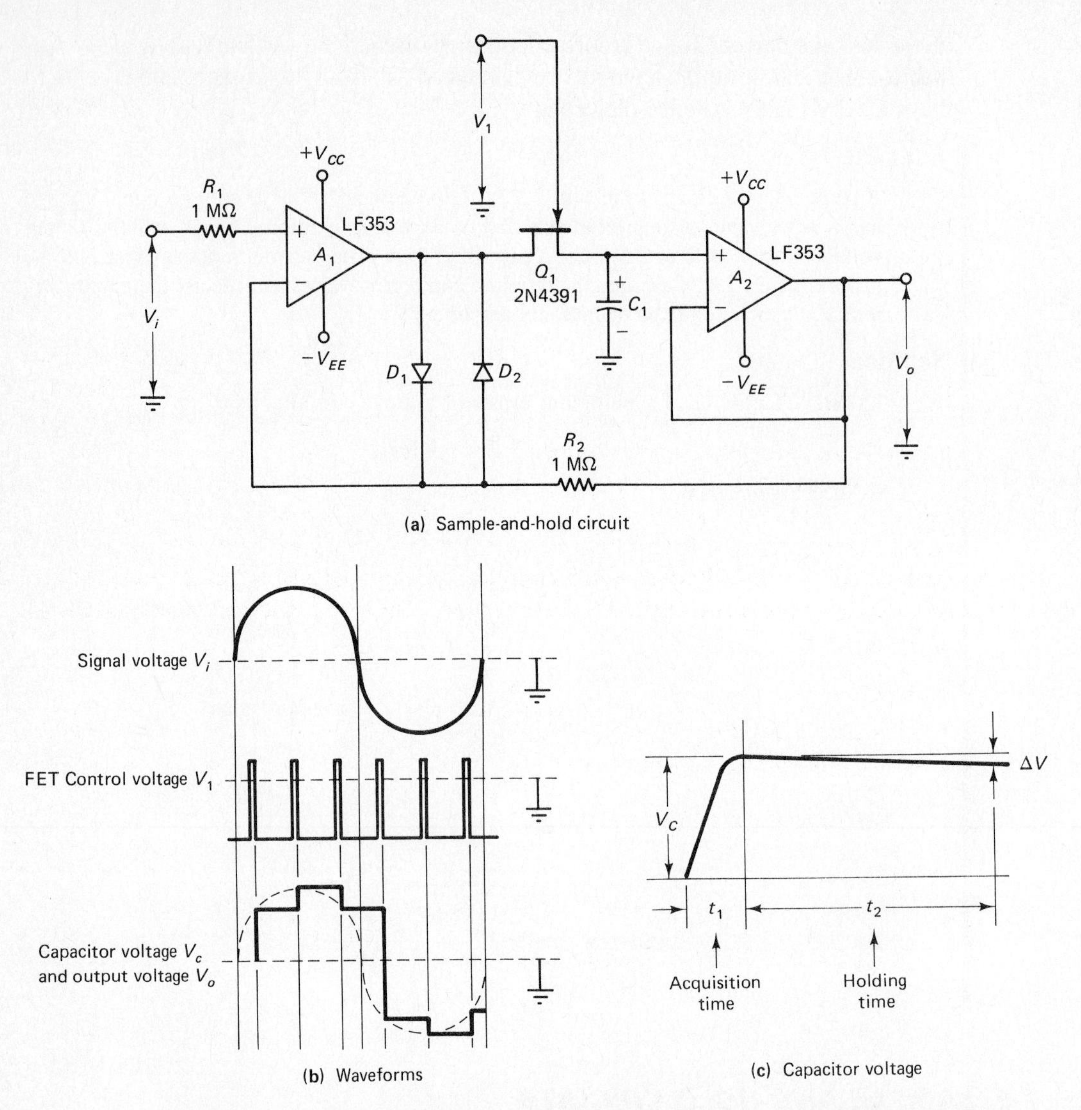

Figure 7-13 Sample-and-hold circuit which functions similarly to the voltage follower-type peak detector (Fig. 7-12), but with the addition of switching FET Q_1 to periodically switch V_i to C_1.

The waveforms in Fig. 7-13(b) illustrate the circuit operation. Q_1 is repeatedly switched *on* and *off* by the pulse waveform (control voltage V_1) applied to its gate terminal. V_1 must go sufficiently negative to drive the FET gate voltage below its pinch-off voltage. It should also go to a positive level approximately equal to the capacitor voltage to ensure complete turn *on* of the FET. If input V_i becomes larger than capacitor voltage V_C while Q_1 is *off*, C_1 rapidly charges to the level of V_i when Q_1 switches *on*. If V_C is initially greater than V_i, C_1 is discharged to the level of V_i when Q_1 is *on*. When Q_1 is *off*, only the input bias current to A_2 and the FET gate-source reverse leakage current are effective in discharging the capacitor. So, C_1

holds the sampled voltage constant until the next sampling instant, giving the step-type capacitor voltage (and output) illustrated in Fig. 7-13(b).

During the *sampling time,* or *acquisition time* (t_1 in Fig. 7-13(c)), C_1 is charged via the FET channel resistance $R_{D(\text{on})}$. If the sampling time is

$$t_1 = 5\,C\,R_{D(\text{on})} \tag{7-6}$$

the capacitor is charged to 0.993 of the input voltage, resulting in a 0.7% error in the sample amplitude.

If,

$$t_1 = 7\,C\,R_{D(\text{on})} \tag{7-7}$$

is used, the error is 0.1%. During the holding time (t_2 in Fig. 7-13(c)), the capacitor is partially discharged, exactly like the case of the peak detector circuit. So, Eq. 7-3 can be used to calculate the capacitor value and either Eqs.7-6 or 7-7 can be employed to determine the minimum acquisition time.

Op-amp A_2 should have a very low input bias current, Q_1 should have a very low gate-source reverse leakage current, and the capacitor should have a low leakage dielectric. Q_1 should also have a low channel resistance for rapid charge and discharge of C_1. The FET gate-source capacitance (C_{GS}) can be one more source of error. C_{GS} charges when the FET is switched *off* and this charge is removed from capacitor C_1.

Example 7-9

A sample-and-hold circuit as in Fig. 7-13 has a signal amplitude of 1 V which is to be sampled with an accuracy of approximately 0.2%. The holding time is to be 500 μs. Design the circuit using LF353 BIFET op-amps and a 2N4391 FET. Determine the minimum acquisition time.

Solution

For the LF 353 op-amp, $I_{B(\text{max})} \approx 50\text{ pA}$

For the 2N4391 FET, the gate-source reverse current is

$$I_{GS} \approx 200\text{ nA}$$

and the channel resistance when *on* is

$$R_{D(\text{on})} = 30\ \Omega$$

let $R_1 = R_2 = 1\text{ M}\Omega$

capacitor discharge current $I_d \approx I_{GS} \approx 200\text{ nA}$

For a 0.2% total error, allow 0.1% due to capacitor discharge and a 0.1% charging error. For 0.1% error due to discharge during the holding time,

let $\Delta V = 0.1\%\text{ of }V_i = 0.1\%\text{ of }1\text{ V}$

$$= 1\text{ mV}$$

Eq. 7-3,

$$C_1 = \frac{I_d t_h}{\Delta V} = \frac{200\text{ nA} \times 500\ \mu\text{s}}{1\text{ mV}}$$

$$= 0.1\ \mu\text{F} \qquad \text{(standard value)}$$

For the 2N4391, $V_{GS(\text{off})} = 10$ V maximum

Therefore, $V_{1(-)} = -10$ V

and $V_{1(+)} \simeq V_o$

$= +1$ V

For 0.1% error due to acquisition time,

Eq. 7-7, $t_{i(\min)} = 7CR_{D(\text{on})} = 7 \times 0.1\ \mu\text{F} \times 30\ \Omega$

$= 21\ \mu\text{s}$

IC Sample-and-Hold

The LH0053 integrated sample-and-hold circuit, shown in Fig. 7-14, is different from the sample-and-hold circuit of Fig. 7-13. This IC requires only the connection of an external capacitor and the provision of a control voltage to switch Q_1 *on* and *off*.

The noninverting input terminal of the output op-amp in Fig. 7-14 is grounded, so negative feedback from the output via C_F keeps the inverting input terminal close to ground. When Q_1 is switched *on*, the two op-amps combined with resistors R_1 and R_2 behave as a single inverting amplifier circuit. And since $R_1 = R_2$, the output voltage is

$$V_o = -V_i$$

Because the inverting input terminal of the output op-amp is held close to ground, the voltage across capacitor C_F equals the output voltage, which is the sampled input voltage. When Q_1 is *on*, C_F is rapidly charged or discharged to V_i. When

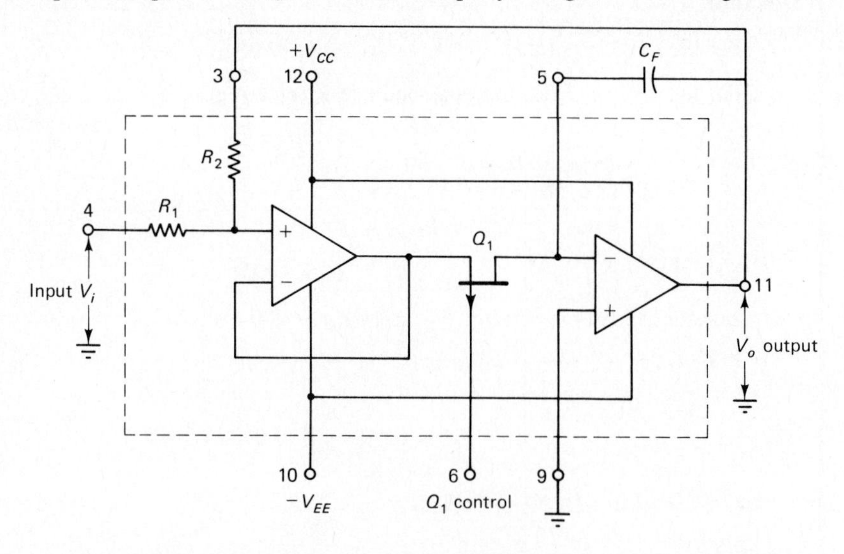

Figure 7-14 LH0053 IC sample-and-hold. With Q_1 *on*, the whole circuit behaves as an inverting amplifier with $V_o = -V_i$. When Q_1 is *off*, C_F holds the output at $-V_i$.

Q_1 is switched *off*, C_F holds its voltage. Feedback keeps the inverting input terminal of the output op-amp close to ground level, so the output is the voltage across the capacitor.

With $C_F = 1000$ pF, the sample acquisition time is typically 4 μs with an accuracy of 0.1%. The error due to capacitor discharge between sample periods is defined in terms of a *drift rate*. This is the capacitor discharge rate, which is specified as typically 6 mV/s.

REVIEW QUESTIONS

7-1. Sketch the circuit of a saturating-type half-wave precision rectifier. Draw the input and output waveforms and explain the circuit operation.

7-2. Discuss the advantages of a precision rectifier over an ordinary diode circuit and show how voltage gain can be achieved with a saturating precision rectifier.

7-3. Sketch the circuit of a non-saturating half-wave precision rectifier. Draw the input and output waveforms, explain the circuit operation, and discuss its advantage over the saturating-type circuit.

7-4. Briefly discuss design procedure and component selection for a half-wave precision rectifier circuit.

7-5. Sketch the circuit of a two-output half-wave precision rectifier. Draw the input and output waveforms and explain the circuit operation.

7-6. Show how a half-wave precision rectifier can be combined with a summing circuit to produce a full wave precision rectifier. Draw the voltage waveforms throughout the circuit and write equations to show that full-wave rectification is performed.

7-7. Draw the circuit of a high input impedance full-wave precision rectifier. Draw the voltage waveforms throughout the circuit and write the appropriate equations to show that full-wave rectification is performed.

7-8. Show how Zener diodes can be used to limit the output voltage of an op-amp circuit. Briefly explain.

7-9. Sketch a Zener diode peak clipper circuit with an adjustable output voltage limit. Explain the circuit operation and write the equations for the upper and lower limits of output voltage.

7-10. Explain what a dead zone circuit does. Sketch an op-amp dead zone circuit, show the waveforms throughout, and explain its operation.

7-11. List the design procedure for an op-amp dead zone circuit.

7-12. Show how a dead zone circuit can be combined with a summing circuit to produce precision limiting on the positive half-cycle of the output waveform. Draw the voltage waveforms throughout the circuit and explain its operation.

7-13. Show how two op-amp dead zone circuits can be combined with a summing circuit to produce precision limiting on both positive and negative half-cycles of the output waveform. Draw the voltage waveforms throughout the circuit, and explain its operation.

7-14. Sketch an op-amp precision clamping circuit, draw the input and output waveforms, and carefully explain the circuit operation. Show how the output voltage can be biased to any desired level.

7-15. Discuss the design procedure for a precision clamping circuit and write the equations for determining the value of each component.

7-16. Sketch a precision rectifier peak detector circuit, draw the input and output waveforms, and explain the circuit operation. Write the equation for calculating the capacitor value for a peak detector circuit.

7-17. Sketch the circuit of a voltage follower-type peak detector. Explain the circuit operation. Write the equations for calculating the capacitor value and the op-amp minimum slew rate.

7-18. Draw an op-amp sample-and-hold circuit. Sketch the signal, control, and output voltage waveforms. Carefully explain the circuit operation.

7-19. Discuss the holding time and acquisition time for a sample-and-hold circuit, and write the equations for determining capacitor size and minimum acquisition time.

7-20. Sketch the circuit of a LH0053 IC sample-and-hold, and briefly explain its operation.

PROBLEMS

7-1. A nonsaturating precision half-wave rectifier using a BIFET op-amp with $V_{CC} = \pm 20$ V is to produce a 3 V peak output. The input signal has a 1 V peak amplitude and a frequency of 50 kHz. Calculate the resistor values and specify the diode reverse recovery time.

7-2. Using a bipolar op-amp with $V_{CC} = \pm 18$ V, design a two-output precision half-wave rectifier as in Fig. 7-3, to give peak outputs of 2 V when the applied input has a 2 V peak amplitude. Specify the diodes to suit a maximum signal frequency of 100 kHz.

7-3. A precision full-wave rectifier circuit with a 50 kHz, 1 V peak input, is to produce a 1 V peak output. Using bipolar op-amps, design the circuit as in Fig. 7-4.

7-4. A 300 kHz signal with a peak amplitude of 500 mV is to be full-wave rectified and amplified to 5 V peak. Design a suitable circuit using BIFET op-amps.

7-5. Design a high input impedance precision full-wave rectifier circuit as in Fig. 7-5. The signal has a 3 V peak amplitude.

7-6. A precision full-wave rectifier has a signal with a 1.5 V peak amplitude. The output peak voltage is to be adjustable from 1.5 V to 4.5 V. Design a suitable circuit using bipolar op-amps.

7-7. An op-amp peak clipper circuit is to amplify its input signal by a factor of 10, and to limit the output voltage to a maximum of ± 7 V. Design a suitable circuit.

7-8. Design an op-amp peak clipper circuit to have unity gain and an output voltage limit adjustable from 6 V to 7.5 V.

7-9. An op-amp dead zone circuit with a 5 V peak input is to have unity gain, and is to pass the upper half of the positive half-cycle of output waveform. Design a suitable circuit using bipolar op-amps. Specify the op-amp and the diodes if the signal frequency ranges from 1 kHz to 33 kHz.

7-10. If a ± 15 V supply is used with the circuit in Problem 7-9, modify the circuit to derive the reference voltage from the supply.

7-11. Further modify the circuit in Problems 7-9 and 7-10 to make the dead zone adjustable between 1/3 and 2/3 of the peak output voltage.

7-12. Design a summing circuit to add to the dead zone circuit in Problem 7-9, to create a precision clipping circuit as in Fig. 7-8.

7-13. Design a plus-minus precision clipping circuit as in Fig. 7-9 to clip the output at a level adjustable from 2 V to 5 V. The op-amp ±12 V supply is to be used instead of reference voltages. Specify suitable op-amps and the diodes for a signal frequency of 60 kHz.

7-14. A 3.3 kHz, ±2 V square wave with a 600 Ω source resistance is to have its negative peak clamped at ground level. Using a bipolar op-amp, design a suitable precision clamping circuit. The tilt on the output is not to exceed 2%.

7-15. Modify the circuit designed for Problem 7-14 to make the clamped level of the output adjustable by ±0.5 V around ground level. Use a potential divider from the ±12 V supply for the bias voltage source.

7-16. Using bipolar op-amps, design the peak detector circuit in Fig. 7-11 to handle a peak input voltage of 1 V. The minimum rise time of the input is 30 μs, the holding time is to be 300 μs, and the maximum output error is to be 5%.

7-17. Design a voltage follower peak detector circuit as in Fig. 7-12 to meet the specification in Problem 7-16. Use BIFET op-amps.

7-18. A ±1.5 V signal is to be sampled by a sample-and-hold circuit as in Fig. 7-13(a). The holding time is to be 700 μs, and the output accuracy is to be approximately 0.3%. LM108 op-amps are to be used, and the available FET has a 25 Ω channel resistance and a 300 nA gate-source reverse leakage current. Design the circuit, and determine the minimum acquisition time.

7-19. A sample-and-hold circuit as in Fig. 7-13(a) has the following components: $R_1 = R_2 = 100\ \text{k}\Omega$, $C_1 = 0.06\ \mu\text{F}$, $A_1 = A_2 = 741$, $Q_1 =$ 2N4391. Analyze the circuit to determine the maximum acquisition time and the holding time. The total error is not to exceed 0.2% of a 3 V input.

COMPUTER PROBLEMS

7-20. Write a computer program to design a precision full-wave rectifier circuit as in Fig. 7-4. Given: $V_{in(peak)}$, $V_{out(peak)}$, f, and V_{CC}.

7-21. Write a computer program to design an adjustable peak clipping circuit as in Fig. 7-6(b). Given: V_{in} and V_{out} range.

7-22. Write a computer program to design a peak detector circuit as in Fig. 7-12. Given: input rise time, holding time error, and V_o.

LABORATORY EXERCISES

7-1 Precision Half-Wave Rectifiers

1. Construct the saturating precision rectifier circuit shown in Fig. 7-1(a). Use a supply of ±9 V to ±15 V, a 741 op-amp, and a 1N914 diode.
2. Apply a 1 kHz sine wave signal input, adjusting it for a peak of 500 mV as displayed on an oscilloscope. Observe the output waveform and carefully measure the peak amplitude of the output voltage.
3. Construct the nonsaturating precision rectifier circuit shown in Fig. 7-2, as designed in Example 7-1. Use a supply of ±9 V to ±15 V, IN914 diodes, and an LF353 op-amp, or a 715 with correct compensation.

4. Repeat Procedure 2.
5. Increase the signal frequency to 1 MHz and check to see if the circuit is still operating correctly without any distortion in the output.

7-2 Clipping and Clamping Circuits.

1. Construct the op-amp clipping circuit shown in Fig. 7-6(b) as designed in Example 7-4. Use a 741 op-amp with a supply of ±15 V.
2. Apply a 1 kHz sine wave input signal, adjusting it to give a peak input of 7 V.
3. Observing the output waveform on an oscilloscope, slowly adjust the moving contact of R_4 from one extreme to the other. Measure the clipped peaks of the output at each extreme.
4. Construct the op-amp clamping circuit shown in Fig. 7-10(b) as designed in Example 7-7. Use a correctly compensated 715 or other fast slew rate op-amp with a supply of ±15 V.
5. Apply a ±5 V, 10 kHz square wave input. Monitor the output voltage waveform on an oscilloscope, noting the peak relationship to ground level.

CHAPTER 8

Differentiating and Integrating Circuits

Objectives

- Sketch op-amp *differentiating* and *integrating* circuits, and explain their operation.
- Design op-amp differentiating and integrating circuits to meet required specifications using bipolar and BIFET op-amps together with transistors and FETs where appropriate.
- Write the equations for the *characteristic frequencies* of differentiating and integrating circuits.
- Sketch the output waveforms produced by differentiating and integrating circuits in response to sine, rectangular, and triangular inputs. Discuss the distortion that can occur.

INTRODUCTION

Differentiation and integration can be performed by op-amp circuits. A differentiating circuit is employed to produce an output amplitude proportional to the rate of change of an input voltage. The output of an integrating circuit must be proportional to the area under each half-cycle of the input waveform. Op-amp differentiating and integrating circuits are inverting amplifiers with appropriately placed capacitors. Differentiating circuits are usually designed to respond to triangular and rectangular input waveforms, while integrating circuits are most often designed to produce a triangular wave output from a square wave input. Both circuits have frequency limitations when processing sine waves.

8-1 DIFFERENTIATING CIRCUIT

A differentiating circuit produces an output voltage which is proportional to the rate of change of the input voltage. Differentiation can be performed by a simple CR series circuit in which the output is taken as the voltage across the resistor. However, the op-amp differentiating circuit in Fig. 8-1 has a much lower output resistance and higher output voltage than is possible with a simple CR circuit.

For dc considerations, C_1 in Fig. 8-1 is an open circuit. The noninverting input terminal is grounded via R_3 and the inverting input is connected to the output via R_2. Thus, the circuit behaves as a voltage follower with the noninverting input grounded. While no signal is present, the output of the circuit remains at ground level.

When the input signal is a positive-going voltage, a current I_1 flows into C_1, as illustrated. With I_1 much larger than the maximum input bias current to the op-amp, effectively all of I_1 flows through R_2. The left side of R_2 at the op-amp inverting input remains close to ground and, as for an inverting amplifier, the output voltage is

$$V_o = -I_1 R_2$$

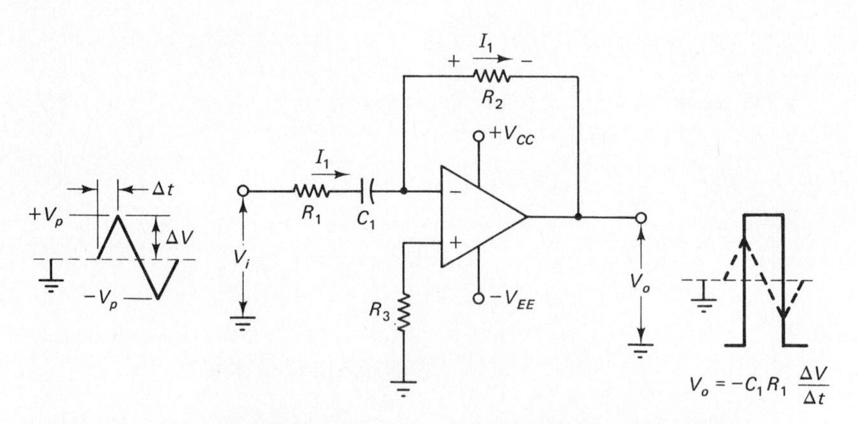

Figure 8-1 Op-amp differentiating circuit. The ramp input causes C_1 to charge at a constant rate of current I_1. The output voltage ($I_1 R_2$) is directly proportional to the rate of change of the input.

Consider the case of the triangular waveform illustrated in Fig. 8-1. When positive-going, it has a rate of change easily identified as $\Delta v/\Delta t$. Assume that the capacitor voltage is zero when the input is zero and positive-going. Also, assume that resistor R_1 has no effect on the capacitor charging current. As the input increases from zero to $+V_p$, the capacitor charges to $+V_p$. So, the capacitor voltage changes by Δv over time interval Δt, as illustrated. During this time, the capacitor charging current is constant as long as the input voltage rate of change remains constant. With a constant charging current,

$$C_1 = \frac{I_1 \, \Delta t}{\Delta v} \tag{8-1}$$

giving

$$I_1 = \frac{C_1 \, \Delta v}{\Delta t}$$

Therefore,

$$V_o = -\frac{C_1 \, \Delta v}{\Delta t} \times R_2$$

or

$$V_o = -C_1 R_2 \frac{\Delta v}{\Delta t} \tag{8-2}$$

This means the output voltage is directly proportional to the rate of change of the input; or the output is the *derivative* of the input. Because the differentiated output of the circuit is inverted, the output voltage is a positive dc level during the time the input is negative-going, and a negative voltage level while the input is positive-going; that is, a square wave.

Resistor R_1 in series with capacitor C_1 is included to help prevent the circuit from oscillating at high frequencies. Capacitor C_1 and resistor R_2 constitute a phase lag circuit in the op-amp feedback loop, which tends to make the circuit unstable. The inclusion of R_1 corrects the phase lag (see Section 5-3) and usually allows the circuit to be compensated for a closed-loop gain of R_2/R_1. R_1 also has some effect on the circuit performance as a differentiator.

8-2 DIFFERENTIATOR DESIGN

To determine suitable values for R_2 and C_1 in Fig. 8-1, the current I_1 is made much larger than the op-amp input bias current. When using a BIFET op-amp (see Section 3-1), it is best to start by selecting a value for C_1 much larger than any stray capacitance that might occur. Note that C_1 should have a low leakage dielectric. The relationship between I_1 and C_1 is given by Eq. 8-1. The resistance of R_2 is calculated from the required output voltage amplitude and the level of I_1,

$$R_2 = \frac{V_o}{I_1}$$

The minimum supply voltage for the op-amp should be chosen approximately 3 V greater than the maximum desired output voltage so that the output stage of the op-amp does not saturate. Thus, for a maximum output of ± 9 V, the supply should be at least ± 12 V.

The resistance of R_1 should be much larger than X_{C1} at the possible oscillation frequency of the circuit (the frequency at which $M\beta = 1$) to minimize the loop phase shift. R_1 should also be much smaller than R_2, to keep the voltage gain high for compensation purposes. A simple rule of thumb is to make $R_1 = R_2/20$. Division by any convenient factor up to 100 might be satisfactory. If the signal source resistance R_s is substantial, it should be considered part of R_1. The voltage gain for determining the op-amp compensating components is then $A_v \simeq R_2/(R_1 + R_s)$.

Resistor R_3 is included in the circuit (as always) to equalize the dc $I_B R_B$ voltage drops at the op-amp input terminals. Since R_1 is dc open-circuited by the presence of C_1, the total resistance in series with the op-amp inverting input terminal is the resistance of R_2. Therefore, R_3 should equal R_2.

Example 8-1

Design a differentiating circuit to give an output of 5 V when the input changes by 1 V in a time of 100 μs. Use an op-amp with a bipolar input stage.

Solution

$$I_1 \gg I_{B(max)}$$

let

$$I_1 = 500\ \mu A$$

$$R_2 = \frac{V_o}{I_1} = \frac{5\ V}{500\ \mu A}$$

$$= 10\ k\Omega \qquad \text{(standard value)}$$

Eq. 8-1,

$$C_1 = \frac{I_1 \times \Delta t}{\Delta v} = \frac{500\ \mu A \times 100\ \mu s}{1\ V}$$

$$= 0.05\ \mu F \qquad \text{(standard value)}$$

$$R_1 = \frac{R_2}{20} = \frac{10\ k\Omega}{20}$$

$$= 500\ \Omega \qquad \text{(use 470 } \Omega \text{ standard value)}$$

$$R_3 = R_2 = 10\ k\Omega$$

$$V_{CC} \geq \pm(V_o + 3\ V) = \pm(5\ V + 3\ V)$$

$$\geq \pm 8\ V$$

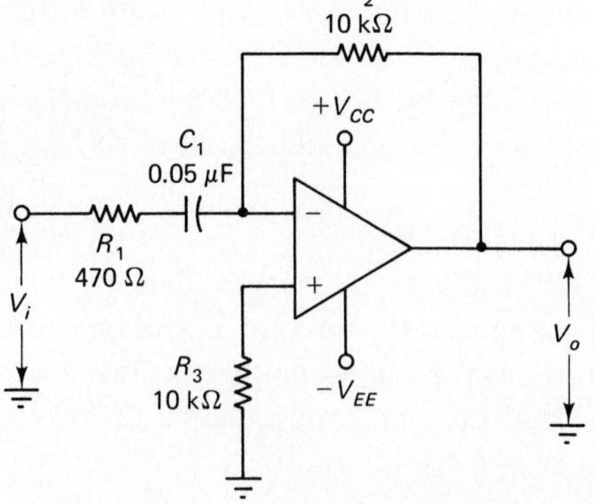

Figure 8-2 Differentiating circuit designed in Example 8-1.

Compensate the op-amp for

$$A_v \simeq \frac{R_2}{R_1} = \frac{10\ \text{k}\Omega}{470\ \Omega}$$
$$\simeq 21$$

8-3 DIFFERENTIATING CIRCUIT PERFORMANCE

Triangular and Rectangular Wave Response

Consider the effect of a triangular input waveform input once again, as illistrated in Fig. 8-3(a). As already explained, the resultant output is a square wave. However, instead of being a perfect square wave, the waveform shown has rounded corners and definite rise and fall times (t_{ro} and t_{fo}). This distortion is a result of the op-amp not having a high enough upper cutoff frequency. Similar distortion occurs on the differentiated output of the rectangular wave in Fig. 8-3(b). In this case, the input voltage is changing during the rise and fall times of the input and remaining constant in the intervals between the leading and lagging edges of the input waveform. Thus, the output is a negative dc voltage level proportional to the positive rate of input change during the rise time, a zero level during the time the input is not changing, and a positive dc level proportional to the negative rate of change at the fall time of the input.

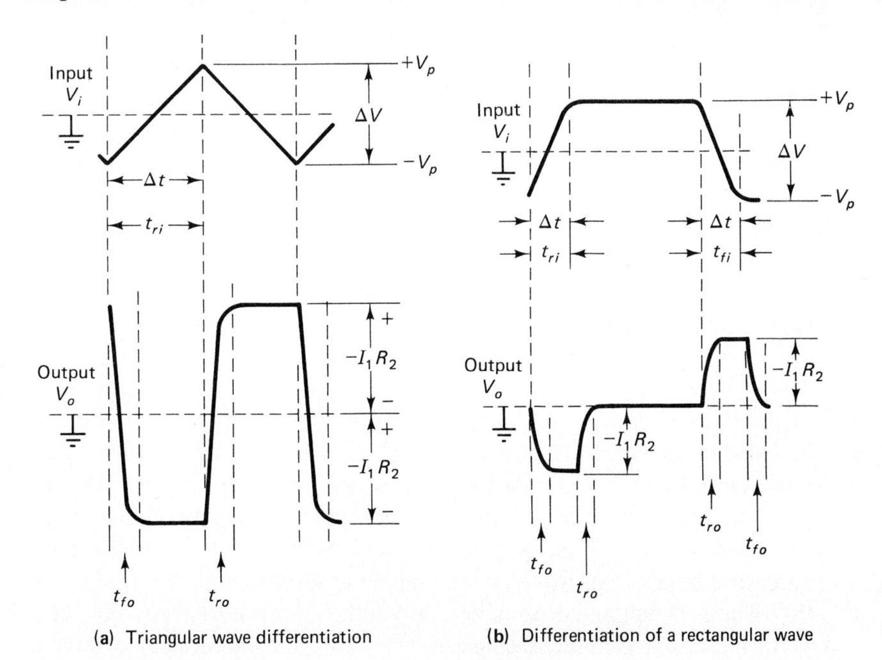

Figure 8-3 Differentiating circuit rise and fall times produce distortion in the output waveforms. The distortion may not be noticeable when the circuit output rise and fall times are less than 10% of the input rise and fall times.

If the rise and fall times of the output waveforms are reduced to 10% of the output pulse width, the outputs would look reasonably undistorted. Since the output pulses are the same width as the rise and fall times of the input waveforms, the output rise times should be approximately 10% of the input rise times. It can be shown that the rise time (t_{ro}) of the output of any linear circuit is related to the circuit upper cutoff frequency (f_2) by the equation (see Section 5-6)

$$f_2 = \frac{0.35}{t_{ro}} \tag{8-3}$$

At this frequency, X_{C1} should normally be much smaller than R_1. So, f_2 is the upper cutoff frequency for a closed-loop gain of R_2/R_1. A waveform with an input rise time of 10 μs would require an output rise time of 1 μs to produce a reasonably undistorted differentiated output wave. So, at $A_v = R_2/R_1$, the circuit upper cutoff frequency would be

$$f_2 = \frac{0.35}{1\ \mu\text{s}} = 350\ \text{kHz}$$

The differentiated waveforms would look even better if the output rise times were less than 10% of the input rise times. However, this would require an op-amp with a much higher cutoff frequency. In some circumstances, greater distortion of the output may be acceptable. For example, if only the output amplitude is to be measured to determine the input rate of change, an output rise time of 50% of the input might be satisfactory. In this case, an op-amp with a much lower cutoff frequency would suffice.

The above method of determining the circuit cutoff frequency applies only when the op-amp is used in a *small signal* situation; that is when the output is only around 1 V or less. For large output voltage swings, the op-amp slew rate must be taken into account. An output voltage amplitude of 10 V with a rise time of 1 μs, requires an op-amp with a minimum slew rate of 10 V/μs.

Sine Wave Response

Now look at the sinusoidal input waveform and the resultant differentiated output in Fig. 8-4. When a sine wave is correctly differentiated, a cosine waveform is produced. However, the inverting effect of the op-amp circuit in Fig. 8-1 results in an inverted cosine wave, as illustrated. An inverting amplifier with unity gain could be included at the circuit output to reinvert the waveform, if desired. The differentiator input circuit consists of resistor R_1 in series with capacitor C_1 (Fig. 8-1) which is connected to the virtual ground at the op-amp inverting input terminal. So, at signal frequencies where R_1 is very much smaller than X_{C1}, the input circuit can be regarded as purely capacitive and the input current (I_1) leads the applied voltage (v_i) by 90°. The $I_1 R_2$ voltage drop is in phase with I_1, so it leads v_i by 90°, and this is what gives the cosine output. Because $V_o = -I_1 R_1$, the output is a negative cosine wave.

At high signal frequencies, when X_{C1} is very much smaller than R_1, the circuit merely behaves as an inverting amplifier with a voltage gain of $-R_2/R_1$. At the fre-

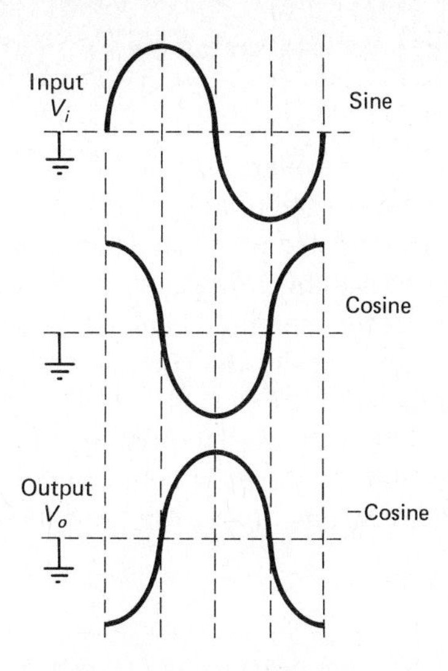

Figure 8-4 A cosine waveform is produced when a sine wave is correctly differentiated. The inversion in the op-amp differentiating circuit results in a -cosine wave.

quency at which $X_{C1} = R_1$, I_1 leads v_i by 45°. So here again, the circuit will not correctly perform differentiation.

The *characteristic frequency* (f_c) of the circuit is defined as the frequency at which X_{C1} equals R_2,

$$\frac{1}{2\pi f_c C_1} = R_2$$

giving,

$$f_c = \frac{1}{2\pi R_2 C_1} \tag{8-4}$$

With $R_1 = R_2/20$ (see Section 8-2) and $X_{C1} = R_2$ at f_c,

$$X_{C1} = 20R_1 \text{ at } f_c$$

In this case,

$$I_1 \approx \frac{v_i}{X_{c1}}$$

and the circuit output voltage amplitude is,

$$|v_o| \approx |v_i| \times \frac{R_2}{X_{c1}} \tag{8-5}$$

With $X_{C1} = R_2$ at f_c,

$$|v_o| \approx |v_i|$$

The phase shift of the output with respect to the input depends on the relationship between X_{C1} and R_1.

$$\phi = \arctan \frac{X_{C1}}{R_1} \tag{8-6}$$

With $X_{C1} = R_2 = 20R_1$ at f_c,

$$\phi = \arctan 20 \approx 87.1°$$

So, the current I_1 leads the applied voltage v_i by 87.1° at f_c. This is 2.9° less than the 90° phase shift for correct differentiation; a 3.2% error.

At signal frequencies above f_c the phase angle error progressively increases. At frequencies below f_c, X_{C1} becomes much larger than R_1, the phase angle error progressively decreases, and the output amplitude falls off in proportion to frequency. It is seen that differentiation occurs at the characteristic frequency of the differentiating circuit and at signal frequencies below f_c.

Fast noise spikes at the input of a differentiating circuit can produce unwanted large amplitude outputs. This effect is usually combatted by connecting a phase lead capacitor across resistor R_2 to further reduce the high frequency voltage gain (see Section 4-6).

Example 8-2

Determine the required minimum slew rate for the op-amp in the differentiating circuit designed in Example 8-1 if an output rise time equal to 30% of the input rise time is acceptable. Also, calculate the maximum sine wave input frequency at which differentiation occurs.

Solution

$$t_{ro} = 30\% \text{ of } t_{ri} = 30\% \text{ of } 100\ \mu s = 30\ \mu s$$

$$S_{min} = \frac{V_o}{t_{ro}} = \frac{5\ V}{30\ \mu s} \approx 0.17\ V/\mu s$$

Eq. 8-4,

$$f_c = \frac{1}{2\pi R_2 C_1} = \frac{1}{2\pi \times 10\ k\Omega \times 0.05\ \mu F} \approx 320\ Hz$$

8-4 INTEGRATING CIRCUIT

Circuit Operation

An integrating circuit produces an output voltage which is proportional to the area (amplitude multiplied by time) contained under the input waveform. Like differentiation, integration can be performed by a simple CR circuit. But use of an op-amp provides a much lower output resistance and higher output voltage than is possible with a CR circuit.

The integrating circuit in Fig. 8-5(a) resembles an inverting amplifier with capacitor C_1 connected between the output and the op-amp inverting input terminal. Because the noninverting input terminal is grounded via resistor R_3, the inverting in-

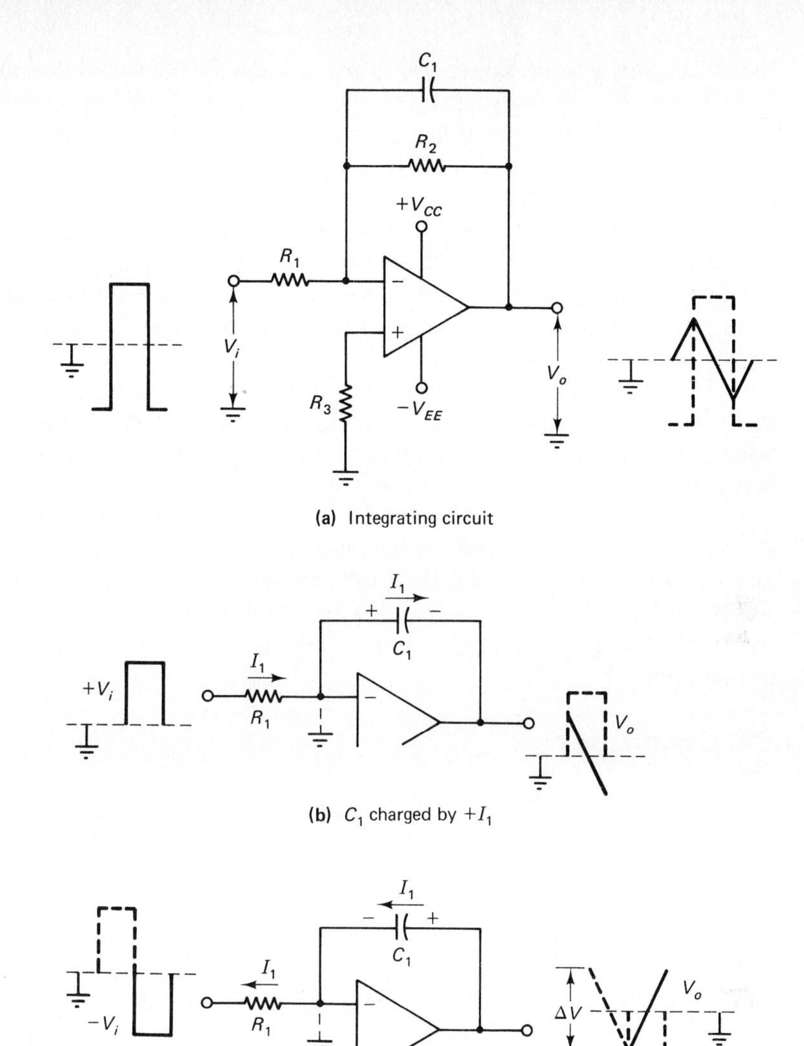

Figure 8-5 Op-amp integrating circuit. A constant input voltage results in a constant charging current for C_1. This generates a capacitor (and output) voltage which is directly proportional to V_i and Δt; that is, proportional to the $(V_i \Delta t)$ area of the input.

put terminal remains close to ground level. Thus, the input voltage appears across R_1 and the input current is simply V_i/R_1.

During the positive half-cycle of a square wave input, current I_1 is a constant quantity flowing into R_1, as illustrated in Fig. 8-5(b). With I_1 much larger than the op-amp input bias current, effectively all of I_1 flows to R_2 and C_1. As will be explained, the resistance of R_2 is so high that virtually all of I_1 flows through the capacitor. So, I_1 charges the capacitor with the polarity positive on the left side and negative on the right as illustrated. With the left side of the capacitor connected to the virtual ground at the op-amp inverting input terminal, the capacitor voltage is the circuit output voltage. Since the input current is constant, the capacitor is charged linearly and the output is a negative-going ramp.

During the negative half-cycle of the square wave input, the direction of I_1 is reversed, as shown in Fig. 8-5(c). The capacitor is now linearly charged negative on the left and positive on the right. This produces a positive-going output ramp. Thus, the square wave input to the integrating circuit generates a triangular output waveform which is negative-going when the input is positive and positive-going when the input is negative.

Disregarding the inversion produced by the op-amp circuit, the peak amplitude of the output is dependant upon I_1 and Δt. Since I_1 is proportional to the input peak voltage v_p and Δt is the time duration of the input pulse, the output voltage is directly proportional to the area under each half-cycle of the input. Doubling the square wave amplitude would double the area of each input half-cycle. It would also double the capacitor charging current and thus increase the output peak voltage by a factor of two. Similarly, increasing the time duration of each half-cycle of the input would increase the capacitor charging time and proportionately increase the output peak amplitude. So, the output peak voltage is the area summation, or *integration,* of the input.

DC Conditions

For dc conditions, C_1 in Fig. 8-5 is an open circuit. So, if resistor R_2 was not present, the op-amp output might be saturated at either its positive or negative dc voltage levels, approximately $+(V_{CC} - 1\text{ V})$ and $-(V_{EE} - 1\text{ V})$. The result is that, instead of being symmetrical above and below ground level, the output waveform would have a dc offset of several volts. With R_2 included, the circuit behaves as an inverting amplifier for dc conditions, giving a zero level output voltage when the input is at ground. Usually, resistor R_2 is selected by the simple rule-of-thumb that it should be twenty times the resistance of R_1.

8-5 INTEGRATOR DESIGN

As for so many other circuits already discussed, the first step in the design of an integrating circuit is to select current I_1 to be much larger than the op-amp input bias current. In the case of a BIFET op-amp, it is best to start by selecting the capacitance of C_1 much larger than any stray capacitance. C_1 should also have a low leakage dielectric. For a triangular output waveform, the relationship between I_1, C_1, and the output voltage amplitude is given by Eq. 8-1.

$$C_1 = \frac{I_1\,\Delta t}{\Delta v}$$

where Δt is the ramp time and Δv is the peak-to-peak amplitude [Fig. 8-5(c)]. So, if I_1 is first selected, the value of C_1 can be readily calculated. If C_1 is chosen first, I_1 is determined from its capacitance value.

Resistor R_1 is calculated as

$$R_1 = \frac{V_i}{I_1}$$

Then $$R_2 = 20R_1$$

and $$R_3 = R_2 \parallel R_1 \approx R_1$$

The supply voltages should, once again, be selected at least 3 V greater than the positive and negative peak output voltages.

At high frequencies, capacitor C_1 behaves as a short circuit, giving 100% feedback. So, for frequency compensation purposes the circuit should be treated as a voltage follower.

Example 8-3

Using a BIFET op-amp, design an integrating circuit to produce a triangular output waveform with a peak-to-peak amplitude of 4 V. The input is a ±5 V square wave with a frequency of 500 Hz.

Solution

$$C_1 \gg \text{stray capacitance}$$

let $$C_1 = 0.1\ \mu\text{F (standard value)}$$

$$\Delta t = \frac{T}{2} = \frac{1}{2f}$$

$$= \frac{1}{2 \times 500\ \text{Hz}} = 1\ \text{ms}$$

$$\Delta v = 4\ \text{V}$$

from Eq. 8-1, $$I_1 = \frac{C_1\,\Delta v}{\Delta t} = \frac{0.1\ \mu\text{F} \times 4\ \text{V}}{1\ \text{ms}}$$

$$= 400\ \mu\text{A}$$

$$R_1 = \frac{V_i}{I_1} = \frac{5\ \text{V}}{400\ \mu\text{A}}$$

$$= 12.5\ \text{k}\Omega$$ (use a 12 kΩ standard value with a 470 Ω connected in series)

$$R_2 = 20R_1 = 20 \times 12.5\ \text{k}\Omega$$

$$= 250\ \text{k}\Omega$$ (use a 270 kΩ standard value)

$$R_3 \approx R_1 = 12.5\ \text{k}\Omega$$ (use a 12 kΩ standard value)

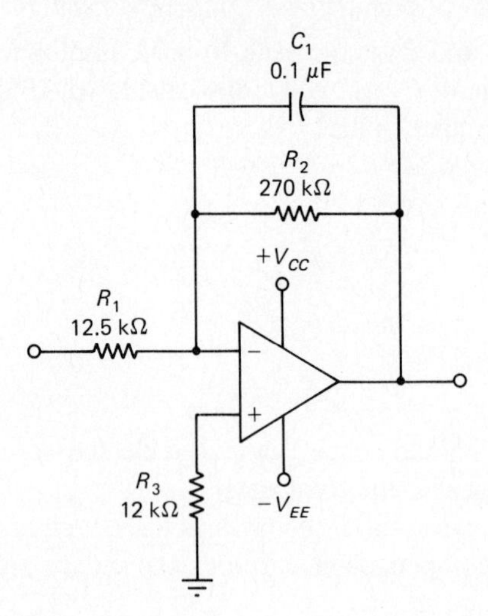

Figure 8-6 Integrating circuit designed in Example 8-3.

8-6 INTEGRATING CIRCUIT PERFORMANCE

Ramp Output Distortion

The ramp output waveform produced by an integrator with a square wave input is redrawn in Fig. 8-7(a). With resistor R_2 in parallel with C_1 (in Fig. 8-5), an increasing portion of I_1 flows through R_2 as the capacitor voltage grows. This is current that should have flowed into C_1 to charge the capacitor linearly and it results in a nonlinearity of the ramp waveform, as illustrated.

For the integrator circuit designed in Example 8-3, the current through R_2 at the peak of the capacitor voltage is

$$I_2 = \frac{V_P}{R_2} = \frac{(4\ \text{V}/2)}{220\ \text{k}\Omega}$$

$$\approx 9.1\ \mu\text{A}$$

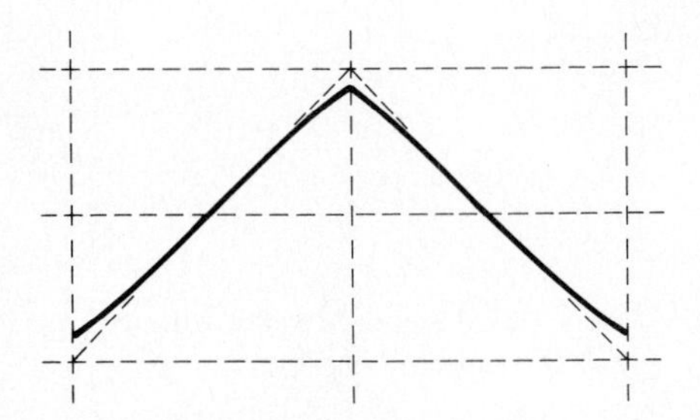

(a) Nonlinearity due to current flow through resistor R_2

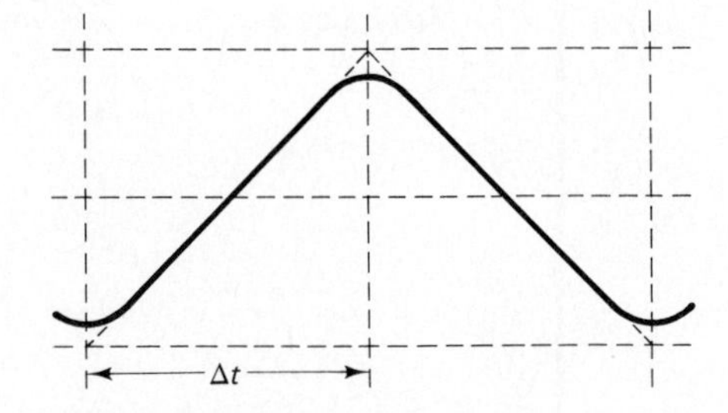

(b) Rounding of the corners of the output ramp due to the op-amp frequency response not being high enough

Figure 8-7 Distortion of the output waveform from an integrating circuit occurs because of the current flow through the resistor in parallel with the capacitor, and because of a limited op-amp frequency response.

This constitutes a 2.3% reduction in the charging current to C_1. When precise linearity is required, R_2 must be omitted and another means of preventing dc output voltage saturation must be employed (see Section 10-1).

Figure 8-7(b) shows the kind of corner-rounding distortion that occurs on the ramp output from an integrating circuit when the op-amp does not have a high enough frequency response. If distortion is to be unnoticeable, the output rise time of the op-amp should not exceed approximately 10% of the ramp time. So, the op-amp minimum slew rate should be

$$S_{(min)} \approx \frac{V_{o(p\text{-to-}p)}}{(\Delta t/10)} \tag{8-7}$$

Sine Wave Response

A sinusoidal input waveform and the resultant integrating circuit output are shown in Fig. 8-8. When a sine wave is correctly integrated, a negative cosine waveform is produced. Here again, as in the case of the differentiating circuit, the inverting effect of the op-amp alters the result. The actual output is a positive cosine wave, as illustrated. This could be inverted if desired. The integrator input circuit consists of resistor R_1 connected to the virtual ground at the op-amp inverting input terminal [Fig. 8-5(a)]. Thus, the input current I_1 is always in phase with the input voltage v_i. The output voltage is developed across capacitor C_1 and resistor R_2 in parallel. At high signal frequencies, where X_{C1} is much smaller than R_2, the parallel circuit can be regarded as purely capacitive and its voltage drop lags its current (I_1) by 90° to give the −cosine output. Because $V_o = -I_1 X_{C1}$, the output is a positive cosine wave.

At low signal frequencies, when X_{C1} is much larger than R_2, the circuit will not integrate sine waves, but instead behaves as an inverting amplifier with a voltage gain of $-R_2/R_1$. At the frequency at which $X_{C1} = R_2$, v_o lags v_i by 45°. So here again, the circuit will not correctly perform integration. The *characteristic frequency*

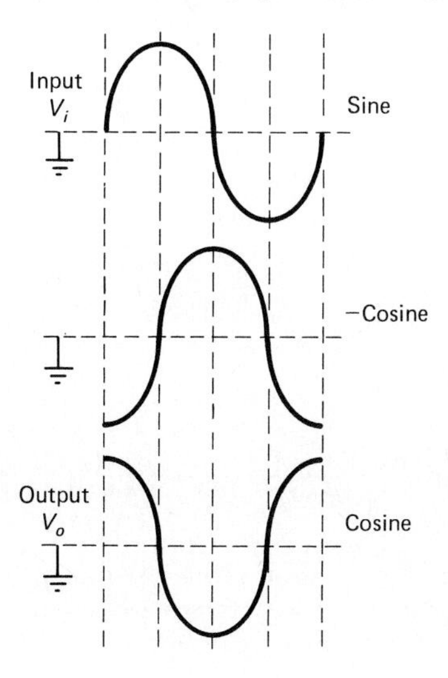

Figure 8-8 A −cosine waveform is produced when a sine wave is correctly integrated. The inversion in the op-amp integrating circuit results in a cosine wave.

(f_c) of the integrator is defined as the frequency at which X_{C1} equals *input resistance* R_1.

$$\frac{1}{2\pi f_c C_1} = R_1$$

giving
$$f_c = \frac{1}{2\pi R_1 C_1} \tag{8-8}$$

With $X_{C1} = R_1$ at f_c, and $R_2 = 20\ \mathrm{R}_1$ (see Section 8-5) the integrator output voltage amplitude is

$$|\mathrm{v}_o| \approx |\mathrm{v}_i| \times \frac{X_{C1}}{R_1} \tag{8-9}$$

or,
$$|\mathrm{v}_o| \approx |\mathrm{v}_i|$$

The phase shift of the output with respect to the input depends upon the relationship between X_{C1} and R_2.

$$\phi = \arctan \frac{R_2}{X_{C1}} \tag{8-10}$$

with $R_2 = 20R_1 = 20X_{C1}$ at f_c,
$$\phi = \arctan 20$$
$$\approx 87.1°$$

As in the case of the differentiating circuit at its characteristic frequency, this is 2.9° less than the 90° phase shift for correct integration, a 3.2% error.

At frequencies below f_c, X_{C1} becomes larger, thus making the parallel circuit more resistive and causing the phase angle error to progressively increase. At signal frequencies above f_c, X_{C1} becomes much smaller than R_2, the phase angle error decreases, and the output amplitude falls off in proportion to frequency. It is seen that integration occurs at f_c and at signal frequencies higher than f_c.

Example 8-4

Determine the required minimum slew rate for the op-amp in the integrator circuit designed in Example 8-3 if the ramp output is to appear distortion-free. Also, calculate the minimum sine wave input frequency at which integration occurs.

Solution

Eq. 8-7,
$$S_{\min} = \frac{v_{o(\text{p to p})}}{\Delta t/10} = \frac{4\ \text{V}}{1\ \text{ms}/10}$$
$$= 0.04\ \text{V}/\mu\text{s}$$

Eq. 8-8,
$$f_c = \frac{1}{2\pi R_1 C_1} = \frac{1}{2\pi \times 12.5\ \text{k}\Omega \times 0.1\ \mu\text{F}}$$
$$= 127\ \text{Hz}$$

REVIEW QUESTIONS

8-1. Sketch an op-amp differentiating circuit. Carefully explain the circuit dc and ac operation and the function of every component.

8-2. List the design procedure for a differentiating circuit (a) using a bipolar op-amp and (b) using a BIFET op-amp.

8-3. Sketch the output waveforms produced by an op-amp differentiating circuit with triangular and rectangular inputs. Explain each output waveshape. Discuss the distortion that occurs and how it can be minimized.

8-4. Explain how the minimum cutoff frequency and slew rate should be determined for an op-amp in a differentiating circuit.

8-5. Sketch the output waveform that should occur and the actual output waveform when a sine wave is applied to an op-amp differentiating circuit. Explain.

8-6. Write the equation for the characteristic frequency of an op-amp differentiating circuit and discuss the output produced by a sine wave input with the characteristic frequency. Explain what happens when the input frequency is (a) above the characteristic frequency and (b) below the characteristic frequency.

8-7. Sketch an op-amp integrating circuit. Carefully explain the circuit dc and ac operation, and the function of every component.

8-8. List the design procedure for an integrating circuit (a) using a bipolar op-amp and (b) using a BIFET op-amp.

8-9. Sketch the output waveform produced by an op-amp integrating circuit with a rectangular input. Explain the output waveshape. Discuss the distortion that occurs and how it can be minimized.

8-10. Show how the minimum slew rate should be determined for an op-amp in an integrating circuit.

8-11. Sketch the output waveform that should occur and the actual output waveform when a sine wave is applied to an op-amp integrating circuit. Explain.

8-12. Write the equation for the characteristic frequency of an op-amp integrating circuit and discuss the output produced by a sine wave input with the characteristic frequency. Explain what happens when the input frequency is (a) above the characteristic frequency and (b) below the characteristic frequency.

PROBLEMS

8-1. A differentiating circuit is to produce an output of 1 V when the input changes by 5 V in a time of 150 μs. Design a suitable circuit using a bipolar op-amp.

8-2. Determine the amplitude of the square wave output from the circuit in Problem 8-1 when a 10 kHz, 2 V peak-to-peak triangular input wave is applied.

8-3. A triangular wave with a frequency of 500 Hz and a peak-to-peak amplitude of 3 V is to be converted into a ± 7 V square wave by means of a differentiating circuit. Design a suitable circuit using a bipolar op-amp.

8-4. Redesign the circuit for Problem 8-3 using a BIFET op-amp.

8-5. Determine the required upper cutoff frequency and minimum slew rate for each of the op-amps in the circuits designed for Problems 8-1 through 8-4.

8-6. Calculate the characteristic frequency for each of the differentiating circuits designed for Problems 8-1 through 8-4.

8-7. Using a bipolar op-amp, design an integrating circuit to produce a triangular wave output with a peak-to-peak amplitude of 5 V, when the input is a square wave with an amplitude of ± 2 V and a frequency of 1 kHz.

8-8. Redesign the circuit for Problem 8-7 to use a BIFET op-amp.

8-9. A triangular wave with a peak-to-peak amplitude of 3 V is to be produced from an integrating circuit when a ± 7 V square wave is applied as input. The square wave frequency is 500 Hz. Design a suitable circuit using a BIFET op-amp.

8-10. Redesign the circuit for Problem 8-9 to use a bipolar op-amp.

8-11. Determine the required slew rate for each of the op-amps in the circuits designed for Problems 8-7 through 8-10.

8-12. Calculate the characteristic frequency for each of the integrating circuits designed for Problems 8-7 through 8-10.

8-13. A differentiating circuit as in Fig. 8-1 has the following components: $R_1 = 680\ \Omega$, $R_2 = 12\ \text{k}\Omega$, $C_1 = 0.1\ \mu\text{F}$, $R_3 = 12\ \text{k}\Omega$. Determine the output produced when a 300 Hz, 4 V peak-to-peak triangular wave is applied as input.

8-14. A 100 Hz, 5 V peak-to-peak square wave with $t_r = 1$ ms and $t_f = 3$ ms is applied as input to the circuit in Problem 8-13. Determine the output waveform that results.

8-15. Calculate the characteristic frequency of the differentiating circuit described in Problem 8-13. Also, determine the required minimum slew rate of the op-amp for satisfactory operation with the input described in Problem 8-14.

8-16. An integrating circuit as in Fig. 8-5 has the following components: $R_1 = 15\ \text{k}\Omega$, $R_2 = 330\ \text{k}\Omega$, $C_1 = 0.03\ \mu\text{F}$, $R_3 = 15\ \text{k}\Omega$. Determine the output produced when a 400 Hz, 3 V peak-to-peak square wave is applied as input.

8-17. Calculate the characteristic frequency of the integrating circuit described in Problem 8-16. Also, determine the required minimum slew rate of the op-amp for satisfactory operation with the specified input.

COMPUTER PROBLEMS

8-18. Write a computer program to design a differentiating circuit as in Fig. 8-1, including calculation of the characteristic frequency and determination of the required op-amp minimum slew rate. Given: a BIFET op-amp, V_o, and input rate of change.

8-19. Write a computer program to design an integrating circuit as in Fig. 8-5, including calculation of the characteristic frequency and determination of the required op-amp minimum slew rate. Given: a bipolar op-amp, V_{in}, f, $V_{o(\text{peak})}$.

LABORATORY EXERCISES

8-1 Differentiating Circuit

1. Construct the op-amp differentiating circuit shown in Fig. 8-2, as designed in Example 8-1. Use a 741 op-amp, with a supply of ± 15 V.

2. Apply a ±0.5 V, 5 kHz triangular wave input.
3. Observe the input and output waveforms on an oscilloscope and measure the positive and negative output voltage peaks. Compare to the waveforms in Fig. 8-3(a) and to the specified output amplitude.
4. Replace the triangular input with a ±0.5 V, 100 Hz square wave. Observe the input and output waveforms and compare to Fig. 8-3(b). If possible, alter the rise and fall times of the square wave, and note the effect on the output.
5. Replace the square wave with a ±0.5 V, 100 Hz sine wave. Observe the input and output waveforms. Slowly increase the frequency to discover the maximum frequency at which sine wave differentiation occurs.

8-2 Integrating Circuit

1. Construct the op-amp integrating circuit shown in Fig. 8-6, as designed in Example 8-3. Use a (correctly compensated) LF 108 with a supply of ±15 V. (A 741 can be used with this particular circuit.)
2. Apply a ±5 V, 500 Hz square wave input.
3. Observe the input and output waveforms on an oscilloscope and measure the positive and negative output voltage peaks. Compare to the waveforms in Fig. 8-5 and to the specified output amplitude.
4. Replace the square wave with a ±0.5 V, 500 Hz sine wave. Observe the input and output waveforms. Slowly reduce the frequency to discover the minimum frequency at which sine wave integration occurs.

CHAPTER 9

Op-Amp Nonlinear Circuits

Objectives

- Explain the effects of *output voltage swing, input voltage range, slew rate,* and *frequency compensation* upon op-amp switching circuits.
- Sketch the following op-amp switching circuits and explain their operation: *direct-coupled zero-crossing detector, capacitor-coupled zero-crossing detector, noninverting Schmitt trigger circuits, inverting Schmitt trigger circuits, astable multivibrator, monostable multivibrator.*
- Sketch and explain the typical input/output characteristics of a Schmitt trigger circuit.
- Design each of the above circuits using bipolar and BIFET op-amps together with transistors and FETs where appropriate.
- Analyze each of the above circuits to determine its performance.

INTRODUCTION

Operational amplifiers are widely used in circuits in which the output is switched between the positive and negative saturation levels. Positive feedback is employed in these circuits and the op-amp normally needs no frequency compensation. In the zero crossing detector, which is the simplest of op-amp switching circuits, one input terminal is grounded and the signal is applied to the other input. The Schmitt trigger circuit is essentially a zero-crossing detector with feedback. It has definite upper and lower input voltage levels that trigger the output voltage to change from one saturation level to the other. Op-amp astable multivibrator and monostable multivibrator circuits are constructed by adding components to crossing detectors or Schmitt trigger cricuits.

9-1 OP-AMPS IN SWITCHING CIRCUITS

Output Voltage Swing

In switching applications the output voltage of an op-amp is normally switched between the extreme positive and negative voltage levels, or output saturation voltages, $+V_{sat}$ and $-V_{sat}$. As already discussed, usually,

$$+V_{sat} \approx (V_{CC} - 1 \text{ V}) \qquad (9\text{-}1)$$

and

$$-V_{sat} \approx (-V_{EE} + 1 \text{ V}) \qquad (9\text{-}2)$$

The 1 V difference is only a rough approximation. It might easily be 2 V or less than 0.5 V, depending on the particular op-amp.

Input Voltage Range

In linear applications, negative feedback keeps the op-amp (noninverting and inverting) input terminal voltages closely equal. In switching applications, the dc voltage level at one input terminal may be several volts different from that at the other input. In fact, switching applications normally employ positive feedback to provide a substantial voltage difference between the two input terminals. This ensures that the output is saturated at either a positive or negative voltage level. So, in switching applications, there is normally a differential voltage at the op-amp input terminals. The minimum differential voltage required to produce output saturation can be calculated as

$$V_{i(\text{dif})} \approx \frac{V_{CC}}{M_{\min}} \qquad (9\text{-}3)$$

where $M_{\min}$ is the minimum open-loop voltage gain of the op-amp. For a 741 op-amp the minimum open-loop voltage gain is 50 000. If a ±15 V supply is used

$$V_{i(\text{dif})} \approx \frac{\pm 15 \text{ V}}{50\,000} = \pm 300 \ \mu\text{V}$$

There is a maximum differential input voltage specified for each op-amp. The op-amp input stage may be damaged if this maximum is exceeded. Most operational amplifiers can accept a differential input voltage equal to twice the supply voltage. For example, with a ±15 V supply, the maximum differential input should not exceed 30 V. However, some op-amps have a more limited differential input voltage range.

Slew Rate

As discussed in Sections 2-5 and 5-6, the slew rate defines the maximum rate of change of the output voltage. In switching applications, an input voltage step is frequently applied and the output that results is often distorted by the effect of the slew rate. An input sine or square wave could produce a near-triangular output because of the op-amp slew rate. This is illustrated in Fig. 9-1. The rise and the fall times (Δt) of the output waveform in Fig. 9-1 are directly related to the slew rate.

Eq. 2-7
$$\Delta t \approx \frac{\Delta V_o}{S}$$

If the rise and fall times are not greater than 10% of the output pulse width (t), the output will at least look like a rectangular wave [Fig. 9-1(b)]. This may be an acceptable amount of distortion is some situations. Other circumstances may require rise and fall times much smaller than 10% of the pulse width. Whatever the specification, the required minimum slew rate can be calculated using Eq. 2-7.

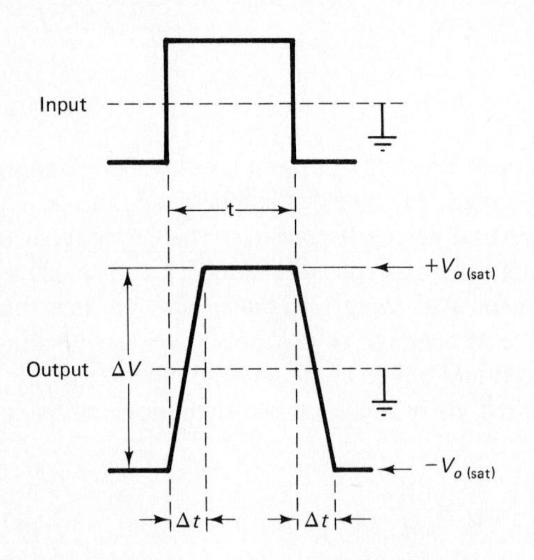

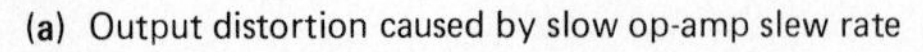
(a) Output distortion caused by slow op-amp slew rate

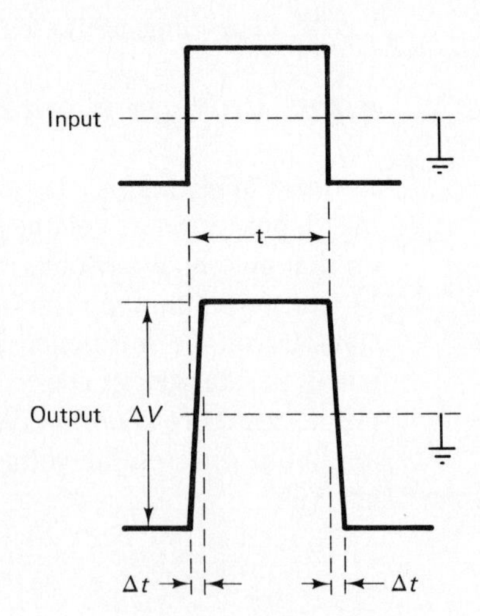

(b) Output distortion minimized if the slew rate gives $\Delta t \ll t$

Figure 9-1 Operational amplifier slew rate can create severe distortion in an output waveform. For minimum distortion, the slew rate must be selected to give rise and fall times much smaller that the output pulse width.

Frequency Compensation

Operational amplifiers employed in nonlinear applications, where they are switching from one output extreme to the other, normally do not require frequency compensation. There is no ac feedback when the output is in a steady-state saturated condition. In switching applications, the feedback is usually a positive dc quantity rather than negative ac. So, the ac stability considerations, which are important in linear applications, do not apply to op-amp switching circuits. However, it should be noted that in some applications often classified as nonlinear, the op-amp is actually functioning as a linear amplifier. This is true in the case of circuits for precision rectification, clamping, differentiation, integration, and sampling. Frequency compensation is required in these situations.

9-2 CROSSING DETECTORS

Zero Crossing Detector

An op-amp connected to function as a *zero crossing detector* is shown in Fig. 9-2(a). The inverting input terminal is grounded and the input voltage is directly connected to the noninverting terminal. When the input is below ground level, the

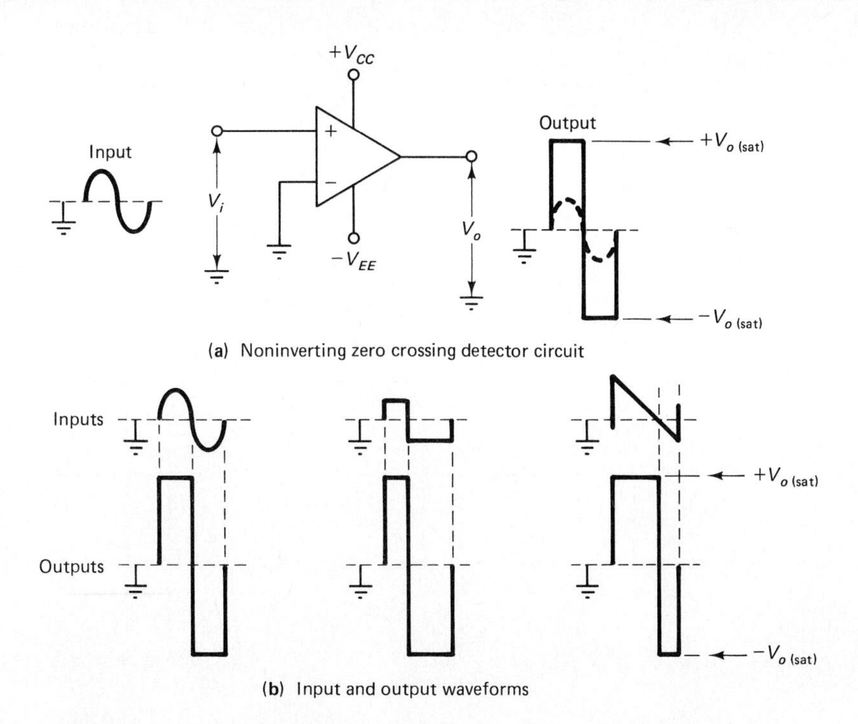

Figure 9-2 A noninverting zero-crossing detector is merely an op-amp with its inverting input terminal grounded. The output switches between $-V_{o(\text{sat})}$ and $+V_{o(\text{sat})}$ each time the input crosses the zero voltage level.

output is saturated at its negative extreme. When the input goes above ground level by about 300 μV (see Eq. 9-3), the output immediately switches to the op-amp positive saturation voltage. Each time the input voltage crosses the zero level, the output switches from one saturation level to the other. Because the output moves in a positive direction when the input crosses zero from negative to positive, the circuit in Fig. 9-2(a) can be classified as a *noninverting* zero-crossing detector.

The output waveforms that result for various inputs to a zero-crossing detector are illustrated in Fig. 9-2(b). Regardless of the input waveshape, the output is always a rectangular waveform.

There are no design calculations involved in using an op-amp as a zero-crossing detector. However, as discussed in Section 9-1, the op-amp input voltage range must be large enough to handle the signal and the slew rate must be fast enough to avoid unacceptable output distortion.

In Fig. 9-3(a), the signal is connected to the op-amp inverting input terminal and the noninverting terminal is grounded. In this case, the output is inverted with respect to the input, as illustrated. When the input voltage crosses the zero line going in a positive direction, the output goes negative, and vice versa. This circuit is an *inverting* zero-crossing detector, sometimes termed an *inverter*.

Instead of being biased to ground level, the noninverting input terminal of the op-amp could be biased to a positive or negative dc voltage level (Fig. 9-3(b)).

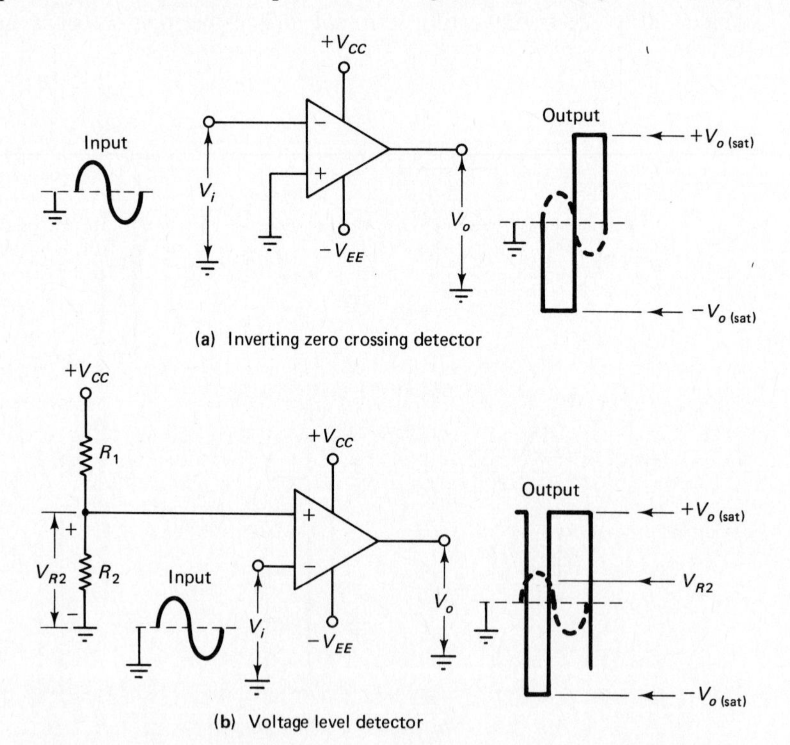

Figure 9-3 An inverting zero-crossing detector has the op-amp noninverting input grounded. In a voltage level detector, one of the op-amp input terminals is biased to the input voltage level at which switching is to occur.

Thus, the circuit output would change when the input arrives at the bias voltage level. This circuit is a *voltage level detector*.

Capacitor-Coupled Crossing Detector

The capacitor-coupled crossing detector in Fig. 9-4 has the op-amp noninverting input terminal connected to ground via resistor R_1 to provide a dc bias current path to the op-amp. Also, the inverting input terminal is biased to a dc voltage level slightly above ground. This is required to ensure that the output is saturated in a negative direction when no input signal is present. The output switches to $+V_{o(sat)}$ when the capacitor-coupled signal drives the noninverting input terminal above V_B, and falls back to $-V_{o(sat)}$ when the input drops below V_B.

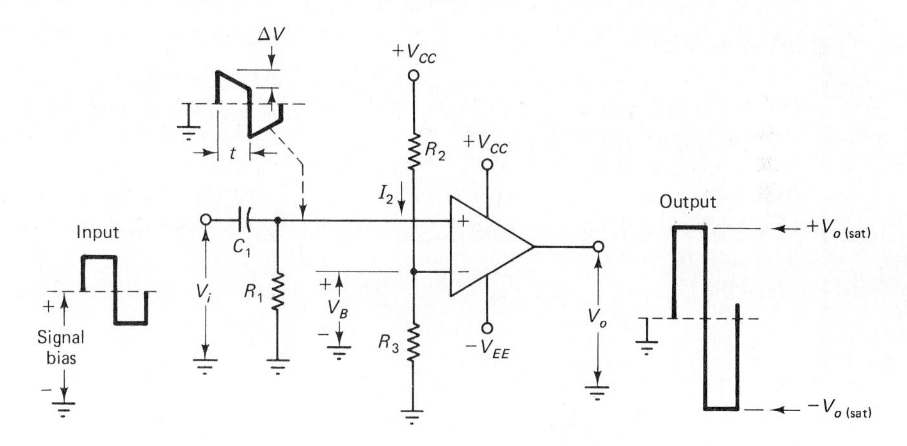

Figure 9-4 A capacitor-coupled crossing detector must have one of the op-amp input terminals biased above or below ground level, to ensure the output remains at the desired saturation level when no input is present.

The potential divider (R_2 and R_3) is designed in the usual way, and resistor R_1 is selected as large as possible using Eq. 3-1. The determination of capacitance C_1 depends upon the type of input waveform involved. With a sinusoidal input, phase shift error is normally to be avoided. So, X_{C1} should be much smaller than R_1 at the minimum signal frequency, assuming that the input impedance of the op-amp is much larger than R_1. With $X_{C1} = R_1/20$ at the lowest signal frequency, there will be approximately a 2.9° phase shift. This will decrease with increasing signal frequency. So, for less than 3° of phase shift error

$$X_{C1} = \frac{R_1}{20} \tag{9-4}$$

The input voltage to capacitor C_1 may not be symmetrical above and below ground level, but it is always symmetrical on the op-amp side of C_1. When a square wave is applied as an input to a capacitor-coupled crossing detector, the waveform at the op-amp input terminal can develop considerable tilt, as illustrated. As long as the op-amp output remains saturated at the required level, the tilt at the input is not important. A suitable capacitance for C_1 can be determined by first estimating an acceptable tilt voltage ΔV and noting the time duration t of the input. Then, the input current is calculated as

$$I_1 = \frac{V_i}{R_1}$$

and assuming that I_1 remains approximately constant,

$$C_1 = \frac{I_1 t}{\Delta V} \tag{9-5}$$

When a BIFET op-amp is to be used in a crossing detector, it is best to commence by selecting a capacitance value much larger than stray capacitances. I_1 can then be calculated from Eq. 9-5 and R_1 can be determined from V_i and I_1.

Example 9-1

A capacitor-coupled zero-crossing detector is to handle a 1 kHz square wave input with a peak-to-peak amplitude of 6 V. Design a suitable circuit as in Fig. 9-4, using a 741 op-amp with a ±12 V supply.

Solution

$$I_2 \gg I_{B(max)}$$

let

$$I_2 = 100 \times 500 \text{ nA}$$
$$= 50 \ \mu\text{A}$$

Let

$$V_B \approx 0.1 \text{ V}$$
$$V_{R2} = V_{CC} - V_B = 12 \text{ V} - 0.1 \text{ V}$$
$$= 11.9 \text{ V}$$
$$R_2 = \frac{V_{R2}}{I_2} = \frac{11.9 \text{ V}}{50 \ \mu\text{A}}$$
$$= 238 \text{ k}\Omega$$

(use 220 kΩ standard value and recalculate I_2)

$$I_2 = \frac{V_{R2}}{R_2} = \frac{11.9 \text{ V}}{220 \text{ k}\Omega}$$
$$= 54.1 \ \mu\text{A}$$
$$V_{R3} = V_B = 0.1 \text{ V}$$
$$R_3 = \frac{V_{R3}}{I_2} = \frac{0.1 \text{ V}}{54.1 \ \mu\text{A}}$$
$$= 1.85 \text{ k}\Omega$$

(use a 1.8 kΩ standard value)

Eq. 3-1,

$$R_1 = \frac{0.1 V_{BE}}{I_{B(max)}} = \frac{0.1 \times 0.7 \text{ V}}{500 \text{ nA}}$$
$$= 140 \text{ k}\Omega$$

(use 120 kΩ standard value)

$$V_{i(\text{peak})} = \frac{6\text{ V}}{2} = 3\text{ V}$$

$$I_1 = \frac{V_i}{R_1} = \frac{3\text{ V}}{120\text{ k}\Omega}$$

$$= 25\ \mu\text{A}$$

let $\quad \Delta V = 1\text{ V}$

$$t = \frac{1}{2f} = \frac{1}{2 \times 1\text{ kHz}}$$

$$= 500\ \mu\text{s}$$

Eq. 9-5, $\quad C_1 = \dfrac{I_1 t}{\Delta V} = \dfrac{25\ \mu\text{A} \times 500\ \mu\text{s}}{1\text{ V}}$

$$= 0.0125\ \mu\text{F}$$ (use a 0.015 μF standard value to give $\Delta V < 1$ V)

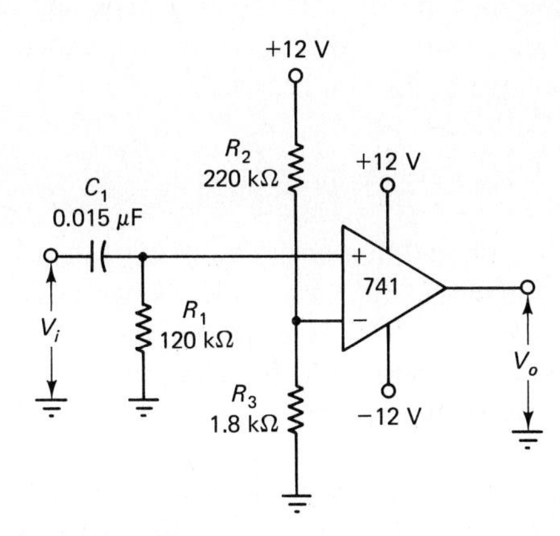

Figure 9-5 Capacitor-coupled level detector designed in Example 9-1.

Example 9-2

For the circuit designed in Example 9-1, estimate the minimum op-amp slew rate to give a reasonably undistorted output. Also, calculate the lowest sine wave input frequency that can be applied without the phase shift error exceeding 3°.

Solution

$$\Delta V_o = +V_{o(\text{sat})} - [-V_{o(\text{sat})}] \approx (V_{CC} - 1\text{ V}) - (V_{EE} + 1\text{ V})$$

$$\approx (12\text{ V} - 1\text{ V}) - (-12\text{ V} + 1\text{ V})$$

$$\approx 22\text{ V}$$

$$\Delta t \approx 0.1 \times t \approx 0.1 \times 500\ \mu\text{s}$$

$$\approx 50\ \mu\text{s}$$

From Eq. 2-7, $S_{\min} = \dfrac{\Delta V_o}{\Delta t} = \dfrac{22\text{ V}}{50\ \mu\text{s}}$

$$= 0.44\text{ V}/\mu\text{s}$$

For a maximum phase shift of 2.9° with a sine wave input,

Eq. 9-4,
$$X_{C1} = \frac{R_1}{20} = \frac{120\ \text{k}\Omega}{20}$$
$$= 6\ \text{k}\Omega$$
$$f_{(\text{min})} = \frac{1}{2\pi X_{C1} C_1} = \frac{1}{2\pi \times 6\ \text{k}\Omega \times 0.015\ \mu\text{F}}$$
$$\approx 1.8\ \text{kHz}$$

9-3 INVERTING SCHMITT TRIGGER CIRCUIT

Circuit Operation

An operational amplifier connected to function as a *Schmitt trigger circuit* is shown in Fig. 9-6(a). The circuit looks like a noninverting amplifier except for the important difference that the input is applied to the op-amp inverting terminal instead of the noninverting input terminal and that feedback from the output is connected to the noninverting input. The voltage at the noninverting input is

$$V_{R2} = \frac{V_o \times R_2}{R_1 + R_2}$$

When the output voltage is saturated in a positive direction at $+V_{o(\text{sat})}$, V_{R2} is a positive quantity. When the output is at $-V_{o(\text{sat})}$, V_{R2} is negative. For example, if $V_o = +14$ V and $R_1 = R_2$, $V_{R2} = +7$ V. With the circuit input voltage at zero, the noninverting input terminal is 7 V positive with respect to the inverting input terminal. Consequently, the op-amp output remains at its positive saturation voltage.

The output will switch from the positive saturation level to the negative saturation voltage only when the inverting input terminal is raised above the voltage at the noninverting input terminal (V_{R2}). When the input level is raised just slightly (microvolts) above V_{R2}, the op-amp output begins to move in a negative direction, causing V_{R2} to decrease, thus making the inverting input terminal more positive with respect to the noninverting input terminal. The differential input voltage causing the output to switch from positive saturation to negative saturation is increased by the movement of the output in the negative direction. This is positive feedback and it causes the output to move rapidly from one saturation level to the other.

Continuing the above use of a 14 V output level and $R_1 = R_2$ as an example, $V_o = -14$ V and $V_{R2} = -7$ V when the output goes negative. Now, while the input voltage remains above -7 V, the output will remain at its negative saturation level. To cause the output to switch to the positive saturation level once again, the input voltage must be decreased to just below the -7 V level of V_{R2}. So, the $+7$ V and -7 V input levels required to change the output voltage are referred to as the *trigger points* of the the circuit. The *upper trigger point* (*UTP*) is $+7$ V and the *lower trigger point* (*LTP*) is -7 V in this particular case. A Schmitt trigger circuit can be designed for virtually any desired trigger point voltages.

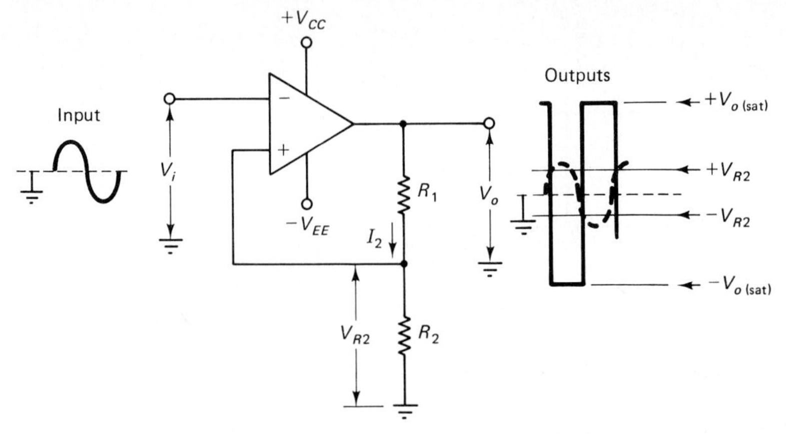

(a) Op-amp inverting Schmitt trigger circuit

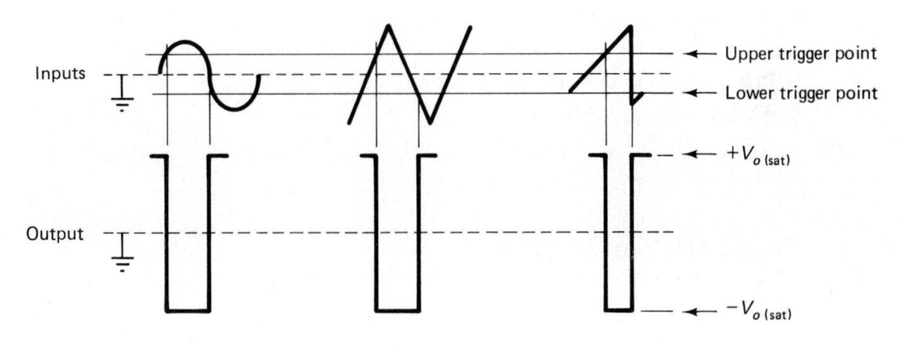

(b) Input and output waveforms

Figure 9-6 The output of an inverting Schmitt trigger circuit switches from $+V_{o(sat)}$ to $-V_{o(sat)}$ when the input voltage rises to the upper trigger point, (the voltage V_{R2}). The output switches back again from $-V_{o(sat)}$ to $+V_{o(sat)}$ when the input falls to the lower trigger point.

The voltage waveforms in Fig. 9-6(b) show the Schmitt trigger circuit response to various inputs. As illustrated, the output switches from positive to negative when the input voltage reaches the upper trigger point and from negative to positive when the input falls to the lower trigger point. So, the output waveform is always rectangular, except, as discussed in Section 9-1, where op-amp slew rate produces distortion.

Input/Output Characteristics

Typical input/output characteristics of an op-amp inverting Schmitt trigger circuit are illustrated in Fig. 9-7. Commencing with the output at $+V_{o(sat)}$ and the input at zero, when V_i is raised to the UTP, the output switches from $+V_{o(sat)}$ to $-V_{o(sat)}$, (point *a* to point *b*). Any further increase in V_i above the UTP maintains the output at $-V_{o(sat)}$, (point *b* to point *c*). While the input is being reduced from UTP to the

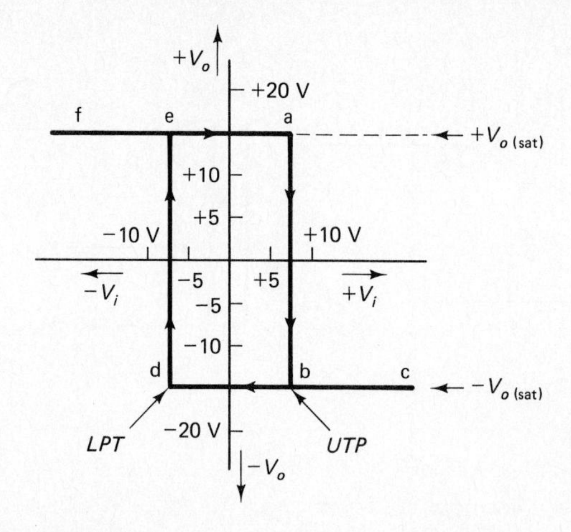

Figure 9-7 Output/input characteristics for an inverting Schmitt trigger circuit. V_o switches from $+V_{o(sat)}$ to $-V_{o(sat)}$ when V_i rises to the *UTP*, and from $-V_{o(sat)}$ to $+V_{o(sat)}$ when V_i falls to the *LTP*.

LTP (point *b* to point *d*), the output remains at $-V_{o(sat)}$. When V_i equals the LTP, the output rapidly switches from $-V_{o(sat)}$ to $+V_{o(sat)}$, (point *d* to point *e*). Now, any further decrease in V_i below the LTP maintains the output voltage at $+V_{o(sat)}$.

The voltage difference between the upper and lower trigger points is referred to as *hysteresis*. The zero-crossing detectors discussed in Section 9-2 have UTP and LTP equal, so they have zero hysteresis.

Schmitt Circuit Design

To design an op-amp Schmitt trigger circuit using a bipolar op-amp, the current through resistors R_1 and R_2 is first selected to be much larger than the op-amp input bias current. R_2 is then calculated as the trigger point voltage divided by I_2.

$$R_2 = \frac{\text{trigger voltage}}{I_2} \tag{9-6}$$

and R_1 is

$$R_1 = \frac{V_o - (\text{trigger voltage})}{I_2} \tag{9-7}$$

When using a BIFET op-amp, it is best to follow the usual procedure of selecting the largest of the two resistors as 1 MΩ, and then calculate the resistance of the other one.

Example 9-3

Using a 741 op-amp with a supply of ±12 V, design an inverting Schmitt trigger circuit to have trigger points of ±2 V.

Solution

$$I_2 \gg I_{B(max)}$$

let $\quad I_2 = 50\ \mu\text{A}$

$$V_{R2} = \text{UTP} = 2\text{ V}$$

$$R_2 = \frac{V_{R2}}{I_2} = \frac{2\text{ V}}{50\ \mu\text{A}}$$

$$= 40\text{ k}\Omega \qquad \text{(use 39 k}\Omega\text{ standard value and recalculate } I_2\text{)}$$

$$I_2 = \frac{V_{R2}}{R_2} = \frac{2\text{ V}}{39\text{ k}\Omega}$$

$$\approx 51.3\ \mu\text{A}$$

$$V_{R1} = V_{o(\text{sat})} - V_{R2} \approx (12\text{ V} - 1\text{ V}) - 2\text{ V}$$

$$= 9\text{ V}$$

$$R_2 = \frac{V_{R2}}{I_2} = \frac{9\text{ V}}{51.3\ \mu\text{A}}$$

$$= 175\text{ k}\Omega \qquad \text{(use 180 k}\Omega\text{ standard value)}$$

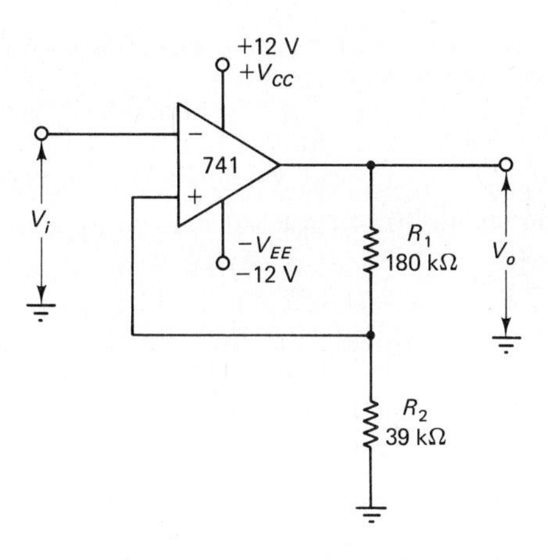

Figure 9-8 Inverting Schmitt trigger circuit designed in Example 9-3.

Adjusting the Trigger Points

Upper and lower trigger points of equal magnitude are not always desirable. One common requirement is to have some positive level of UTP combined with a zero voltage LTP. This can easily be achieved by including a diode in series with R_1, as illustrated in Fig. 9-9(a). When the output is positive, D_1 is forward-biased, and the UTP is the voltage drop across R_2. When the output goes negative, D_1 is reverse-biased, only the op-amp input bias current flows in R_2, and the op-amp noninverting input terminal is held close to ground level. The output will go positive once again when the input voltage is reduced below ground level. Thus, the lower trigger point is effectively zero volts.

Diode D_1 must be selected to have a maximum reverse voltage greater than the circuit supply voltage. Its maximum reverse recovery time (t_{rr}) should be much smaller than the minimum pulse width of the input signal. Typically,

$$t_{rr} \le \frac{\text{min pulse width}}{10}$$

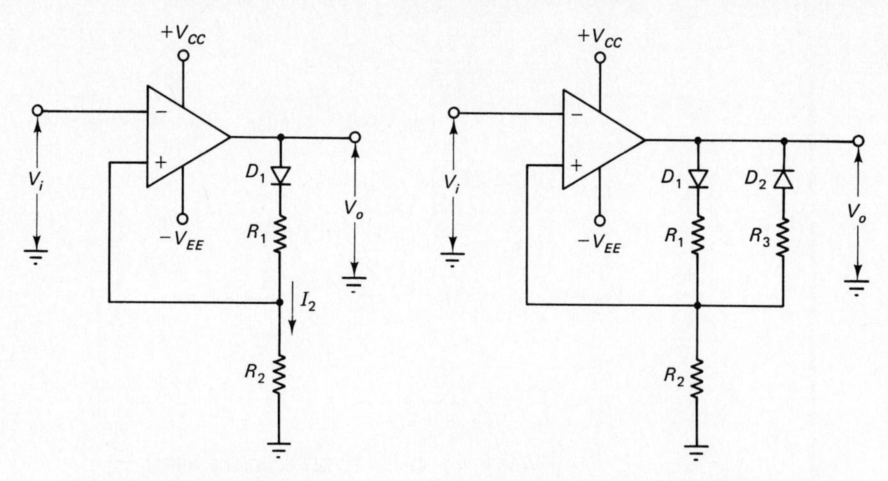

(a) Diode D_1 gives $LTP = 0$

(b) Two diodes give different UTP and LTP levels

Figure 9-9 Inverting Schmitt trigger circuits with different UTP and LTP voltages. Inclusion of diode D_1 in (a) gives $I_2 = 0$ when V_o is negative. Thus, $LTP = 0$. With D_1 and D_2 in (b), and different resistances for R_1 and R_3, the UTP and LTP can be any desired voltage levels.

The diode forward current should be at least 100 μA, although a 500 μA minimum might be more satisfactory.

The circuit in Fig. 9-9(b) has two different-level trigger points. When the output is positive, D_1 is forward biased and D_2 is reversed. The upper trigger point is

$$\text{UTP} = V_{R2} = \frac{(|V_o| - V_F) \times R_2}{R_1 + R_2} \tag{9-8}$$

where V_F is the forward voltage drop of D_1. When the output is negative, D_2 is forward-biased and D_1 is reversed, giving the lower trigger point as

$$\text{LTP} = V_{R2} = \frac{(|V_o| - V_F) \times R_2}{R_3 + R_2} \tag{9-9}$$

Different resistance values for R_1 and R_3 give different trigger point voltages.

9-4 NONINVERTING SCHMITT TRIGGER CIRCUIT

Circuit Operation

The noninverting Schmitt trigger circuit shown in Fig. 9-10(a) resembles the circuit of an inverting amplifier. However, once again, there is the important difference that the op-amp input terminals are interchanged. Suppose that the input voltage is at ground level and that the output voltage is at its negative saturation level. The voltage across resistor R_1 is

$$V_{R1} = \frac{V_o \times R_1}{R_1 + R_2}$$

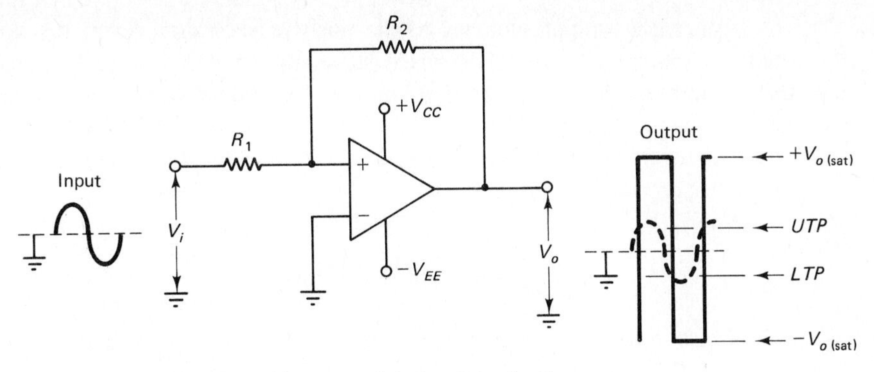

(a) Noninverting op-amp Schmitt trigger circuit

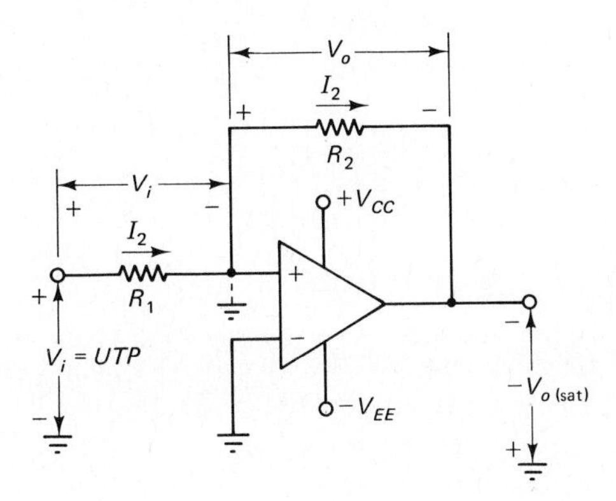

(b) Voltage and current conditions when V_i = UTP

Figure 9-10 Noninverting op-amp Schmitt trigger circuit. The output switches from $-V_{o(\text{sat})}$ to $+V_{o(\text{sat})}$ when the input voltage rises to the upper trigger point, and back again when the input falls to the lower trigger point. At the trigger points, $V_{R2} = V_o$, and $V_{R1} = V_i$.

With V_o negative, V_{R1} is a negative quantity. Thus, the noninverting input terminal is negative with respect to the (grounded) inverting input terminal and the output voltage is held at the negative saturation level.

To cause the output to switch from the negative saturation level to positive saturation, the input voltage must be raised until the voltage at the noninverting input terminal is just slightly (microvolts) above ground level. The situation is illustrated in Fig. 9-10(b). With the noninverting input terminal at ground level, the output voltage V_o is developed across resistor R_2 and the input voltage V_i appears across R_1. Therefore, the UTP for this circuit is

$$\text{UTP} = V_i = I_2 R_1$$

where

$$I_2 = \frac{|V_o|}{R_2}$$

giving

$$\text{UTP} = \frac{|V_o| \times R_1}{R_2} \tag{9-10}$$

When the output switches to the positive saturation level, the voltage at the noninverting input terminal is raised substantially above ground, thus maintaining the output at its positive saturation voltage. To return the output to the negative saturation voltage, the input has to be made sufficiently negative to pull the noninverting terminal down to ground level. By the same reasoning that derived Eq. 9-10, the lower trigger voltage is

$$\text{LTP} = \frac{|V_o| \times R_1}{R_2} \tag{9-11}$$

The circuit is designed by first selecting I_2 to be much larger than $I_{B(\max)}$.

Then,
$$R_1 = \frac{\text{trigger voltage}}{I_2}$$

and
$$R_2 = \frac{|V_o|}{I_2}$$

Adjusting the Trigger Points

As in the case of the inverting Schmitt circuit, the trigger points of the noninverting Schmitt circuit can be adjusted by the use of diodes. In Fig. 9-11(a), diode D_1 in series with R_2 is reverse-biased when the output is positive. This makes the voltage at the noninverting input terminal equal to the input voltage V_i. The output switches

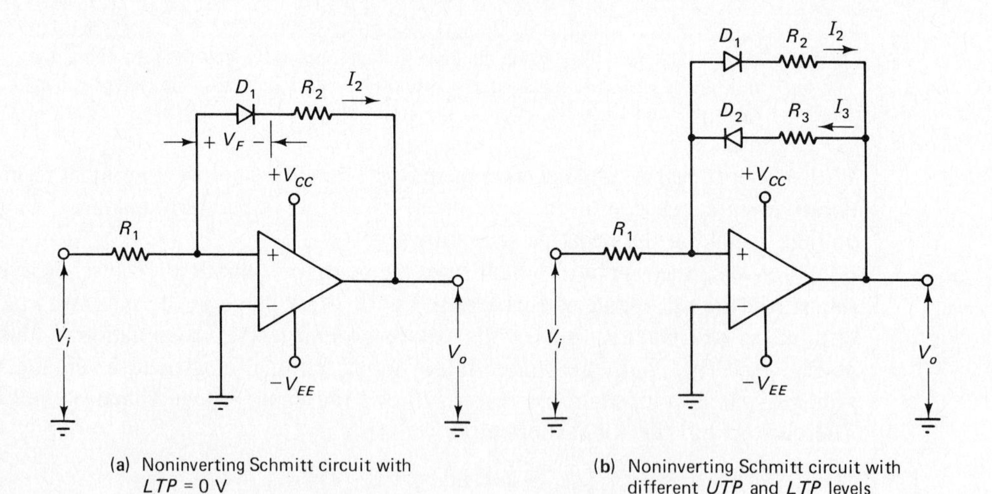

(a) Noninverting Schmitt circuit with *LTP* = 0 V

(b) Noninverting Schmitt circuit with different *UTP* and *LTP* levels

Figure 9-11 Noninverting Schmitt trigger circuits with different *UTP* and *LTP* voltages. With diode D_1 in (a), $I_2 = 0$ when V_o is positive, giving $LTP = 0$. With D_1 and D_2 in (b), and different resistances for R_2 and R_3, the *UTP* and *LTP* can be any desired voltage levels.

from positive to negative saturation when V_i goes just below ground level, so the lower trigger point is zero. The upper trigger point is

$$\text{UTP} = \frac{(|V_o| - V_F) \times R_1}{R_2} \tag{9-12}$$

where V_F is the forward voltage drop of the diode.

Figure 9-11(b) shows a noninverting Schmitt circuit which has different UTP and LTP voltage levels determined by the resistances of R_2 and R_3. The upper trigger point is given by Eq. 9-12 and the lower trigger point is

$$\text{LTP} = \frac{(|V_o| - V_F) \times R_1}{R_3} \tag{9-13}$$

Example 9-4

Design a noninverting Schmit trigger circuit as in Fig. 9-11(b), to have UTP = +3 V, and LTP = −5 V. Use a 741 op-amp with $V_{CC} = \pm 15$ V.

Solution

Design first for the UTP.

For adequate diode forward current, let

$$I_2 = 500\ \mu\text{A}$$

$$V_{R1} = \text{UTP} = 3\text{ V}$$

$$R_1 = \frac{V_{R1}}{I_2} = \frac{3\text{ V}}{500\ \mu\text{A}}$$

$$= 6\text{ k}\Omega \qquad \text{(use 5.6 k}\Omega\text{ standard value and recalculate } I_2\text{)}$$

$$I_2 = \frac{V_{R1}}{R_1} = \frac{3\text{ V}}{5.6\text{ k}\Omega}$$

$$\approx 536\ \mu\text{A}$$

$$V_{R2} = |V_o| - V_F \approx (15\text{ V} - 1\text{ V}) - 0.7\text{ V}$$

$$= 13.3\text{ V}$$

$$R_2 = \frac{V_{R2}}{I_2} = \frac{13.3\text{ V}}{536\ \mu\text{A}}$$

$$= 24.8\text{ k}\Omega \qquad \text{(Use series-connected 22 k}\Omega\text{ and 2.7 k}\Omega\text{ standard value resistors)}$$

Now design for the LTP, using the already selected resistance for R_1,

$$V_{R1} = LTP = 5\text{ V}$$

$$I_3 = \frac{V_{R1}}{R_1} = \frac{5\text{ V}}{5.6\text{ k}\Omega}$$

$$\approx 893\ \mu\text{A}$$

$$V_{R3} = |V_o| - V_F \approx (15\text{ V} - 1\text{ V}) - 0.7\text{ V}$$
$$= 13.3\text{ V}$$
$$R_3 = \frac{V_{R3}}{I_3} = \frac{13.3\text{ V}}{893\ \mu\text{A}}$$
$$= 14.9\text{ k}\Omega \qquad \text{(use 15 k}\Omega\text{ standard value)}$$

Select the diodes.

Minimum reverse voltage, $V_R > V_{CC} = 15$ V

$$t_{rr} \leq \frac{\text{min pulse width}}{10}$$

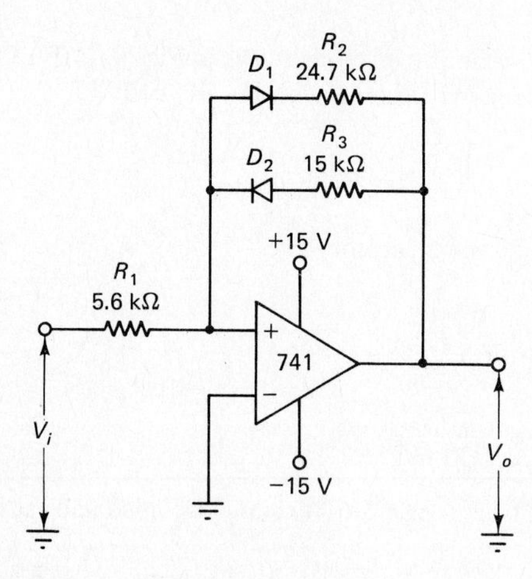

Figure 9-12 Noninverting Schmitt trigger circuit designed in Example 9-4.

Example 9-5

Using the selected standard resistance values, calculate the actual UTP and LTP for the circuit designed in Example 9-4.

Solution

$$I_2 = \frac{|V_o| - V_F}{R_2} = \frac{14\text{ V} - 0.7\text{ V}}{22\text{ k}\Omega + 2.7\text{ k}\Omega}$$
$$\approx 538\ \mu\text{A}$$
$$UTP = I_2 \times R_1 = 538\ \mu\text{A} \times 5.6\text{ k}\Omega$$
$$\approx 3.02\text{ V}$$
$$I_3 = \frac{|V_o| - V_F}{R_3} = \frac{14\text{ V} - 0.7\text{ V}}{15\text{ k}\Omega}$$
$$\approx 887\ \mu\text{A}$$
$$LTP = -I_3 \times R_1 = 887\ \mu\text{A} \times 5.6\text{ k}\Omega$$
$$\approx -5\text{ V}$$

9-5 ASTABLE MULTIVIBRATOR

Circuit Operation

An *astable multivibrator* is a circuit that is continuously switching its output voltage between high and low levels. It has no stable state. An astable multivibrator using an operational amplifier is shown in Fig. 9-13. The op-amp together with resistors R_2 and R_3 constitute an inverting Schmitt trigger circuit, exactly as in Fig. 9-6(a). The input voltage to the Schmitt circuit is the voltage across capacitor C_1, which is charged from the op-amp output via resistor R_1.

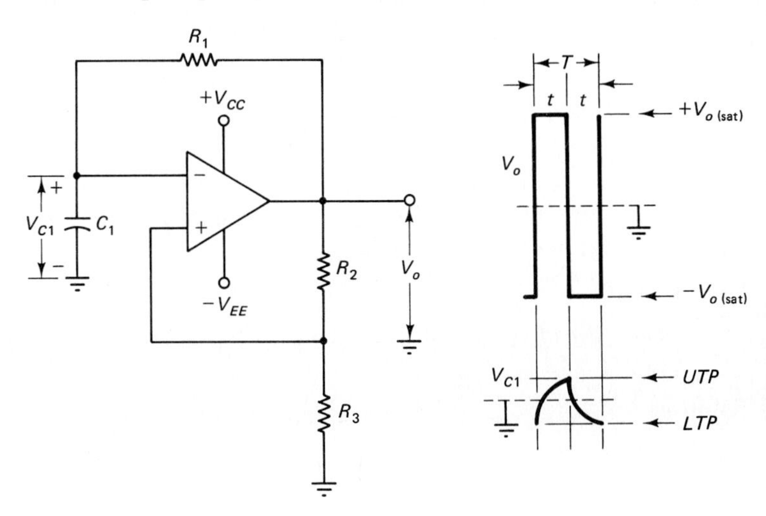

Figure 9-13 Astable multivibrator consisting of an inverting Schmitt trigger circuit, (op-amp, R_2 and R_3), and a *CR* circuit. When $V_o = +V_{o(sat)}$, C_1 charges to the *UTP*, causing V_o to switch to $-V_{o(sat)}$. When $V_o = -V_{o(sat)}$, C_1 charges to the *LTP*, causing V_o to switch to $+V_{o(sat)}$.

When the circuit output is at the positive saturation level, current flows into the capacitor, charging it positive at the top, as illustrated, until V_{C1} reaches the upper trigger point of the Schmitt circuit. The output then rapidly switches to the op-amp negative saturation level. Now current flows from the capacitor, removing its positive charge and recharging it with the opposite polarity. This continues until V_{C1} arrives at the Schmitt lower trigger point. Then, the op-amp output rapidly switches back to the positive saturation level and the cycle starts again.

It is seen that the circuit is a square wave generator with an output that swings between the op-amp positive and negative saturation levels. The frequency of the output depends on the capacitance of C_1 and the resistance of R_1.

Design

The Schmitt portion of the circuit can be designed in the usual way for any convenient trigger point levels. The minimum current through R_1 is selected to be much larger than the op-amp input bias current. R_1 is calculated as

$$R_1 = \frac{|V_o| - \text{UTP}}{I_1} \tag{9-14}$$

Once R_1 is determined, C_1 can be calculated from the capacitor charging equation and the desired half-cycle time of the output. If the Schmitt UTP and LTP are selected to be much smaller than the op-amp output voltage, the voltage drop across R_1 will not change very much. Consequently, the capacitor charging current (I_1) can be treated as constant and the simple constant current capacitor equation can be used to determine the capacitance of C_1.

$$C_1 = \frac{I_1 \times t}{\Delta V}$$

or

$$C_1 = \frac{I_1 \times t}{\text{UTP} - \text{LTP}} \tag{9-15}$$

If a BIFET op-amp is used, the capacitance of C_1 should be first selected to be much larger than stray capacitance. Then, I_1 is determined from Eq. 9-15 and R_1 is calculated using Eq. 9-14.

Example 9-6

Using a BIFET op-amp, design an astable multivibrator to have a ±9 V output with a frequency of 1 kHz.

Solution

For $V_o \approx \pm 9$ V,

$$V_{CC} \approx \pm(V_o + 1\text{ V}) = \pm(9\text{ V} + 1\text{ V})$$

$$\approx \pm 10\text{ V}$$

select UTP and LTP $\ll V_o$

Let $|\text{UTP}| = |\text{LTP}| = 0.5$ V

Let $R_2 = 1\text{ M}\Omega$

$$I_3 = \frac{|V_o| - \text{UTP}}{R_2} = \frac{9\text{ V} - 0.5\text{ V}}{1\text{ M}\Omega}$$

$$= 8.5\ \mu\text{A}$$

$$R_3 = \frac{\text{UTP}}{I_3} = \frac{0.5\text{ V}}{8.5\ \mu\text{A}}$$

$$\approx 59\text{ k}\Omega \quad \text{(use 56 k}\Omega\text{ standard value)}$$

Let $C_1 = 0.1\ \mu\text{F}$

$$t = \frac{1}{2f} = \frac{1}{2 \times 1\text{ kHz}}$$

$$= 500\ \mu\text{s}$$

From Eq. 9-15,

$$I_1 = \frac{C_1(\text{UTP} - \text{LTP})}{t} = \frac{0.1\ \mu\text{F}[0.5\text{ V} - (-0.5\text{ V})]}{500\ \mu\text{s}}$$

$$= 200\ \mu\text{A}$$

the noninverting input terminal toward ground level. When the noninverting terminal goes slightly above ground, the op-amp output immediately switches back to the positive saturation level once again, and the circuit is returned to its original state. The circuit produces a negative-going output pulse each time it is triggered. The pulse width (PW) of the output depends on the capacitance of C_2, the bias voltage V_{R2}, and the resistance of R_1 and R_2.

Figure 9-15(b) shows an adjustable resistor R_4 included in the circuit. The total resistance in series with the capacitor is now $(R_4 + R_1 \parallel R_2)$. Usually, R_4 is selected to be much larger than $R_1 \parallel R_2$, so that the capacitor charge/discharge resistance is effectively R_4. Adjusting R_4 alters the capacitor charge/discharge time, thus controlling the output pulse width.

Design

Resistors R_1 and R_2 are determined in the usual way to give V_{R2} typically around 0.5 V to 1 V. Equation 3-1 can be used to calculate the resistance of R_3 as large as possible. To generate spikes from the input pulse, the time constant $C_1 R_3$ should approximately equal one-tenth of the input pulse width.

$$C_1 R_3 \approx 0.1t$$

or,

$$C_1 = \frac{0.1t}{R_3} \tag{9-16}$$

C_1 might be first selected much larger than stray capacitance, then R_3 can be calculated from Eq. 9-16. Note that $0.1t$ must be long enough for the op-amp output to commence changing.

To calculate the required capacitance value for C_2, the capacitor charging equation must be employed.

$$e_c = E - (E - E_o)\epsilon^{-t/CR}$$

Rewriting the equation in the form for calculation of capacitance,

$$C_2 = \frac{PW}{(R_1 \parallel R_2) \ln [(E - E_o)/(E - e_c)]} \tag{9-17}$$

In Eq. 9-17, PW is the desired output pulse width. E is the capacitor charging voltage, that is, the voltage that it would charge to after triggering if it were allowed to continue charging without the output switching from $-V_{o(\text{sat})}$. So

$$E = V_{R2} - [-V_{o(\text{sat})}] \tag{9-18}$$

E_o is the initial capacitor voltage before triggering. Taking E as a positive quantity, E_o must be assigned a negative polarity.

$$E_o = -[+V_{o(\text{sat})} - V_{R2}] \tag{9-19}$$

The final capacitor voltage e_c is the voltage at which the op-amp output switches back from $-V_{o(\text{sat})}$ to $+V_{o(\text{sat})}$. This is the voltage across C_2 when the noninverting input terminal is at ground level and the output is still at $-V_{o(\text{sat})}$. Therefore,

$$e_c = 0 - (-V_{O(\text{sat})})$$

or,

$$e_c = V_{o(\text{sat})} \tag{9-20}$$

Example 9-7

Design a monostable multivibrator circuit, as in Fig. 9-15(a), to have an output pulse width of 1 ms when triggered by a 2 V, 100 μs input pulse. Use a 741 op-amp with a $\pm$12 V supply.

Solution

$$I_2 \gg I_{B(\text{max})}$$

let

$$I_2 = 50\ \mu\text{A}$$

$$V_{R2} < V_i$$

let

$$V_{R2} = 0.5\ \text{V}$$

$$R_2 = \frac{V_{R2}}{I_2} = \frac{0.5\ \text{V}}{50\ \mu\text{A}}$$

$$= 10\ \text{k}\Omega \qquad \text{(standard value)}$$

$$R_1 = \frac{V_{CC} - V_{R2}}{I_2} = \frac{12\ \text{V} - 0.5\ \text{V}}{50\ \mu\text{A}}$$

$$= 230\ \text{k}\Omega \qquad \text{(use 220 k}\Omega\text{ standard value)}$$

Eq. 9-18,

$$E = V_{R2} - [-V_{o(\text{sat})}] \approx 0.5\ \text{V} - (-12\ \text{V} + 1\ \text{V})$$

$$\approx 11.5\ \text{V}$$

Eq. 9-19,

$$E_0 = -(+V_{o(\text{sat})} - V_{R2}) \approx -(12\ \text{V} - 1\ \text{V} - 0.5\ \text{V})$$

$$\approx -10.5\ \text{V}$$

Eq. 9-20,

$$e_c = V_{o(\text{sat})} \approx 12\ \text{V} - 1\ \text{V}$$

$$\approx 11\ \text{V}$$

Eq. 9-17,

$$C_2 = \frac{PW}{(R_1 \parallel R_2) \ln [(E - E_o)/(E - e_c)]}$$

$$= \frac{1\ \text{ms}}{(220\ \text{k}\Omega \parallel 10\ \text{k}\Omega) \ln \dfrac{11.5\ \text{V} - (-10.5\ \text{V})}{11.5\ \text{V} - 11\ \text{V}}}$$

$$= 0.027\ \mu\text{F} \qquad \text{(standard value)}$$

Eq. 3-1,

$$R_{3(\text{max})} = \frac{0.1 V_{BE}}{I_{BE(\text{max})}} = \frac{0.1 \times 0.7\ \text{V}}{500\ \text{nA}}$$

$$= 140\ \text{k}\Omega \qquad \text{(use 120 k}\Omega\text{ standard value)}$$

Eq. 9-16, $$C_1 = \frac{0.1t}{R_3} = \frac{0.1 \times 100\ \mu\text{s}}{120\ \text{k}\Omega}$$

$$= 83\ \text{pF} \qquad \text{(use 91 pF standard value)}$$

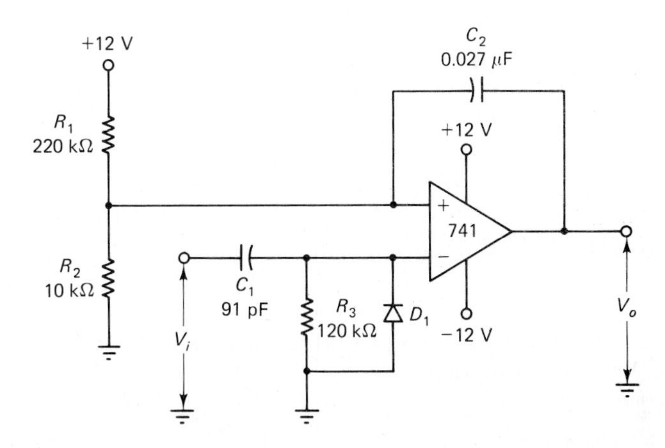

Figure 9-16 Monostable multivibrator circuit designed in Example 9-7.

REVIEW QUESTIONS

9-1. For operational amplifiers employed in switching circuits, briefly discuss output voltage swing, input voltage range, slew rate and frequency compensation.

9-2. Sketch the circuit of an op-amp employed as a noninverting zero crossing detector. Also, sketch typical input and output waveforms. Briefly explain.

9-3. Draw the circuits of an op-amp inverting zero-crossing detector and a voltage level detector. Briefly explain each circuit.

9-4. Sketch the circuit of a capacitor-coupled zero-crossing detector. Show the waveforms at various points in the circuit, and explain its operation.

9-5. Write the equation for calculating each component value in a capacitor-coupled zero-crossing detector. Briefly discuss the design process when using a bipolar op-amp and when using a BIFET op-amp.

9-6. Draw an op-amp inverting Schmitt trigger circuit. Sketch typical input and output waveforms. Explain the circuit operation and the shape of the waveforms.

9-7. Sketch typical input/output characteristics for an inverting Schmitt tigger circuit and carefully explain their shape.

9-8. Discuss the design process for an op-amp inverting Schmitt trigger circuit, and write equations for calculating each component value.

9-9. Draw circuits to show how diodes may be used to select the trigger points of an inverting Schmitt trigger circuit. Explain.

9-10. Draw an op-amp noninverting Schmitt trigger circuit and explain its operation.

9-11. Show how diodes may be used to select the trigger points of a noninverting Schmitt trigger circuit. Explain.

9-12. Sketch the circuit of an op-amp astable multivibrator. Show the voltage waveforms at various points in the circuit and explain its operation.

9-13. Discuss the design procedure for an op-amp astable multivibrator, and write the equations for calculating each component value.

9-14. Draw the circuit of an op-amp monostable multivibrator. Show the voltage waveforms throughout the circuit and carefully explain its operation.

9-15. Discuss the design process for an op-amp monostable multivibrator and write the equations for calculating each component value.

PROBLEMS

9-1. A voltage level detector, as in Fig. 9-3(b), is to switch its output between approximately -13 V and $+13$ V when the input exceeds 1.5 V. Design a suitable circuit to use a 741 op-amp.

9-2. Redesign the capacitor-coupled zero-crossing detector in Example 9-1 to use a BIFET op-amp.

9-3. A capacitor-coupled zero-crossing detector is to provide an output voltage of approximately ± 17 V when a 3 kHz, ± 2 V square wave input is applied. Design a suitable circuit to use a bipolar op-amp.

9-4. For the circuit designed for Problem 9-3, calculate the lowest sine wave signal frequency that can be applied without serious phase shift error. Also, determine the minimum op-amp slew rate to give a reasonably undistorted output for the 3 kHz square wave input.

9-5. Redesign the circuit in Problem 9-3 to use a BIFET op-amp.

9-6. Using a bipolar op-amp, design an inverting Schmitt trigger circuit as in Fig. 9-6(a), to trigger at ± 0.5 V and to produce an output of approximately ± 11 V.

9-7. Redesign the circuit in Problem 9-6 to use a BIFET op-amp.

9-8. An inverting Schmitt trigger circuit is to have UTP = 0 and LTP = 1 V. Design a suitable circuit using a bipolar op-amp and a ± 15 V supply.

9-9. Using a bipolar op-amp with a ± 18 V supply, design an inverting Schmitt trigger circuit to have UTP = 1.5 V and LTP = -3 V.

9-10. Modify the circuit in Problem 9-9 to make each trigger point adjustable by ± 0.5 V.

9-11. Using a bipolar op-amp, design a noninverting Schmitt trigger circuit as in Fig. 9-10(a), to trigger at ± 0.5 V, and to produce an output of approximately ± 14 V.

9-12. Redesign the circuit in Problem 9-11 to use a BIFET op-amp.

9-13. A noninverting Schmitt trigger circuit is to have UTP = 0 and LTP = 2.5 V. Design a suitable circuit using a bipolar op-amp and a ± 18 V supply.

9-14. Using a bipolar op-amp with a ± 15 V supply, design a noninverting Schmitt trigger circuit to have UTP = 1 V and LTP = -1.5 V.

9-15. Modify the circuit designed for Problem 9-14 to make each trigger point adjustable by ± 0.25 V.

9-16. An op-amp astable multivibrator is to have a 3 kHz output with an amplitude of ±11 V. Using a bipolar op-amp, design a suitable circuit.

9-17. Determine the minimum op-amp slew rate required for the circuit designed for Problem 9-16.

9-18. Redesign the circuit in Problem 9-16 to use a BIFET op-amp.

9-19. Design an op-amp astable multivibrator to have an output frequency of 400 Hz. Use a 741 op-amp with a supply of ±18 V.

9-20. Modify the circuit designed for Problem 9-19 to make the frequency adjustable by ±20%.

9-21. Redesign the astable multivibrator circuit in Problem 9-19 to use a BIFET op-amp.

9-22. A monostable multivibrator circuit is to have a 650 μs, ±9 V, output pulse when triggered by a 50 μs, 3 V, input pulse. Using a bipolar op-amp, design a suitable circuit.

9-23. Modify the circuit designed for Problem 9-22 to make the output pulse width adjustable from 650 μs to 1.3 ms.

9-24. Determine suitable minimum slew rates for the op-amps used in the circuits designed for Problems 9-19 and 9-22.

COMPUTER PROBLEMS

9-25. Write a computer program to design a noninverting Schmitt trigger circuit as in Fig. 9-11(b). Given: UTP and LTP.

9-26. Write a computer program to analyze the astable multivibrator circuit in Fig. 9-13 to determine output amplitude and frequency, and capacitor voltage amplitude. Given: supply voltage and all component values.

9-27. Write a computer program to analyze the monostable multivibrator circuit in Fig. 9-15 to determine output amplitude and pulse width, and the required triggering voltage. Given: supply voltage and all component values.

LABORATORY EXERCISES

9-1 Inverting Schmitt Trigger

1. Construct the op-amp inverting Schmitt trigger circuit shown in Fig. 9-8, as designed in Example 9-3.
2. Apply ±5 V, 1 kHz triangular wave input.
3. Monitor the input and output waveforms on an oscilloscope and investigate the upper and lower trigger points. Compare to the designed levels.
4. Modify the circuit by including a diode, as in Fig. 9-9(a).
5. Once again investigate the upper and lower trigger points.

9-2 Noninverting Schmitt Trigger

1. Construct the op-amp noninverting Schmitt trigger circuit shown in Fig. 9-12, as designed in Example 9-4.
2. Apply a ±7 V, 1 kHz triangular wave input.
3. Monitor the input and output waveforms on an oscilloscope and investigate the upper and lower trigger points. Compare to the designed levels.

9-3 Astable Multivibrator

1. Construct the astable multivibrator circuit shown in Fig. 9-14, as designed in Example 9-6. The circuit is designed for a BIFET op-amp but a 741 can also be used.
2. Monitor the output and capacitor waveforms on an oscilloscope. Compare the frequency and amplitudes of each to the designed quantities.

9-4 Monostable Multivibrator

1. Construct the monostable multivibrator circuit shown in Fig. 9-16, as designed in Example 9-7.
2. Apply a 200 Hz pulse waveform with a pulse amplitude of 2 V and a pulse width of 100 μs as input.
3. Monitor the input and output waveforms on an oscilloscope. Measure the output pulse width and compare to the designed value.

CHAPTER 10

Signal Generators

Objectives

- Sketch the following op-amp signal generating circuits, explain the operation of each, and show the various waveforms throughout each circuit: *triangular/rectangular waveform generators, phase shift oscillator, Wein bridge oscillator, amplitude stabilization circuits for oscillators, output and dc level control circuits for signal generators*.
- Design each of the above circuits to meet a given specification.
- Analyze each of the above circuits to determine its performance.

INTRODUCTION

Operational amplifiers are widely used in circuits for generating wavefroms of all shapes. A triangular/rectangular waveform generator can be constructed simply by using an integrating circuit and a Schmitt trigger circuit. Most sinusoidal oscillators consist of only an amplifier and an RC feedback network. Amplitude stabilization of sinusoidal oscillators is necessary to prevent the signal amplitude from approaching the op-amp saturation voltages. An output stage that provides amplitude and dc voltage level control is also necessary.

10-1 TRIANGULAR/RECTANGULAR WAVE GENERATOR

The circuit in Fig. 10-1 consists of a integrator (discussed in Section 8-4) and a non-inverting Schmitt trigger circuit (see Section 9-4). The Schmitt output is applied as an input to the integrator and the integrator output is the Schmitt circuit input. As illustrated by the waveform diagrams in Fig. 10-1, the integrator produces a triangular

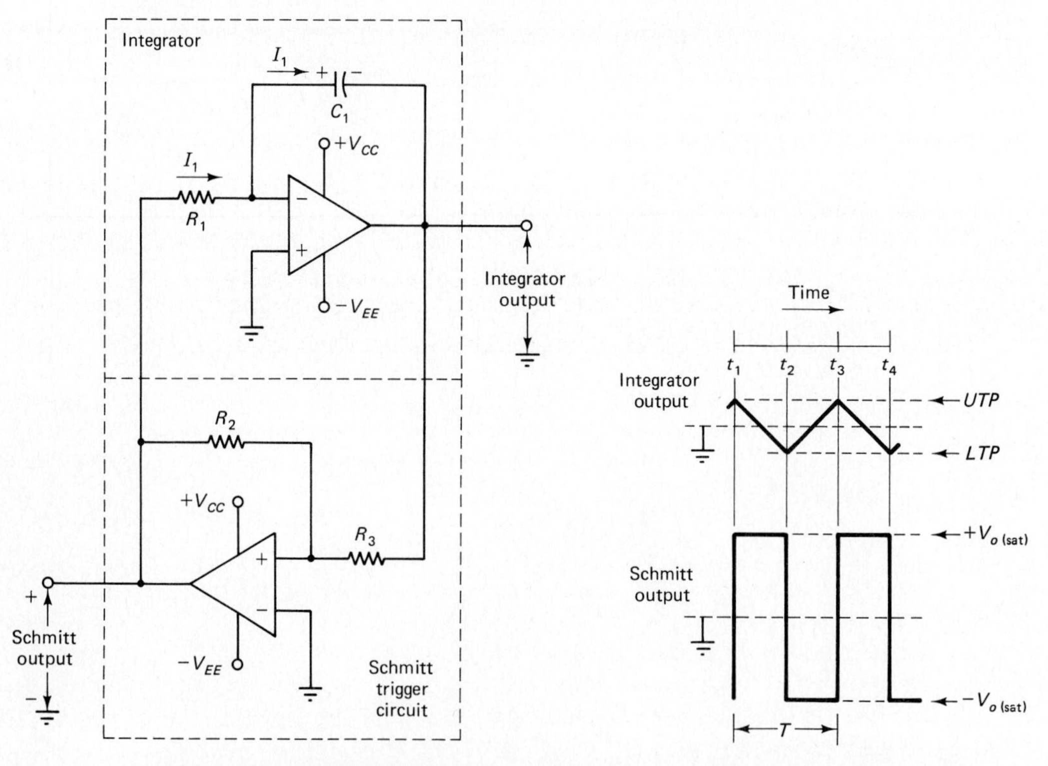

Figure 10-1 Triangular/rectangular waveform generator using an integrator and a Schmitt trigger circuit. When the Schmitt output is positive, the integrator output goes negative until it reaches the Schmitt *LTP*. The Schmit output switches to negative and the integrator output then moves in a positive direction to the *UTP* of the Schmitt.

output waveform when it has a square wave input, and the Schmitt output changes from one saturation voltage level to the other each time the integrator output arrives at the Schmitt upper or lower trigger point.

At time instant t_1, the integrator output is at the UTP and the Schmitt output is at $+V_{o(sat)}$. The positive input voltage to the integrator causes current I_1 to flow through R_1 and C_1, as illustrated. I_1 charges C_1 positive on the left and negative on the right, thus producing a negative-going ramp output from the integrator during the time interval t_1 to t_2.

At t_2, the ramp voltage arrives at the Schmitt LTP. The Schmitt output immediately switches from $+V_{o(sat)}$ to $-V_{o(sat)}$, and reverses the direction of I_1. C_1 is now discharged and recharged with the opposite polarity, generating a positive-going ramp output voltage. The positive-going ramp continues during time interval t_2 to t_3 until it arrives at the Schmitt UTP. At this point, the Schmitt output switches to $+V_{o(sat)}$ once again and the cycle recommences.

The circuit is a free-running signal generator producing triangular and square output waveforms. Note that, unlike the integrating circuit in Fig. 8-5(a), there is no resistor in parallel with capacitor C_1 in Fig. 10-1. Because the integrator output cannot go above the Schmitt UTP or below the LTP, no resistor is required to avoid output voltage saturation.

Frequency and Duty Cycle Adjustment

The frequency of the triangular and square wave outputs from the circuit in Fig. 10-1 can be adjusted by including a variable resistor in series with R_1. Increasing the resistance reduces the level of capacitor charging current, thus charging C_1 more slowly and increasing the time intervals t_1 to t_2 and t_2 to t_3. This constitutes an increase in time period T and a reduction in output frequency. Decreasing the total resistance at the integrator input reduces the capacitor charging times and increases the output frequency. In Fig. 10-2(a), resistor R_4 is connected in series with the integrator input to provide frequency adjustment.

Along with resistor R_4, resistors R_5, R_6, and R_7 in Fig. 10-2(a) have an effect on the capacitor charging times. Ignoring R_5 for the moment, it is seen that when the Schmitt output voltage is positive, current I_1 flows through diode D_1 and resistor R_6 to charge capacitor C_1. When the Schmitt output is negative, the capacitor current is I_2, flowing through D_2 and R_7.

Now note that I_1 also flows through R_{5a} (the top portion of potentiometer R_5) and that I_2 flows through R_{5b} (the bottom portion of R_5). With the moving contact exactly at the center of R_5, and R_6 equal to R_7, I_1 will equal I_2. Consequently, the time intervals t_1 to t_2 and t_2 to t_3, will be equal, as in Fig. 10-1. With the moving contact adjusted to make R_{5a} smaller than R_{5b}, I_1 will be larger than I_2, and C_1 will charge faster during the time interval t_1 to t_2 than during t_2 to t_3. This gives a sawtooth output waveform from the integrator and a pulse output from the Schmitt, as illustrated in Fig. 10-2(b).

When the R_5 moving contact is adjusted to make R_{5a} larger than R_{5b}, I_2 is larger than I_1. This makes the interval t_1 to t_2 longer than t_2 to t_3 and gives the kind of sawtooth and pulse outputs illustrated in Fig. 10-2(c).

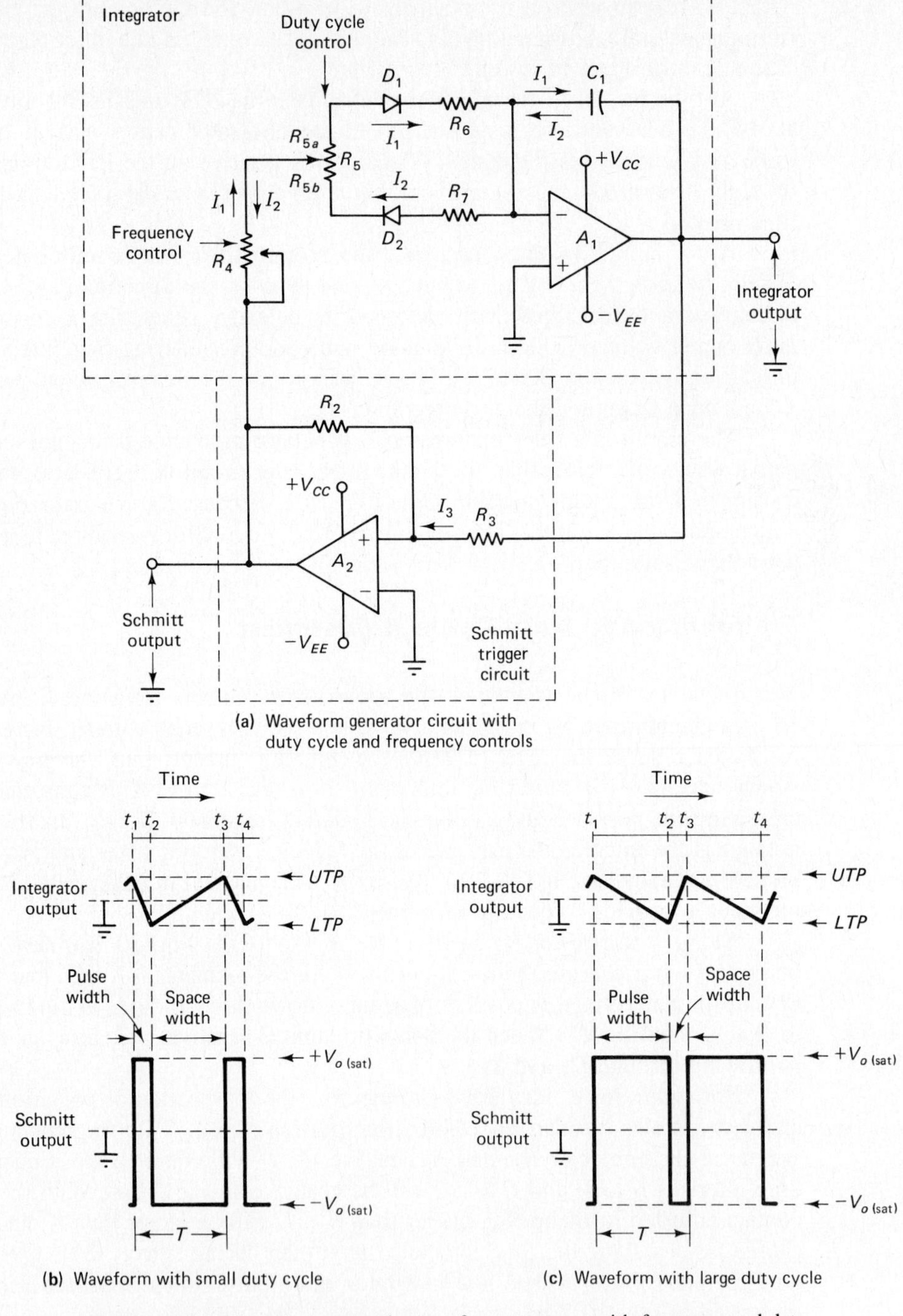

Figure 10-2 Triangular/rectangular waveform generator with frequency and duty cycle controls. R_4 controls the level of current to capacitor C_1 and thus controls the output frequency. R_5 simultaneously adjusts the charge and discharge current levels for C_1, thus adjusting the duty cycle.

The *duty cycle* of the pulse waveform is the ratio of the positive pulse width PW to the time period T. So, the duty cycle of the output waveform is controlled by adjustment of potentiometer R_5. The circuit in Fig. 10-2(a) is a ramp and pulse waveform generator with frequency adjustment afforded by resistor R_4 and duty cycle adjustment provided by R_5. Output amplitude and dc level controls are discussed in Section 10-6.

10-2 WAVEFORM GENERATOR DESIGN

The performance specification for the signal generator must be known before any design calculations can be made. In particular, the output frequency and duty cycle ranges must be specified. From the frequency and duty cycle, the time durations of the maximum and minimum pulse widths at the high and low frequencies can be determined. A 70% duty cycle means that the positive pulse width must be 70% of the time period of the the output waveform.

The noninverting Schmitt circuit is designed exactly as discussed in Section 9-4. The integrator design is a little more complicated than the procedure explained in Section 8-5. Start by selecting the minimum level of capacitor charging current to be much larger than the op-amp input bias current. Then use Eq. 8-1 to calculate the capacitance of C_1.

Eq. 8-1,
$$C_1 = \frac{I_1\,\Delta t}{\Delta v}$$

In this case, $I_1 = I_{1(\mathrm{min})}$, $\Delta t = PW_{(\mathrm{max})}$ at the lowest output frequency, and $\Delta v = \mathrm{UTP} - \mathrm{LTP}$. Alternatively, if using a BIFET op-amp begin by selecting C_1 to be much larger than stray capacitance. Then calculate $I_{1(\mathrm{min})}$ from Eq. 8-1.

$I_{1(\mathrm{min})}$ and $PW_{(\mathrm{max})}$ at the lowest frequency, are used in the initial calculation because the minimum level of capacitor charging current will give the longest charging time, and hence, the longest output pulse width. The longest pulse width occurs when the output frequency is at its lowest. Shorter pulse widths occur with higher charging current levels, so the current is adjusted up, above the selected minimum level.

Minimum I_1 level requires that R_4 be a maximum and that all of R_5 be in series with resistor R_6. Therefore,

$$R_4 + R_5 + R_6 = \frac{+V_{o(\mathrm{sat})} - V_F}{I_{1(\mathrm{min})}} \tag{10-1}$$

where V_F is the forward voltage drop of diode D_1.

The output frequency is at its highest when R_4 is adjusted to zero. Also, the ratio of charging currents for maximum pulse width at the low frequency f_1 and maximum pulse width at the high frequency f_2 is

$$\frac{I_{f1}}{I_{f2}} = \frac{f_1}{f_2}$$

or,
$$\frac{I_{1(\min)}}{I_{f2}} = \frac{f_1}{f_2} \tag{10-2}$$

Using I_{f2} determined from Eq. 10-2,

$$R_5 + R_6 = \frac{+V_{o(\text{sat})} - V_F}{I_{f2}} \tag{10-3}$$

The resistance of R_4 is calculated from the results of Eqs. 10-1 and 10-3.

The ratio of R_5 and R_6 depends on the maximum and minimum pulse widths at any given frequency. For maximum (positive) pulse width, all of R_5 must be in series with R_6. Minimum pulse width occurs when no part of R_5 is in series with R_6. Therefore,

$$\frac{R_5 + R_6}{R_6} = \frac{PW_{(\max)}}{PW_{(\min)}} \tag{10-4}$$

It is important to note that for Eq. 10-4, $PW_{(\max)}$ and $PW_{(\min)}$ must be calculated at the same frequency. The resistances of R_5 and R_6 are determined from Eqs. 10-3 and 10-4. For equality in the output pulse width and space width adjustments (see Fig. 10-2(b) and (c)), the resistance of R_7 should equal R_6.

The required op-amp slew rates are determined from the usual considerations of output waveform distortion. The diode reverse recovery times must be much smaller than the minimum pulse width at the highest frequency. Recall from Section 8-5 that the integrator op-amp (A_1) must be frequency compensated as a voltage follower.

Example 10-1

A triangular/rectangular signal generator as in Fig. 10-2(a) is to be designed to have a 5 V peak-to-peak triangular output, a frequency ranging from 200 Hz to 2 kHz, and a duty cycle adjustable from 20% to 80%. Bipolar op-amps with a supply of ±15 V are to be used. Determine suitable component values.

Solution

Schmitt circuit design

$$I_3 \gg I_{B(\max)}$$

let $\quad I_3 = 50\ \mu\text{A}$

$$R_2 = \frac{V_{o(\text{sat})}}{I_3} \approx \frac{15\text{ V} - 1\text{ V}}{50\ \mu\text{A}}$$

$$\approx 280\text{ k}\Omega \qquad \text{(use 270 k}\Omega\text{ and recalculate } I_2\text{)}$$

$$I_3 \approx \frac{15\text{ V} - 1\text{V}}{270\text{ k}\Omega} \approx 52\ \mu\text{A}$$

$$R_3 = \frac{\text{UTP}}{I_3} \approx \frac{5\text{ V}/2}{52\ \mu\text{A}}$$

$$= 48\text{ k}\Omega \qquad \text{(use 47 k}\Omega\text{ and 1 k}\Omega\text{)}$$

Integrator design

let C_1 charging current be $\quad I_{1(\min)} = 50\ \mu\text{A}$

at lowest frequency f_1, $\quad PW_{(\max)} = 80\%$ of $T_{(\max)}$

$$= 0.8 \times \frac{1}{f_1} = 0.8 \times \frac{1}{200\ \text{Hz}} = 4\ \text{ms}$$

Eq. 8-1,
$$C_1 = \frac{I_1\ \Delta t}{\Delta v} = \frac{50\ \mu\text{A} \times 4\ \text{ms}}{5\ \text{V}}$$
$$= 0.04\ \mu\text{F} \qquad \text{(standard value)}$$

Eq. 10-1,
$$R_4 + R_5 + R_6 = \frac{+V_{o(\text{sat})} - V_F}{I_{1(\min)}} \approx \frac{14\ \text{V} - 0.7\ \text{V}}{50\ \mu\text{A}}$$
$$\approx 266\ \text{k}\Omega$$

From Eq. 10-2,
$$I_{f2} = \frac{I_{1(\min)} \times f_2}{f_1} = \frac{50\ \mu\text{A} \times 2\ \text{kHz}}{200\ \text{Hz}}$$
$$= 500\ \mu\text{A}$$

Eq. 10-3,
$$R_5 + R_6 = \frac{+V_{o(\text{sat})} - V_F}{I_{f2}} \approx \frac{14\ \text{V} - 0.7\ \text{V}}{500\ \mu\text{A}}$$
$$\approx 26.6\ \text{k}\Omega$$
$$R_4 = (R_4 + R_5 + R_6) - (R_5 + R_6)$$
$$= 266\ \text{k}\Omega - 26.6\ \text{k}\Omega$$
$$\approx 240\ \text{k}\Omega \qquad \text{(use 250 k}\Omega\text{ standard value potentiometer)}$$

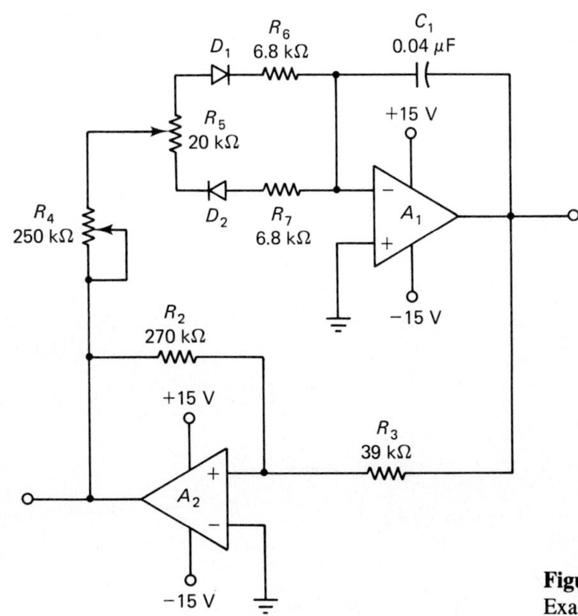

Figure 10-3 Circuit designed for Example 10-1.

at f_1, $PW_{(max)} = 4$ ms

and, $PW_{(min)} = 20\%$ of $T_{max} = 0.2 \times \dfrac{1}{200 \text{ Hz}}$

$= 1$ ms

From Eq. 10-4,

$$R_6 = \frac{(R_5 + R_6) \times PW_{(min)}}{PW_{(max)}} = \frac{26.6 \text{ k}\Omega \times 1 \text{ ms}}{4 \text{ ms}}$$

$$= 6.6 \text{ k}\Omega \qquad \text{(use 6.8 k}\Omega\text{ standard value)}$$

$$R_5 = (R_5 + R_6) - R_6 = 26.6 \text{ k}\Omega - 6.8 \text{ k}\Omega$$

$$\approx 20 \text{ k}\Omega \qquad \text{(standard value potentiometer)}$$

$$R_7 = R_6 = 6.8 \text{ k}\Omega$$

10-3 PHASE SHIFT OSCILLATOR

Circuit Operation

The phase shift oscillator circuit in Fig. 10-4 consists of an inverting amplifier and an RC phase shift network. The RC network feeds a portion of the amplifier ac output back to the amplifier input. The amplifier has an internal phase shift of $-180°$ and the phase shift network provides $+180°$ of phase shift. So, the signal fed back to the input can be amplified to reproduce the output. The circuit is then generating its own input signal, which means it is oscillating.

For oscillations to be sustained in any sinusoidal oscillator, certain conditions, known as the *Barkhausen criteria,* must be fulfilled. These are that *the loop gain around the circuit must be equal to (or greater than) one, and the phase shift around the circuit must be zero*. This is also discussed in Section 5-1.

The RC network connected between the amplifier output and input terminals consists of three resistors and three capacitors. Resistor R_1 functions as the last resistor in the phase shift network and as the amplifier input resistor. The phase shift network illustrated is a *phase lead* network. This produces a $+180°$ phase shift, so the total phase shift around the loop is $(-180° + 180°) = 0°$. A phase lag network would work just as well, giving a total loop phase shift of $(-180° - 180°) = -360°$.

The frequency of the oscillator output depends on the capacitor and resistor values employed in the phase shift network. If three equal-value resistors and three equal-value capacitors are used, the RC circuit can be analyzed to show that the network phase shift is 180° when

$$X_C = \sqrt{6}\,R$$

This gives an oscillation frequency of,

$$f = \frac{1}{2\pi RC\sqrt{6}} \qquad (10\text{-}5)$$

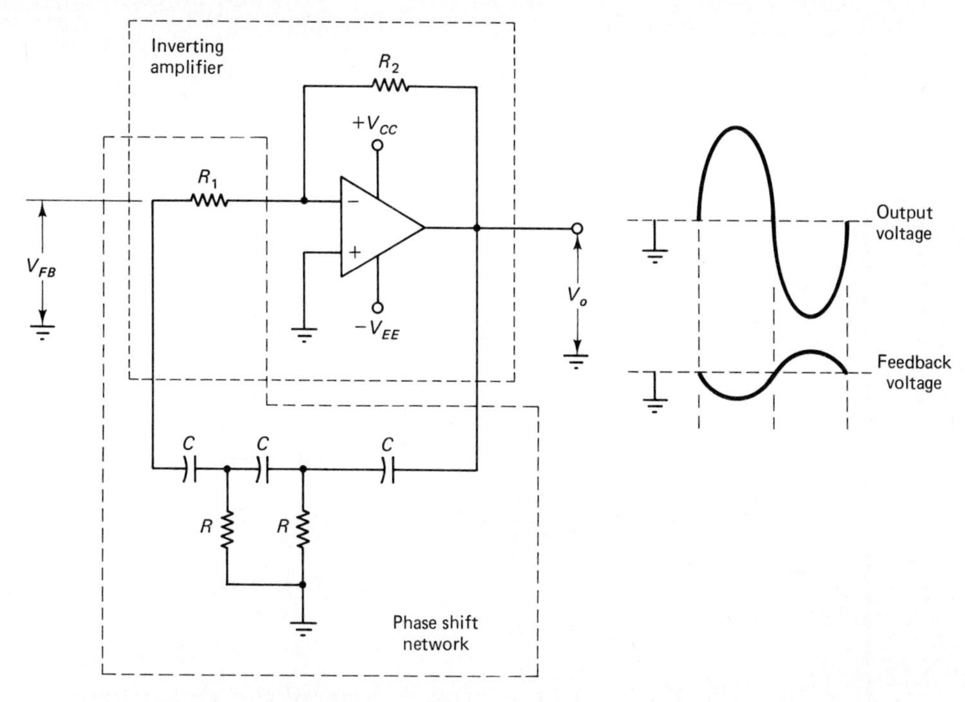

Figure 10-4 Phase shift oscillator consisting of an inverting amplifier and a phase shifting feedback network. The output is attenuated and phase shifted through $+180°$, and then applied to the amplifier which amplifies it and phase shifts it through $-180°$.

In addition to providing network phase shift, the RC network attenuates the amplifier output. Network analysis shows that at the required 180° phase shift, the RC network has an attenuation factor of 29. This means, that the amplifier must have a voltage gain of 29 for the loop gain to equal 1.

If the amplifier gain is less than 29, the circuit will not oscillate. When the gain is substantially greater than 29, the oscillator output waveform is likely to be distorted. A gain just slightly greater than 29 gives a reasonably undistorted sinusoidal waveform. In the absence of amplitude stabilization (discussed in Section 10-4), the output peaks will approach the positive and negative saturation voltages of the op-amp.

Phase Shift Oscillator Design

Design of a phase shift oscillator commences with design of the inverting amplifier to have a voltage gain just slightly greater than 29. The phase shift network resistors are then made equal to the the amplifier input resistor R_1, and the network capacitance is determined from Eq. 10-5. Where an uncompensated op-amp is used, frequency compensating components should be included. If the desired oscillating frequency is much lower than the frequency at which instability might occur, the circuit can usually be heavily over-compensated.

Example 10-2

Using a 741 op-amp with a supply of ±12 V, design a phase shift oscillator to have an output frequency of 3.5 kHz.

Solution

$$I_1 \gg I_{B(\max)}$$

let $$I_1 = 50\ \mu\text{A}$$

$$V_o \approx \pm(V_{CC} - 1\text{ V}) \approx \pm(12\text{ V} - 1\text{ V})$$

$$\approx \pm 11\text{ V}$$

$$R_2 = \frac{V_o}{I_1} = \frac{11\text{ V}}{50\ \mu\text{A}}$$

$$= 220\text{ k}\Omega \qquad \text{(standard value)}$$

$$R_1 = \frac{R_2}{A_v} = \frac{220\text{ k}\Omega}{29}$$

$$= 7.6\text{ k}\Omega \qquad (\text{use } 6.8\text{ k}\Omega \text{ to give } A_v > 29)$$

$$R = R_1 = 6.8\text{ k}\Omega$$

From Eq. 10-5, $$C = \frac{1}{2\pi R f \sqrt{6}} = \frac{1}{2\pi \times 6.8\text{ k}\Omega \times 3.5\text{ kHz} \times \sqrt{6}}$$

$$= 2730\text{ pF} \qquad \text{(use 2700 pF standard value)}$$

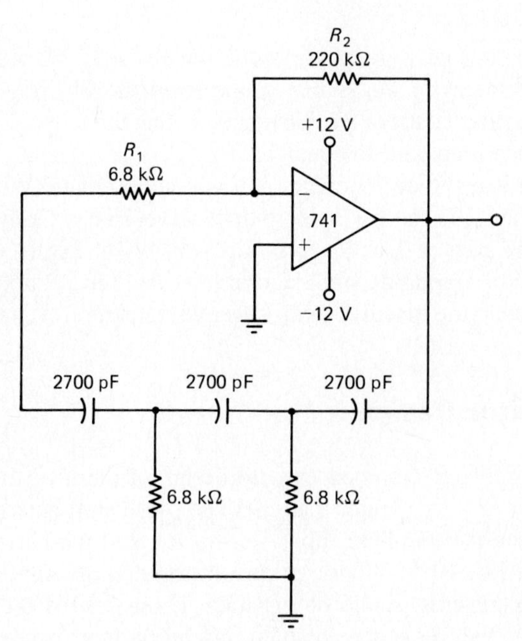

Figure 10-5 Circuit designed for Problem 10-2

10-4 OSCILLATOR AMPLITUDE STABILIZATION

As already mentioned, the output amplitude from an oscillator tends to be the maximum possible output voltage the amplifier can produce. The waveform is also likely to be distorted to some extent because it is limited only by the amplifier saturation voltages. To minimize distortion and reduce the output amplitude to an acceptable level, amplitude stabilization circuitry must be employed. Amplitude stabilization operates by ensuring that oscillation is not sustained if the output exceeds a predetermined level.

For the phase-shift oscillator discussed in Section 10-3, the amplifier gain must always exceed 29 to sustain oscillations. Now consider the inverting amplifier in the phase-shift oscillator circuit in Fig. 10-6(a). Here part of resistor R_2 is bypassed with diodes. When the output amplitude is low, none of the diodes is forward biased and they have no effect on the circuit. The amplifier voltage gain is $A_v = -R_2/R_1$, which is designed to exceed the critical value of 29.

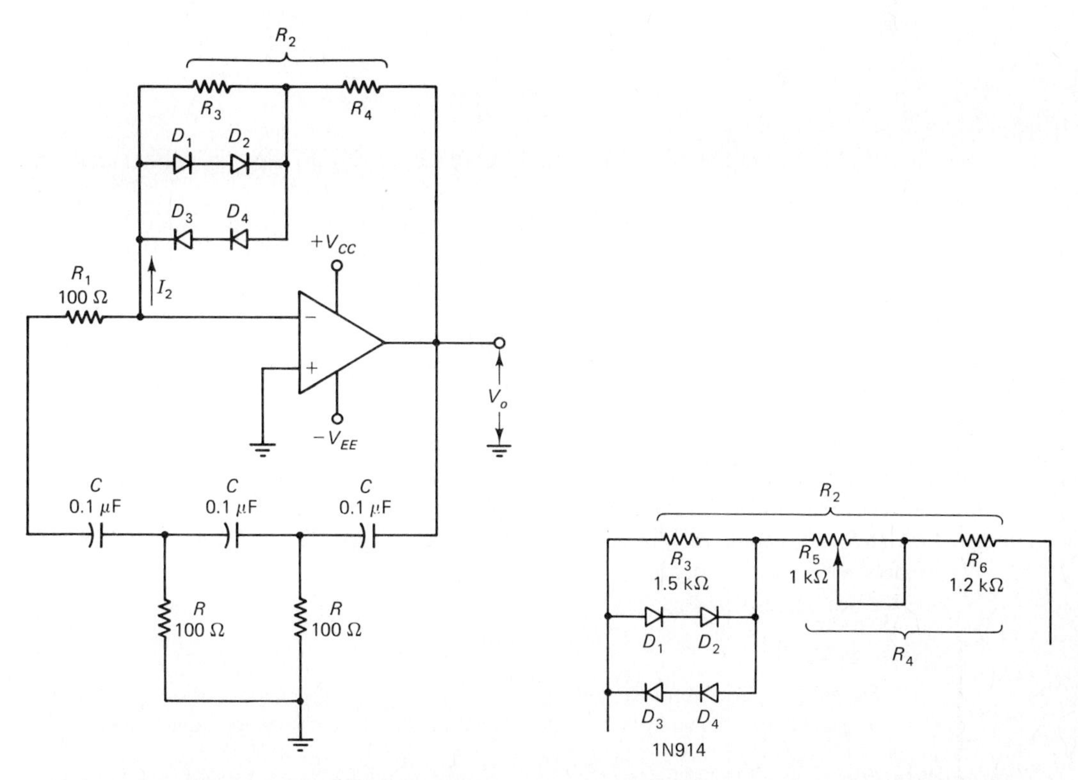

(a) Phase shift oscillator with amplitude stabilization

(b) Adjustable resistor included to control distortion

Figure 10-6 The output amplitude of a phase-shift oscillator can be limited by using diodes to modify the amplifier voltage gain. When the output is high, the diodes short-out part of the feedback network and render the amplifier gain too low to sustain oscillations.

When the output amplitude is large enough to forward bias either D_1 and D_2 or D_3 and D_4, resistor R_3 is short-circuited. The amplifier voltage gain now becomes $A_v = -R_4/R_1$. This is designed to be too small to sustain oscillations. So the circuit cannot oscillate with a high amplitude output. However, it can and does oscillate with a low amplitude output.

To design the amplitude stabilization circuit, the amplifier is first designed in the usual manner with one important difference. The current I_2, used in calculating the resistor for the feedback network, must be large enough for the diodes to be forward-biased into the near-linear region of their characteristics. Resistor R_1 is then determined as

$$R_1 = \frac{V_o/29}{I_2} \tag{10-6}$$

R_2 is calculated from

$$R_2 = 29 \times R_1 \tag{10-7}$$

R_3 is determined from

$$R_3 \approx \frac{2V_F}{I_2} \tag{10-8}$$

and

$$R_4 = R_2 - R_3$$

The resultant component values should give $(R_3 + R_4)/R_1$ slightly greater than 29, and R_4/R_1 less than 29.

The diodes should be low current, switching devices. They should have a reverse breakdown voltage greater than the circuit supply voltage, and their reverse recovery time should be a maximum of one-tenth of the time period of the oscillation frequency.*

$$t_{rr(\max)} = \frac{T}{10} \tag{10-9}$$

Some distortion can still occur in the output waveform if $(R_3 + R_4)/R_1$ is much larger than 29. On the other hand, attempts to make $(R_3 + R_4)/R_1$ exactly equal to 29 may result in the circuit not oscillating. This problem is solved by making a portion of R_4 adjustable as illustrated in Fig. 10-6(b). Typically R_5 should equal approximately 40% of the calculated value of R_4, and R_6 should be 80% of R_4. This gives a ±20% adjustment in R_4.

Example 10-3

Design the phase shift oscillator in Fig. 10-6 to give a maximum output of ±3 V with an oscillation frequency of 6 kHz. Include distortion minimization adjustment.

Solution

Let $I_2 \approx 1$ mA when the diodes are forward-biased, that is, when the peak output is $V_P \approx 3$ V.

* David A. Bell, *Electronic Devices and Circuits,* 3rd ed. (Englewood Cliffs, NJ: Prentice-Hall, Inc., 1986), pp. 49–50.

Eq. 10-6, $R_1 = \frac{V_o/29}{I_2} = \frac{3\text{ V}}{29 \times 1\text{ mA}}$

$= 103\ \Omega$ (use $100\ \Omega$ standard value)

Eq. 10-7, $R_2 = 29 \times R_1 = 29 \times 100\ \Omega$

$= 2.9\text{ k}\Omega$

Eq. 10-8, $R_3 = \frac{2V_F}{I_2} \approx \frac{2 \times 0.7\text{ V}}{1\text{ mA}}$

$= 1.4\text{ k}\Omega$ (use $1.5\text{ k}\Omega$ standard value)

$R_4 = R_2 - R_3 = 2.9\text{ k}\Omega - 1.5\text{ k}\Omega$

$= 1.4\text{ k}\Omega$

$R_5 = 0.4 \times R_4 = 0.4 \times 1.4\text{ k}\Omega$

$= 560\ \Omega$ (use a $1\text{ k}\Omega$ potentiometer)

$R_6 = 0.8 \times R_4 = 0.8 \times 1.4\text{ k}\Omega$

$= 1.12\text{ k}\Omega$ (use $1.2\text{ k}\Omega$ standard value)

$R = R_1 = 100\ \Omega$

$$C = \frac{1}{2\pi R f \sqrt{6}} = \frac{1}{2\pi \times 100\ \Omega \times 6\text{ kHz} \times \sqrt{6}}$$

$\approx 0.1\ \mu\text{F}$ (standard value)

Diodes D_1 through D_4,

Eq. 10-9, $t_{rr(\max)} = \frac{T}{10} = \frac{1}{6\text{ kHz} \times 10}$

$= 16.7\ \mu\text{s}$

$V_{R(\max)} > V_{CC} = \pm 15\text{ V}$ (typically)

The IN914 has $t_{rr} \approx 4$ ns and $V_{R(\max)} = 75$ V. Use IN914 diodes.

10-5 WEIN BRIDGE OSCILLATOR

Circuit Operation

A Wein bridge is an ac bridge in which balance is obtained only when the supply voltage has a particular frequency. In the *Wein bridge oscillator* in Fig. 10-7, the resistive and capacitive components constitute a Wein bridge. The output of the bridge is applied to the op-amp input terminals, and the op-amp output provides the ac supply to the bridge. Another way of looking at the circuit is that the op-amp together with resistors R_3 and R_4 is a noninverting amplifier, and the other components are a feedback network, as illustrated. The output voltage from the amplifier is attenuated

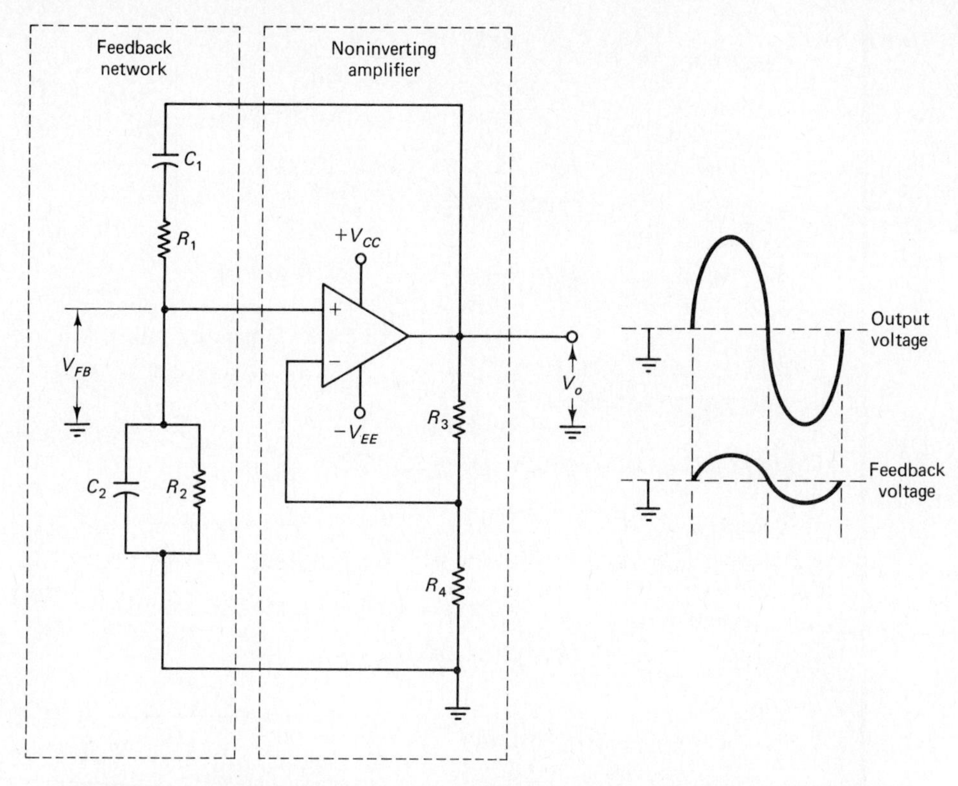

Figure 10-7 Wein bridge oscillator consisting of a noninverting amplifier and a feedback network. The feedback network attenuates, but does not phase shift, the output. The amplifier amplifies the feedback voltage.

but not phase shifted by the feedback network, and the feedback voltage is amplified to produce the output.

The Barkhausen criteria for zero-loop phase shift is fulfilled, in this case, by the amplifier and the feedback network each having zero phase shift. This occurs at only one particular frequency, the resonant frequency of the bridge. At all other frequencies, the bridge output and feedback voltages do not have the correct phase shift or amplitude relationship to sustain oscillations.

Analysis of the bridge circuit reveals that balance is obtained when the following two equations are fulfilled

$$\frac{R_3}{R_4} = \frac{R_1}{R_2} + \frac{C_2}{C_1} \tag{10-10}$$

and

$$2\pi f = \frac{1}{\sqrt{(R_1 C_1 R_2 C_2)}} \tag{10-11}$$

If $R_1 = R_2 = R$ and $C_1 = C_2 = C$, Eq. 10-11 yields

$$f = \frac{1}{2\pi CR} \tag{10-12}$$

and Eq. 10-10 gives,

$$R_3 = 2R_4 \tag{10-13}$$

The amplifier voltage gain is

$$A_v = \frac{R_3 + R_4}{R_4} = 3$$

The feedback network attenuates the amplifier output by a factor of 3, so the voltage gain of the noninverting amplifier must be a minimum of 3 to sustain oscillations.

Wein Bridge Oscillator Design

A Wein bridge oscillator is designed by first determining the noninverting amplifier components in the usual way. Then, R_1 and R_2 are usually made equal to R_4, and the capacitor values are determined from Eq. 10-12. If a BIFET op-amp is used, it is best to start by selecting a capacitance value very much larger than stray capacitance. Then Eq. 10-12 can be used to calculate R.

Example 10-4

Using a BIFET op-amp with a supply of ±12 V, design a Wein bridge oscillator to have an output frequency of 15 kHz.

Solution

Select, $C = C_1 = C_2 = 0.01\ \mu\text{F}$

From Eq. 10-12, $R = \frac{1}{2\pi Cf} = \frac{1}{2\pi \times 0.01\ \mu\text{F} \times 15\ \text{kHz}}$

$= 1.06\ \text{k}\Omega$ (use 1 kΩ standard value)

$R_1 = R_2 = R = 1\ \text{k}\Omega$

Let, $R_4 = R_2 = 1\ \text{k}\Omega$

For $A_v = 3$, $R_3 = 2R_4 = 2 \times 1\ \text{k}\Omega$

$= 2\ \text{k}\Omega$ (use 2.2 kΩ standard value to give $A_v > 3$)

Amplitude Stabilization for the Wein Bridge Oscillator

Figure 10-8 shows two amplitude stabilization methods that can be used with a Wein bridge oscillator. The circuit in Fig. 10-8(a) operates in exactly the same way as the similar circuit already described for use with a phase shift oscillator. Resistor R_5 be-

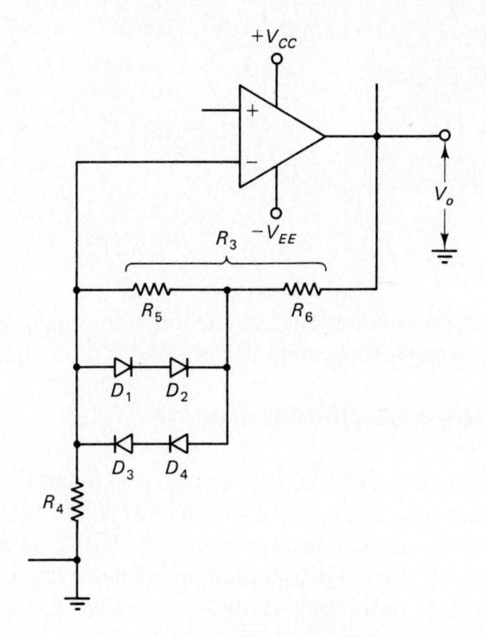

(a) Amplitude stabilization by diodes

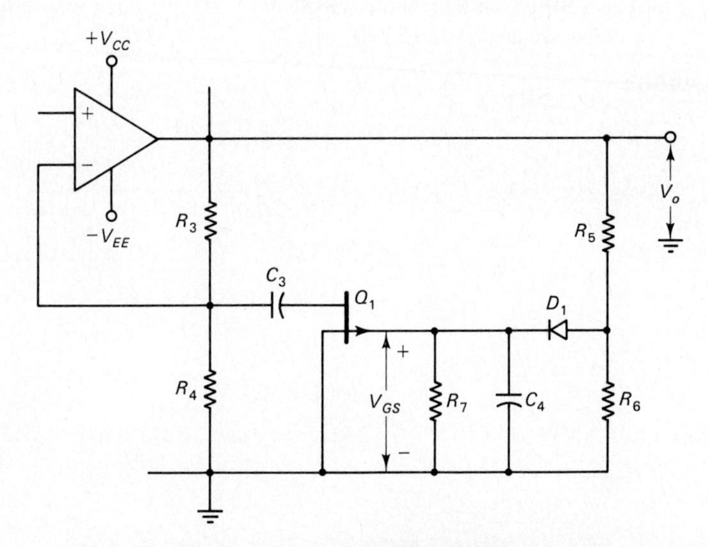

(b) FET circuit for output amplitude stabilization

Figure 10-8 A Wein bridge oscillator may have its output amplitude stabilized by any one of several methods. In all cases, when the output amplitude exceeds a predetermined level, the amplifier voltage gain is reduced below the value required to sustain oscillations.

comes shorted-out by the diodes when the output amplitude exceeds the desired level and the amplifier voltage gain is then not large enough to sustain oscillations.

The circuit in Fig. 10-8(b) once again stabilizes the output voltage by reducing the amplifier gain when the output amplitude is greater than the desired level. The channel resistance of the p-channel field effect transistor (Q_1) is in parallel with resistor R_4 for alternating voltages. Capacitor C_3 ensures that Q_1 has no effect on the dc bias conditions of the circuit. The bias voltage at the FET gate is derived from the amplifier output. The output voltage is potentially divided across resistors R_5 and R_6 and rectified by diode D_1. Capacitor C_4 smoothes out the rectified waveform to give a dc bias at the FET gate. When the output amplitude is low, V_{GS} is low and the FET drain resistance (R_D) remains low. The amplifier voltage gain is

$$A_v = \frac{R_3 + R_4 \parallel R_D}{R_4 \parallel R_D} \tag{10-14}$$

A low value of R_D keeps A_v at a maximum. When V_o goes to a high level V_{GS} is increased and this results in increased FET channel resistance. The increase in R_D reduces A_v and thus prevents the circuit from oscillating with a high V_o level.

To design the FET ampliutude stabilization circuit, a knowledge of the FET channel resistance and how it is affected by the gate-source bias voltage is required. Suppose the circuit is to oscillate when the FET channel resistance is $R_D = 200\ \Omega$ at $V_{GS} = 2$ V and assume that the oscillator peak output is to be $V_p \approx 5$ V. Resistors R_5 and R_6 should be selected to give $V_{GS} = 2$ V when $V_o = 5$ V peak, allowing for V_F across the diode. These can be fairly large resistors (in the 100 kΩ range) because of the high input resistance at the FET gate. Capacitor C_4 smoothes out the half-wave rectified waveform. C_4 is discharged via R_7 during the time interval between peaks of the output waveform. So the value of C_4 is calculated to allow perhaps a 10% discharge during the time period of the oscillator frequency. C_3 is a coupling capacitor, so its impedance at the oscillating frequency should be much smaller than the FET channel resistance. Resistors R_3 and R_4 are determined using Eq. 10-14 to give the required amplifier gain when $R_D = 200\ \Omega$.

10-6 SIGNAL GENERATOR OUTPUT CONTROLS

All signal generators, whether pulse or sine-wave generators, need to have a control for adjusting the output amplitude and (often) another one for altering the dc voltage level of the output. Output amplitude control is easily achieved by means of an amplifier with adjustable gain. Output dc voltage control merely requires the addition of a potential divider to the amplifier.

Figure 10-9 shows an inverting amplifier with an adjustable resistor R_2. When R_2 equals R_1, the input signals are passed to the output unchanged, except for the phase inversion. If R_2 is greater than R_1, the signal is amplified; when R_2 is smaller than R_1, attenuation occurs. Coupling capacitor C_1 ensures that the dc voltage level at the output of the signal generating circuit is not passed to the output stage.

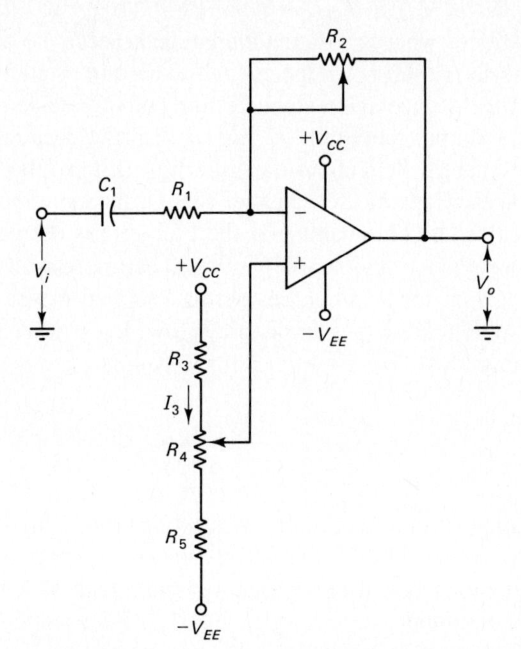

Figure 10-9 Signal generator output stage. Variable resistor R_2 provides output amplitude adjustment and R_4 affords dc voltage level shifting.

Normally with an inverting amplifier, the noninverting input terminal of the op-amp is grounded. This sets the dc output voltage level at ground. In Fig. 10-9, the noninverting terminal is connected to a potential divider (R_3, R_4, and R_5), with R_4 being a potentiometer. With this arrangement, the dc output voltage level can be adjusted above or below ground.

The inverting amplifier component values are calculated in the usual way, with the maximum and minimum resistances of R_2 selected according to the required range of voltage gain/attenuation. For sinusoidal signals, the impedance of capacitor C_1 should be much smaller than R_1 at the lowest operating frequency. When dealing with pulse signals, C_1 must be calculated to give the maximum acceptable tilt on the longest output pulse. The potential divider circuit is designed to have zero volts at the center of R_4 and the desired output voltage extremes at each end of R_4.

The op-amp should have a frequency range higher than the highest frequency of any sinusoidal signal. Also, it should be compensated as a voltage follower, because, when R_2 is adjusted to (or near to) zero resistance, the circuit has the feedback characteristics of a voltage follower. For pulse signals, the slew rate of the op-amp should be such that the output rise time does not exceed 10% of the shortest output pulse width.

Example 10-5

A signal generator output stage as in Fig. 10-9 is to be designed to afford output amplitude adjustment from ±0.1 V to ±5 V and dc voltage level control over a range of ±2.5 V. The signal applied to the output stage has a ±1 V amplitude and a frequency ranging from 50 Hz to 20 kHz. Using a bipolar op-amp with a ±15 V supply, determine suitable component values.

Solution

$$I_1 \gg I_{B(max)}$$

let $I_1 = 50\ \mu A$

$$R_1 = \frac{V_i}{I_1} = \frac{1\ V}{50\ \mu A}$$

$$= 20\ k\Omega \qquad \text{(use 18 k}\Omega\text{ standard value)}$$

$$R_{2(max)} = \frac{V_{o(max)}}{V_i} \times R_1 = \frac{5\ V}{1\ V} \times 18\ k\Omega$$

$$= 90\ k\Omega$$

$$R_{2(min)} = \frac{V_{o(min)}}{V_i} \times R_1 = \frac{0.1\ V}{1\ V} \times 18\ k\Omega$$

$$= 1.8\ k\Omega$$

For R_2, use a 100 kΩ potentiometer in series with a 1.8 kΩ resistor.

$$I_3 \gg I_{B(max)}$$

let $I_3 = 50\ \mu A$

$$R_4 = \frac{V_{R4}}{I_3} = \frac{2.5\ V - (-2.5\ V)}{50\ \mu A}$$

$$= 100\ k\Omega \qquad \text{(standard value potentiometer)}$$

$$R_3 = \frac{V_{R3}}{I_3} = \frac{15\ V - 2.5\ V}{50\ \mu A}$$

$$= 250\ k\Omega \qquad \text{(use 220 k}\Omega\text{ to give larger output adjustment than required)}$$

$$R_5 = R_3 = 220\ k\Omega$$

$$X_{C1} \ll R_1 \text{ at } f_{(min)}$$

let $X_{C1} = R_1/10$ at $f_{(min)}$

$$C_1 = \frac{1}{2\pi f_{(min)} R_1/10} = \frac{1}{2\pi \times 50\ Hz \times 18\ k\Omega/10}$$

$$\approx 1.8\ \mu F \qquad \text{(standard value)}$$

REVIEW QUESTIONS

10-1. Sketch the circuit of a triangular/rectangular waveform generator. Draw the output waveforms from the circuit showing their phase relationship and carefully explain the circuit operation.

10-2. Draw the circuit of a triangular/rectangular waveform generator which has frequency and duty cycle controls. Show all waveforms and explain the circuit operation.

10-3. Discuss the design procedure for a triangular/rectangular waveform generator and write the equations for calculating the component values.

10-4. Draw the circuit of a phase shift oscillator. Sketch the output and feedback voltage waveforms and explain the circuit operation.

10-5. State the Barkhausen criteria and explain how it is fulfilled in the phase shift oscillator. Write the equation for oscillation frequency in a phase shift oscillator.

10-6. Sketch the circuit of a phase shift oscillator that uses diodes for output amplitude stabilization. Explain how the amplitude stabilization circuit operates and show how a distortion control may be included.

10-7. Draw the circuit of a Wein bridge oscillator. Sketch the output and feedback voltage waveforms and explain the circuit operation.

10-8. Discuss the design procedure for a Wein bridge oscillator, and write the equations for output frequency and amplifier voltage gain.

10-9. Sketch a circuit to show how diodes may be used for output amplitude stabilization with a Wein bridge oscillator. Explain the circuit operation.

10-10. Sketch a Wein bridge oscillator circuit which uses a FET for output amplitude stabilization. Explain the operation of the amplitude stabilization circuit.

10-11. Draw the circuit of an output stage for controlling the output amplitude and dc voltage level of a signal generator. Explain how it operates and discuss the selection of component values.

PROBLEMS

10-1. Design a triangular/rectangular waveform generator as in Fig. 10-1 to have an output frequency of 1 kHz, a triangular output amplitude of ±6 V, and a square wave output amplitude of approximately ±10 V. Use bipolar op-amps and estimate a minimum suitable op-amp slew rate.

10-2. Redesign the circuit for Problem 10-1 to use BIFET op-amps.

10-3. Using BIFET op-amps with a ±12 V supply, design a triangular/rectangular waveform generator. The triangular wave is to have an output amplitude of ±3 V, the frequency is to be adjustable from 1 kHz to 6 kHz, and the duty cycle range is to be 40% to 60%. Estimate a minimum suitable op-amp slew rate.

10-4. Modify the circuit designed for Problem 10-1 to make the amplitude of the triangular waveform adjustable over the range ±4 V to ±8 V. Analyze the circuit to determine the effect that this adjustment has on the output frequency.

10-5. Using a BIFET op-amp with a ±9 V supply, design a phase shift oscillator to have an output frequency of 10 kHz.

10-6. Redesign the circuit for Problem 10-5 to use a bipolar op-amp.

10-7. The circuit designed for Problem 10-5 is to have its output frequency changed to 3.3 kHz. (a) calculate new capacitor values to effect this change, (b) determine suitable resistor values that would change the frequency as required.

10-8. Modify the phase shift oscillator circuit designed for Problem 10-6 to include a voltage stabilization circuit as in Fig. 10-6. The output amplitude is to be stabilized at ±7 V and distortion control is to be included.

10-9. Using bipolar op-amps with a ±18 V supply, design a phase shift oscillator to have an output frequency of 15 kHz. The output amplitude is to be stabilized at ±10 V and distortion control is to be included.

10-10. Design a Wein bridge oscillator as in Fig. 10-7 to have an output frequency of 12 kHz. Use a bipolar op-amp with a ±15 V supply.

10-11. The circuit designed for Problem 10-10 is to have its output frequency changed to 9 kHz. (a) calculate suitable capacitor values to change the frequency, (b) determine new resistor values to change the frequency as required.

10-12. Modify the circuit designed for Problem 10-10 to use a diode amplitude stabilization circuit as in Fig. 10-8(a). The output amplitude is to be stabilized to ±8 V.

10-13. Modify the circuit designed for Problem 10-10 to use a FET amplitude stabilization circuit as in Figure 10-8(b). The output amplitude is to be stabilized at ±10 V. The available FET has R_D = 5 kΩ at V_{GS} = 1 V and R_D = 3.3 kΩ at V_{GS} = 2 V.

10-14. Design a signal generator output stage as in Fig. 10-9 to operate with the circuit designed for Problem 10-13. The output amplitude is to be adjustable from ±0.5 V to ±5 V and the dc voltage level is to be adjustable over the range ±3 V. Use a BIFET op-amp.

10-15. The circuit designed for Problem 10-8 is to have an output stage that will give an output amplitude range of ±0.1 V to ±3 V and a dc level adjustable from −2 V to +2 V. Using a bipolar op-amp, design a suitable circuit.

10-16. The square wave output of the circuit designed for Problem 10-3 is to have an output stage that will provide an amplitude range of ±0.3 V to ±5 V and dc level control of ±3 V. The tilt on the output is not to exceed 1% of the peak output amplitude. Using a bipolar op-amp, design a suitable circuit.

COMPUTER PROBLEMS

10-17. Write a computer program to analyze the pulse/triangular waveform circuit in Fig. 10-2 to determine the triangular wave amplitude, the pulse amplitude, and pulse width range. Given: supply voltage and all component values.

10-18. Write a computer program to design a phase shift oscillator as in Fig. 10-6(a). Given: supply voltage, f, and V_o.

10-19. Write a computer program to analyze the Wein bridge oscillator circuit in Fig. 10-7 to determine output amplitude and frequency. Given: supply voltage and all component values.

LABORATORY EXERCISES

10-1 Triangular/Rectangular Waveform Generator

1. Construct the triangular/rectangular waveform generator shown in Fig. 10-3, as designed in Example 10-1. Use 741 op-amps.
2. Monitoring the integrator and Schmitt output waveforms on an oscilloscope, adjust the moving contact of R_5 approximately to the half-way point and adjust R_4 to give a frequency of 200 Hz.

3. Further adjust R_5 and R_4 in turn, to investigate the duty cycle and frequency ranges. Compare the ranges of adjustment and the amplitudes of each output to the designed quantities.

10-2 Phase Shift Oscillator

1. Construct the phase shift oscillator shown in Fig. 10-6(a), as designed in Example 10-3. Use IN914 diodes and a 741 op-amp with a supply of ± 9 V to ± 15 V.
2. Measure the output amplitude and frequency, and compare to the design quantities.

10-3 Wein Bridge Oscillator

1. Construct the Wein bridge oscillator circuit in Fig. 10-7, as designed in Example 10-4. Although the circuit is designed for a BIFET op-amp, a 741 can be used.
2. Measure the output amplitude and frequency, and compare to the design quantities.

CHAPTER 11

Active Filters

Objectives

- Sketch the following op-amp filter circuits and explain the operation of each: *all-pass phase lag, all-pass phase lead, first-order low-pass, first-order high-pass, second-order low-pass, second-order high-pass, third-order low-pass, third-order high-pass, bandpass using low-pass and high-pass, wide-band and narrow-band single-stage bandpass, state-variable bandpass, bandstop using low-pass and high-pass, bandstop using bandpass and summing circuit.*
- Sketch and explain the typical frequency response graphs for each of the above circuits, showing *Butterworth* and *Chebyshev* responses where appropriate.
- Design each of the above circuits to meet a given specification.
- Analyze each of the above circuits to determine its performance.

INTRODUCTION

Filters are circuits that pass only a certain range of signal frequencies; they are typically classified as *low-pass, high-pass, bandpass,* and *bandstop*. Passive filters employ only passive components such as resistors, capacitors, and inductors. *Active filters* use transistors, or most often, operational amplifiers together with passive components. In addition to being classified as low-pass, high-pass and so forth, active filters are further defined in terms of the rate at which the output falls off at the edge of the frequency range. Those which have a fall-off rate of 20 dB per decade are called *first-order* filters. When the rate is 40 dB per decade, they are termed *second-order* filters. A filter with a 60 dB per decade fall-off rate is a *third-order* filter. Analysis of filter circuits can be extremely complex, but the behavior of the most commonly used filters can be readily understood and many simple design techniques are available.

11-1 ALL-PASS PHASE SHIFTING CIRCUITS

Phase Lag Circuit

A circuit that will pass all signal frequencies without attenuation but with varying phase shifts is illustrated in Fig. 11-1(a). Known as an *all-pass circuit,* this circuit is similar to the difference amplifier discussed in Section 3-6 except that its input terminals are connected together and have V_i applied to both, and capacitor C_1 replaces a resistor. The procedure used in the analysis of the difference amplifier can also be employed to investigate the performance of the all-pass circuit.

With terminal 2 grounded and V_i applied to terminal 1,

$$v_{o1} = -\frac{R_2}{R_1} \times v_i$$

with $R_1 = R_2$, $$v_{o1} = -v_i$$

With terminal 1 grounded and v_i applied to terminal 2,

$$v_{o2} = \frac{R_2 + R_1}{R_1} \times v_{C1}$$

with $R_1 = R_2$ $$v_{o2} = 2v_{C1}$$

$$v_{C1} = \frac{-j(1/\omega C_1)}{R_3 - j(1/\omega C_1)} \times v_i$$

$$= \frac{v_i}{1 + j\,\omega C_1 R_3}$$

so, $$v_{o2} = \frac{2v_i}{1 + j\,\omega C_1 R_3}$$

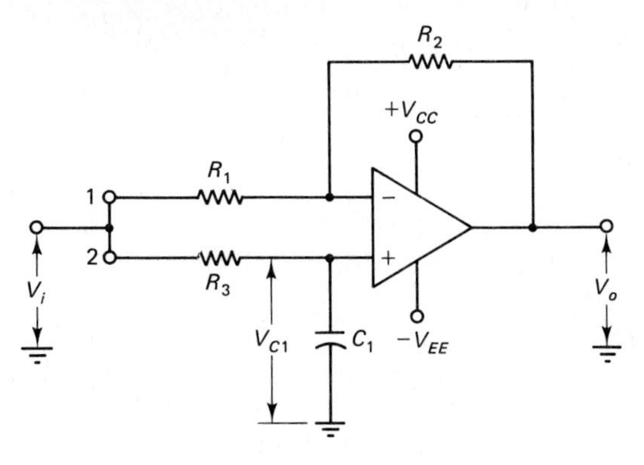

(a) All-pass phase lag circuit

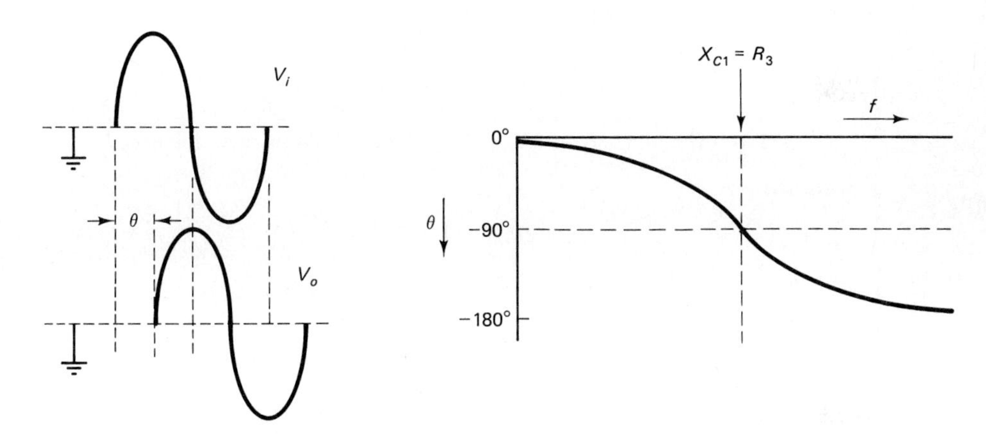

(b) Input and output waveforms, and graph of phase lag versus frequency

Figure 11-1 All-pass circuit which passes a signal without attenuation while introducing a phase lag dependent on the signal frequency. A 90° phase lag occurs when $X_{C1} = R_3$.

and

$$v_o = v_{o1} + v_{o2}$$

$$= -v_i + \frac{2v_i}{1 + j\,\omega C_1 R_3}$$

$$= v_i \times \frac{1 - j\,\omega C_1 R_3}{1 + j\,\omega C_1 R_3}$$

which gives

$$v_o = v_i \angle -2 \arctan (\omega C_1 R_3) \tag{11-1}$$

It is seen that the output voltage amplitude is always equal to the input, regardless of the signal frequency. Also, the output lags the input by an angle of

$$\theta = -2 \arctan (\omega C_1 R_3) \tag{11-2}$$

or,

$$\theta = -2 \arctan (R_3/X_{C1})$$

When $X_{C1} = R_3$, $\theta = -90°$

at high frequency, when $X_{C1} \ll R_3$, $\theta \approx -180°$

at low frequency, when $X_{C1} \gg R_3$, $\theta \approx 0°$

This is illustrated in Fig. 11-1(b) where the waveforms and the shape of the typical phase angle versus frequency graph is shown.

An all-pass circuit is designed by first determining resistors R_1 and R_2 as for an inverting amplifier with a gain of 1. This gives $R_1 = R_2$. Then R_3 is usually made equal to $R_1 \| R_2$ and X_{C1} is equated to R_3 at the frequency at which a 90° phase shift is desired. The phase shift can be adjusted by using a variable resistor for R_3. When using a BIFET op-amp, capacitor C_1 should first be selected; then R_3 equals X_{C1} at the 90° phase shift frequency. If an uncompensated op-amp is used, it should be compensated as an inverting amplifier with a gain of 1. The highest operating frequency of the circuit is limited by the op-amp cutoff frequency.

Example 11-1

Using a 741 op-amp, design an all-pass circuit to have a phase lag adjustable from 80° to 100°. The input signal has a 1 V amplitude and a 5 kHz frequency.

Solution

$$I_1 \gg I_{B(\max)}$$

let

$$I_1 = 50\ \mu\text{A}$$

$$R_1 = \frac{v_i}{I_1} = \frac{1\text{ V}}{50\ \mu\text{A}}$$

$$= 20\text{ k}\Omega \qquad \text{(use 18 k}\Omega\text{ standard value)}$$

$$R_2 = R_1 = 18\text{ k}\Omega$$

$$R_3 \approx R_1 \| R_2 = 9\text{ k}\Omega$$

For a 90° phase shift,

$$X_{C1} = R_3$$

or,

$$C_1 = \frac{1}{2\pi f R_3} = \frac{1}{2\pi \times 5\text{ kHz} \times 9\text{ k}\Omega}$$

$$= 3537\text{ pF} \qquad \text{(use 3600 pF standard value)}$$

For a 80° phase shift,

Eq. 11-2,

$$\theta = 2\arctan(\omega C_1 R_3)$$

or,

$$R_3 = \frac{\tan(\theta/2)}{\omega C_1} = \frac{\tan(80°/2)}{2\pi \times 5\text{ kHz} \times 3600\text{ pF}}$$

$$\approx 7.4\text{ k}\Omega$$

For a 100° phase shift,

$$R_3 = \frac{\tan(\theta/2)}{\omega C_1} = \frac{\tan(100°/2)}{2\pi \times 5\text{ kHz} \times 3600\text{ pF}}$$

$$\approx 10.5\text{ k}\Omega$$

For R_3, use a 6.8 kΩ fixed value resistor in series with a 5 kΩ variable resistor to give a total resistance adjustable from 6.8 kΩ to 11.8 kΩ.

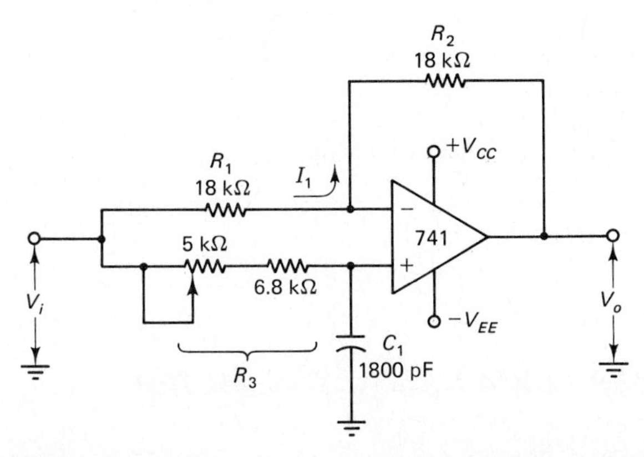

Figure 11-2 All-pass circuit with adjustable phase lag, designed in Example 11-1.

Phase Lead Circuit

The phase lead all-pass circuit in Fig. 11-3 is similar to the phase lag circuit, except that R_3 and C_1 are interchanged. This has the effect of producing an output which leads the input by a phase angle dependent on the relationship between X_{C1} and R_3. The design procedure for this circuit is similar to that for the phase lag circuit.

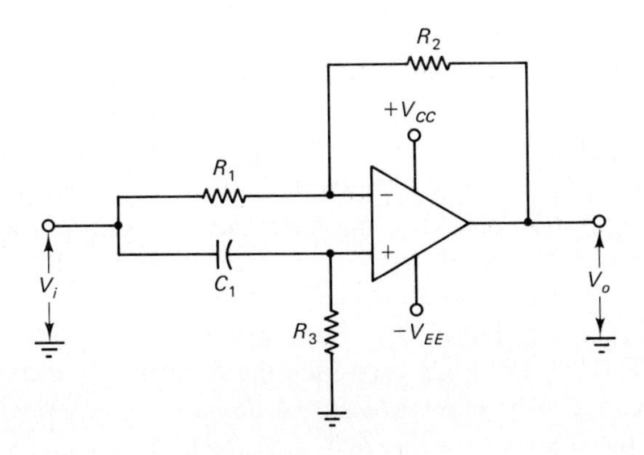

(a) All-pass phase lead circuit

Figure 11-3 All-pass phase lead circuit. The signal is passed without attenuation while a phase lead is introduced dependent on the signal frequency. A 90° phase lead occurs when $X_{C1} = R_3$.

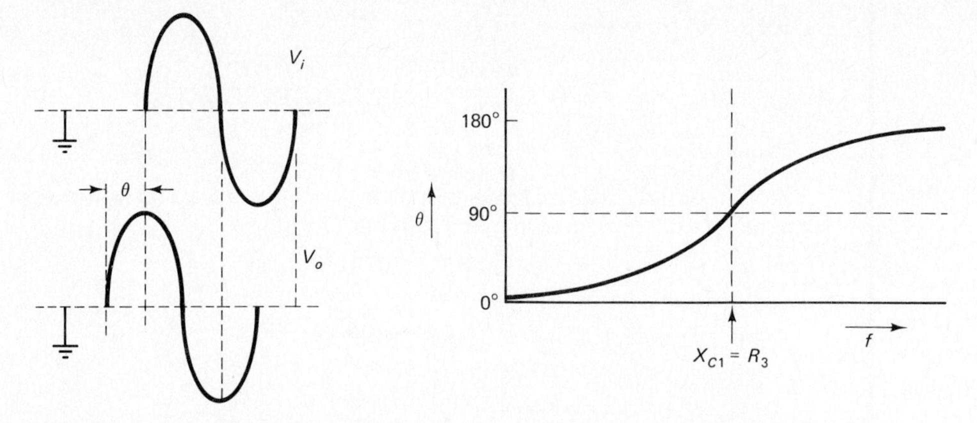

(b) Input and output waveforms, and graph of phase lead versus frequency

Figure 11-3 *(continued)*

11-2 FIRST-ORDER LOW-PASS ACTIVE FILTER

A passive low-pass filter circuit consisting of a resistor and a capacitor is shown in Fig. 11-4(a). The filter load (Z_L) is connected in parallel with capacitor C_1. This illustrates a major problem for passive filters: Unless Z_L is much larger than X_{C1}, the load is likely to affect the filter performance. Connecting a voltage follower to the circuit as illustrated in Fig. 11-4(b) eliminates the load problem and converts the circuit into an *active filter*. This circuit is also known as a *first-order low-pass* filter.

The voltage gain of the active filter is

$$\frac{V_o}{V_i} = \frac{X_{C1}}{\sqrt{(R_1^2 + X_{C1}^2)}} \tag{11-3}$$

The active filter frequency response shown in Fig. 11-4(c) is exactly the same as that of the unloaded passive filter. At low frequencies X_{C1} is much larger than R_1, and so Eq. 11-3 gives the gain as approximately 1. At higher frequencies, as X_{C1} approaches the value of R_1, the gain decreases (there is an attenuation). The frequency at which the output voltage is down by 3 dB from its pass-frequency level is defined as the *cutoff frequency* (f_c) of the filter. As illustrated in Fig. 11-4(c), f_c is the frequency at which X_{C1} equals R_1. This can be proved by substituting $X_{C1} = R_1$ into Eq. 11-3.

At frequencies higher than f_c the gain of the circuit falls off at a rate of *20 dB per decade*, a fall of 20 dB each time the frequency is increased by a factor of 10. This rate can also be expressed as *−6 dB per octave*, which means a fall of 6 dB each time the frequency is doubled.

Design of the basic low-pass filter circuit in Fig. 11-4(b) starts with selection of resistor R_1 small enough that no significant voltage drop occurs across it due to the op-amp input bias current. As explained in Section 3-1, a maximum drop of about 70 mV should be allowed across the resistor. It is also desirable to keep R_1 as

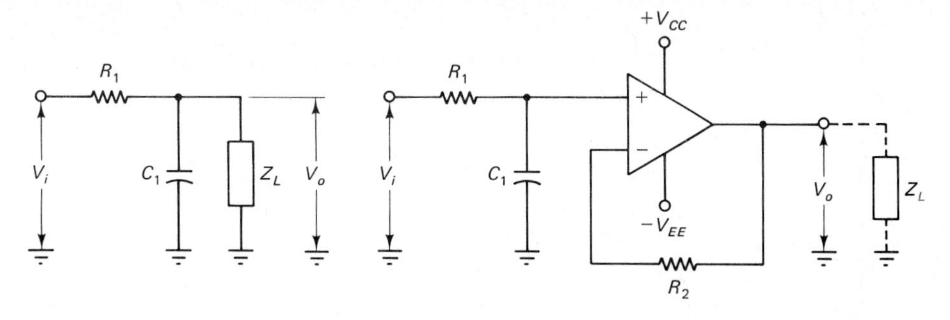

(a) Passive low-pass filter (b) Active low-pass filter

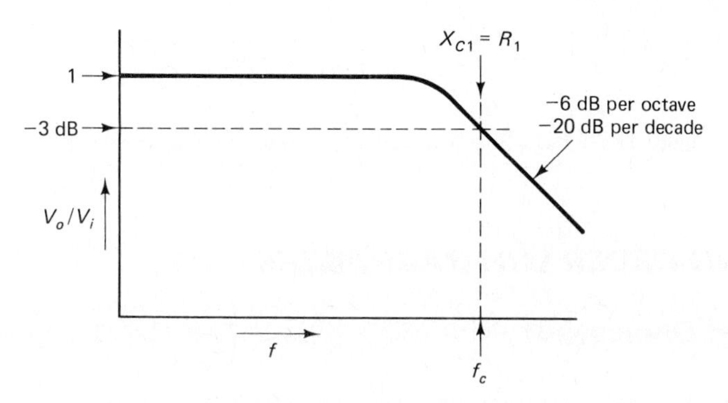

(c) Frequency response of low-pass filter

Figure 11-4 First-order low-pass filters. The active low-pass filter is simply a passive filter circuit with a voltage follower output stage. The output is attenuated by 3 dB when $X_{C1} = R_1$, and the fall-off rate is 20 dB per decade.

large as possible for maximum filter input impedance. (At high frequencies when X_{C1} is very small, $Z_i \approx R_1$). R_2 is made equal to R_1 to equalize the resistance at the op-amp input terminals. Capacitor C_1 is determined from

$$X_{C1} = R_1 \text{ at } f_c \tag{11-4}$$

The capacitance of C_1 must be much larger than stray capacitance. This points to the usual approach of selecting C_1 first when using BIFET op-amps.

Example 11-2

Using a 741 op-amp, design a first-order active low-pass filter to have a cutoff frequency of 1 kHz.

Solution

Eq. 3-1, $$R_1 = \frac{70 \text{ mV}}{I_{B(\max)}} = \frac{70 \text{ mV}}{500 \text{ nA}}$$

$$= 140 \text{ k}\Omega \qquad \text{(use 120 k}\Omega\text{ standard value)}$$

$$R_2 = R_1 = 120 \text{ k}\Omega$$

Eq. 11-4, $$X_{C1} = R_1 \text{ at } = f_c$$

so, $$C_1 = \frac{1}{2\pi f_c R_1} = \frac{1}{2\pi \times 1 \text{ kHz} \times 120 \text{ k}\Omega}$$

$$= 1326 \text{ pF} \qquad \text{(use 1300 pF standard value)}$$

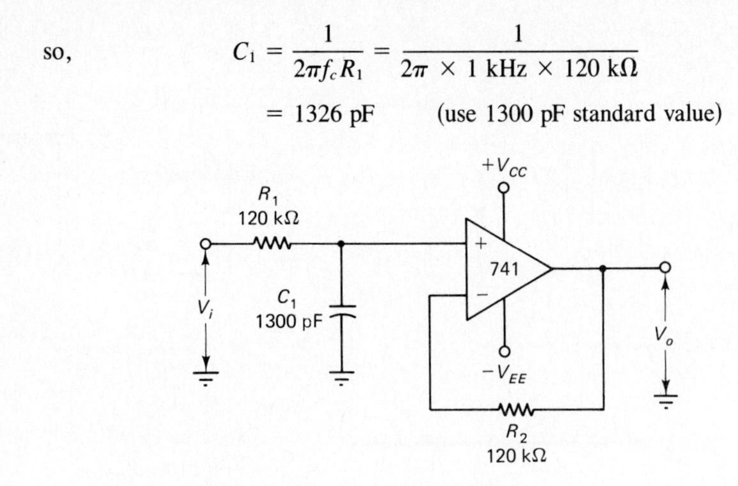

Figure 11-5 First-order low-pass filter designed in Example 11-2; $f_C = 1$ kHz.

11-3 SECOND-ORDER LOW-PASS FILTER

Circuit Operation

The basic low-pass filter circuit discussed in Section 11-2 has a voltage gain which falls off at the rate of 20 dB per decade. Ideally, when the gain of a filter is falling, the slope of the gain/frequency graph should be as steep as possible to severely attenuate unwanted frequencies. Figure 11-6(a) shows a filter circuit, known as a *second-order* low pass filter, which has a frequency response that falls off at the rate of 40 dB per decade above the upper cutoff frequency. This steeper roll-off rate is achieved by using the $C_1 R_2$ section together with feedback from the output via capacitor C_2 to the junction of R_1 and R_2.

At low frequencies, X_{C1} and X_{C2} are much larger than R_1 and R_2 and consequently, they have no significant effect on the circuit. So, the output voltage is equal to the input, giving a voltage gain of 1. At high frequencies, the effect of C_1 and R_2 causes the output to fall off at a rate of 20 dB per decade as the frequency increases. There is no phase shift from the input of the voltage follower to its output. However, there is a phase lag introduced by C_1 and R_2, and there is a phase lead generated by C_2 combined with R_1 and R_2. The result of these phase differences is that the feedback via C_2 produces a further fall off (or *roll off*) of 20 dB per decade. The two gain reducing effects combine to produce a fall-off rate of 40 dB per decade, as illustrated by the frequency response in Fig. 11-6(b).

Circuit Design

The design and analysis of active filter circuits can be very complex. Many equations, design charts, and computer programs have been created to produce optimum designs for various filter performances. The most common active filters are

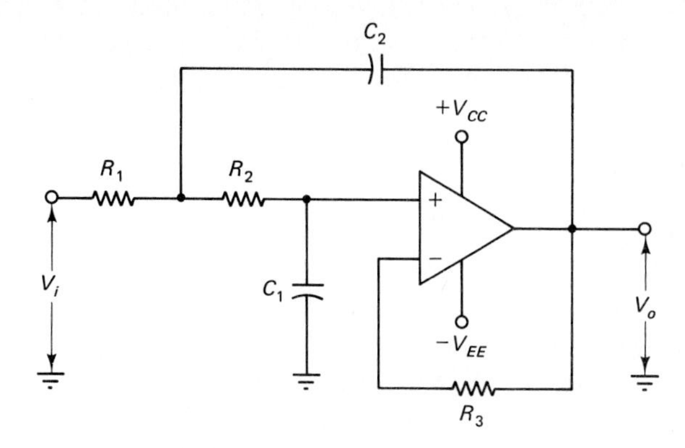

(a) −40 dB per decade low-pass filter circuit

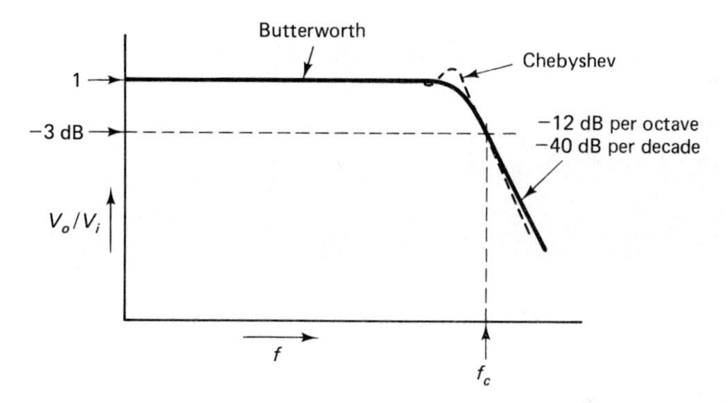

(b) Frequency response of Butterworth and Chebyshev low-pass filters

Figure 11-6 Second-order low-pass filter. The combined effect of the two resistors and two capacitors gives a fall-off rate of 40 dB per decade. The cutoff frequency f_C occurs at $X_{C1} = \sqrt{2}\, R_2$, and $X_{C2} = R_1/\sqrt{2}$.

Butterworth, Chebyshev, and Bessell. In general, a Butterworth filter has a flat response which rolls off at 40 dB per decade. But the initial roll-off rate is quite gradual, as illustrated in Fig. 11-6(b). A Chebyshev filter has a much sharper initial roll-off, but the response curve dips and then peaks at the roll-off point, as illustrated. A Bessel filter has a performance similar to that of a Butterworth, but with a slightly sharper initial roll-off. There are also important differences in the phase responses of the various filters.

A very simple approach can be taken in the design of a −40 dB per decade Butterworth low-pass filter. The resistance of $R_1 + R_2$ is first determined in the usual way. Then, $R_1 = R_2$, and $R_3 = R_1 + R_2$. Capacitor C_1 is calculated from

$$X_{C1} = \sqrt{2}\, R_2 \text{ at } f_c \tag{11-5}$$

and

$$C_2 = 2C_1 \tag{11-6}$$

to give,

$$R_1 = \sqrt{2}\, X_{C2} \text{ at } f_c \tag{11-7}$$

The effect of selecting the capacitors in this way should produce a total attenuation of 3 dB at the cutoff frequency. As always, the capacitor values must be much larger than stray capacitance.

Example 11-3

Design a second-order low-pass filter circuit as in Fig. 11-6 to have a cutoff frequency of 1 kHz.

Solution

The frequency response of the 741 extends to approximately 800 kHz when its voltage gain is 1, [see Fig. 5-13(a)]. So the 741 is suitable.

From Eq. 3-1,
$$R_1 + R_2 = \frac{70\text{ mV}}{I_{B(\max)}} = \frac{70\text{ mV}}{500\text{ nA}}$$
$$= 140\text{ k}\Omega$$
$$R_1 = R_2 = 70\text{ k}\Omega \qquad \text{(use 68 k}\Omega\text{ standard value)}$$
$$R_3 \simeq R_1 + R_2 = 136\text{ k}\Omega \qquad \text{(use 150 k}\Omega\text{ standard value)}$$

Eq. 11-5,
$$X_{C1} = \sqrt{2}\,R_2 \text{ at } f_c$$

or,
$$C_1 = \frac{1}{2\pi f_c \sqrt{2}\,R_2} = \frac{1}{2\pi \times 1\text{ kHz} \times \sqrt{2} \times 68\text{ k}\Omega}$$
$$= 1655\text{ pF} \qquad \text{(use 1600 pF standard value)}$$

Eq. 11-6,
$$C_2 = 2\,C_1 = 2 \times 1600\text{ pF}$$
$$= 3200\text{ pF} \qquad \text{(use 3300 pF standard value)}$$

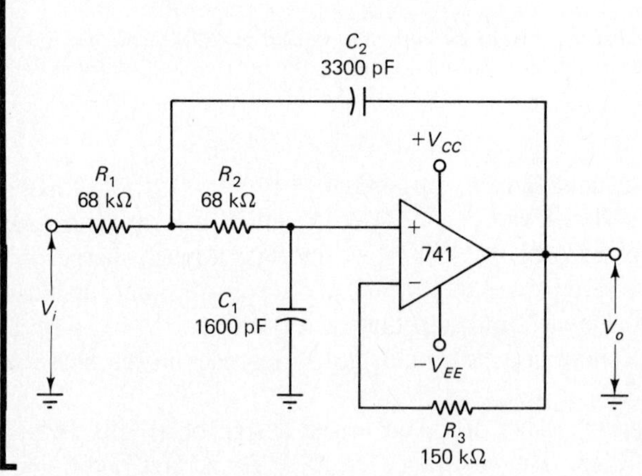

Figure 11-7 Second-order low-pass filter designed in Example 11-3; $f_C = 1$ kHz.

Cutoff Frequency Accuracy

The selected components in Example 11-3 are not exactly as calculated. Consequently, the circuit cutoff frequency may not be exactly as specified. Even if all the calculations gave standard size components, the component tolerances are likely to affect the cutoff frequency. To more precisely achieve a desired cutoff frequency in any filter circuit, it may be necessary to select low tolerance components and to make some of the components partially adjustable.

11-4 FIRST-ORDER HIGH-PASS FILTER

The *first-order high-pass filter* circuit in Fig. 11-8(a) is made up of a passive high-pass circuit, C_1 and R_1, and a voltage follower to isolate the load. At high frequencies, X_{C1} is very much smaller than R_1, there is virtually zero attenuation, and the circuit voltage gain is 1. When the frequency drops to the point at which X_{C1} equals R_1, the voltage gain drops by 3 dB [see Fig. 11-8(b)]. As the frequency decreases further, the voltage gain rolls off at 20 dB per decade.

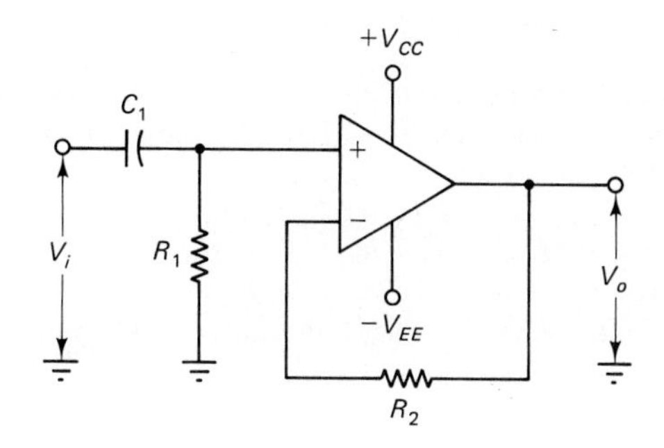

(a) Active high-pass filter

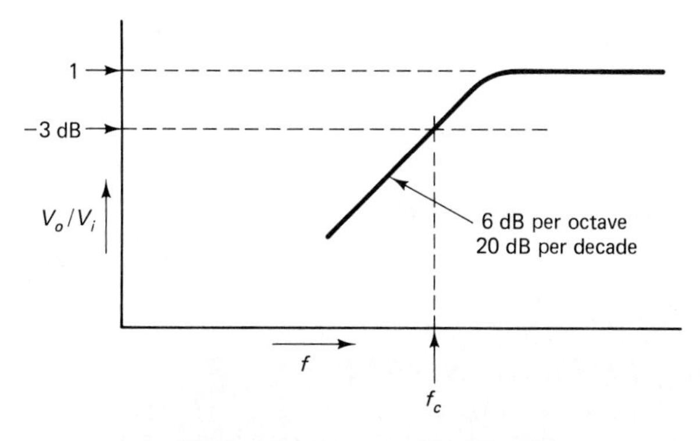

(b) Frequency response of high-pass filter

Figure 11-8 First-order high-pass active filter consisting of a passive high-pass filter circuit with a voltage follower output stage. The output is attenuated by 3 dB when $X_{C1} = R_1$, and the fall-off rate is 20 dB per decade.

The design procedure for the -20 dB per decade (*first-order*) active high-pass filter circuit is the same as for the similar low-pass filter, except that a low cutoff frequency (f_c) is involved instead of a high cutoff frequency. The highest frequency that the op-amp will pass can also be important for a high-pass filter. This is dealt with in Section 5-5.

Example 11-4

Design a first-order high-pass active filter circuit to have a cutoff frequency of 5 kHz. Use an LM108 op-amp and estimate the highest frequency that can be passed.

Solution

Because the LM108 has an extremely low input bias current, it should be treated as a BIFET op-amp. Therefore, select a capacitance value for C_1 very much larger than stray capacitance.

Let $\quad C_1 = 1000\ \text{pF}$

From Eq. 11-4,
$$R_1 = \frac{1}{2\pi f_c C_1} = \frac{1}{2\pi \times 5\ \text{kHz} \times 1000\ \text{pF}}$$
$$= 31.8\ \text{k}\Omega \qquad \text{(use 31.6 k}\Omega \pm 1\%\ \text{standard value)}$$
$$R_2 = R_1 = 31.6\ \text{k}\Omega$$

From the LM108 gain/frequency response (data sheet in Appendix 1-2), the op-amp unity gain frequency is $f_u \approx 1$ MHz. (Maximum frequency compensation must be used for a voltage follower.)

From Eq. 5-6,
$$f_2 = \frac{f_u}{A_v}$$
$$= 1\ \text{MHz}$$

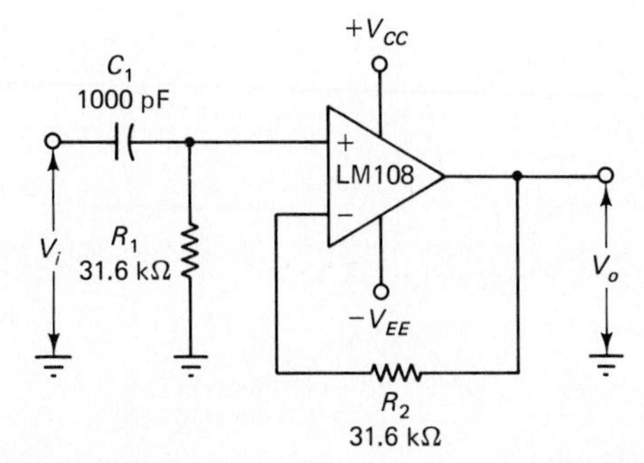

Figure 11-9 First-order high-pass filter designed in Example 11-4; $f_C = 5$ kHz.

11-5 SECOND-ORDER HIGH-PASS FILTER

The *second-order* (-40 *dB per decade*) *high-pass filter* circuit in Fig. 11-10(a) is similar to the low-pass filter in Figure 11-6(a), except that the resistor and capacitor positions are interchanged. At high frequencies, X_{C1} and X_{C2} are much smaller than R_1 and R_2, consequently, C_1, C_2, R_1 and R_2 have no significant effect on the circuit. The output voltage is then equal to the input, giving a voltage gain of 1. At low frequencies, the effect of C_2 and R_2 causes the output to fall off at a rate of 20 dB per decade as the frequency decreases. A phase lead is produced by C_2 and R_2, and a phase lag is generated by R_1 combined with C_1 and C_2. So, feedback via R_1 produces a further roll-off of 20 dB per decade and the total roll-off rate is 40 dB per decade, as illustrated in Fig. 11-10(b).

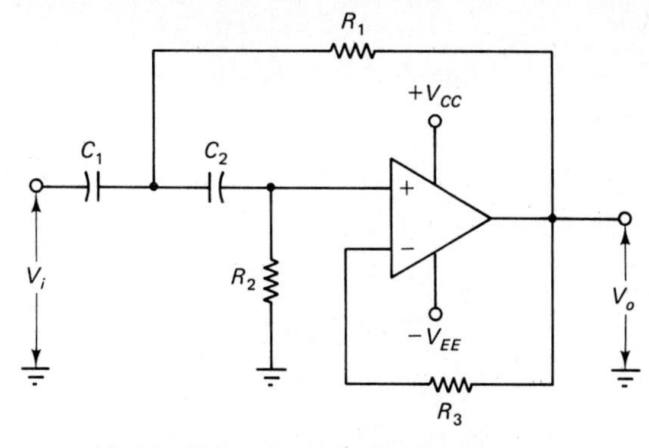

(a) −40 dB per decade high-pass filter circuit

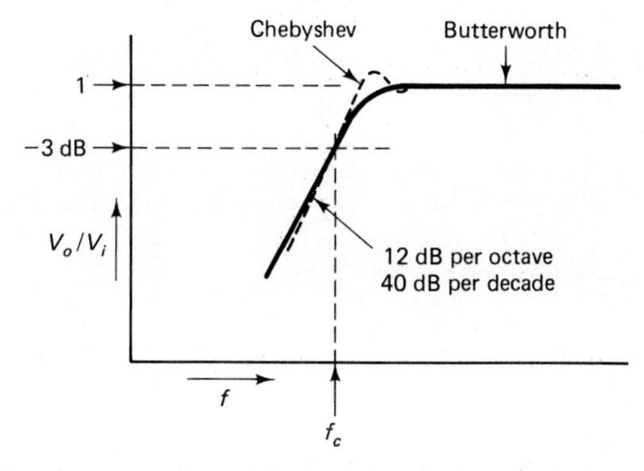

(b) Frequency response of Butterworth and Chebyshev high-pass filters

Figure 11-10 Second-order high-pass filter. The fall-off rate is 40 dB per decade, and the cutoff frequency f_C occurs at $X_{C1} = \sqrt{2}\,R_1$, and $X_{C2} = R_2/\sqrt{2}$.

Like active low-pass filters, high-pass filter circuits can be classified as *Butterworth, Chebyshev, or Bessell,* depending on the design approach employed. Also like the low-pass circuits, a very simple design procedure can be used for a −40 db per decade Butterworth high-pass filter. Capacitor C_2 is selected to give

$$R_2 = \sqrt{2}\,X_{C2} \text{ at } f_c \tag{11-8}$$

then,

$$C_1 = C_2 \tag{11-9}$$

and

$$R_1 = R_2/2 \tag{11-10}$$

to give

$$X_{C1} = \sqrt{2}\,R_1 \text{ at } f_c \tag{11-11}$$

Example 11-5

Design a second-order high-pass active filter to have a cutoff frequency of 12 kHz. Use a 715 op-amp and estimate the highest signal frequency that will be passed.

Solution

From the 715 data sheet in Appendix 1-4,

$$I_{B(max)} = 1.5\ \mu A$$

$$R_2 = \frac{70\ mV}{I_{B(max)}} = \frac{70\ mV}{1.5\ \mu A}$$

$$\approx 47\ k\Omega \qquad \text{(standard value)}$$

$$R_1 = R_2/2 = 23.5\ k\Omega \qquad \text{(use 22 k}\Omega\text{ and 1.5 k}\Omega\text{ in series)}$$

$$R_3 = R_2 = 47\ k\Omega$$

Eq. 11-8, $R_2 = \sqrt{2}\,X_{C2}$ at f_c

or,

$$C_2 = \frac{1}{2\pi f_c(R_2/\sqrt{2})} = \frac{\sqrt{2}}{2\pi \times 12\ kHz \times 47\ k\Omega}$$

$$= 398\ pF \qquad \text{(use 390 pF standard value)}$$

Eq. 11-9, $C_1 = C_2 = 390\ pF$

From the 715 data sheet in Appendix 1-4, the op-amp unity gain cutoff frequency is $f_u \approx 11$ MHz (using maximum frequency compensation).

From Eq. 5-6,

$$f_2 = \frac{f_u}{A_v}$$

$$= 11\ MHz$$

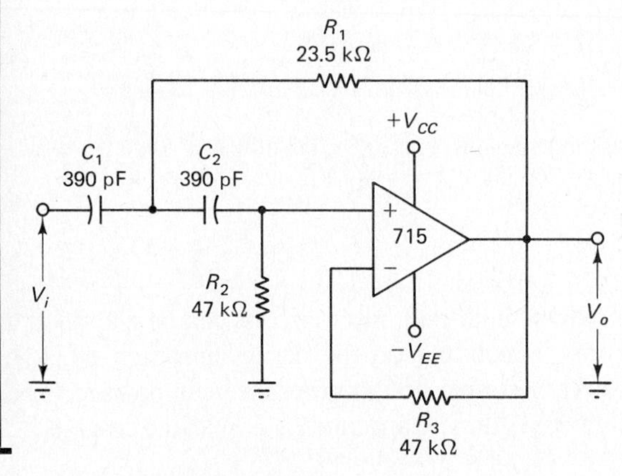

Figure 11-11 Second-order high-pass filter designed in Example 11-5; f_C = 12 kHz.

11-6 THIRD-ORDER LOW-PASS FILTER

The low-pass filter circuit in Fig. 11-12 consists of two stages: A −20 dB per decade (first-order) stage and a −40 dB per decade (second-order) stage. The complete circuit is known as a *third-order low-pass filter*. The overall gain of the circuit falls off at a rate of 60 dB per decade at frequencies above the cutoff frequency. Each stage of the circuit is independently designed in the usual way. However, if the gain of stage 1 is down by 3 dB at f_c and the gain of stage 2 is also down by 3 dB at f_c, then

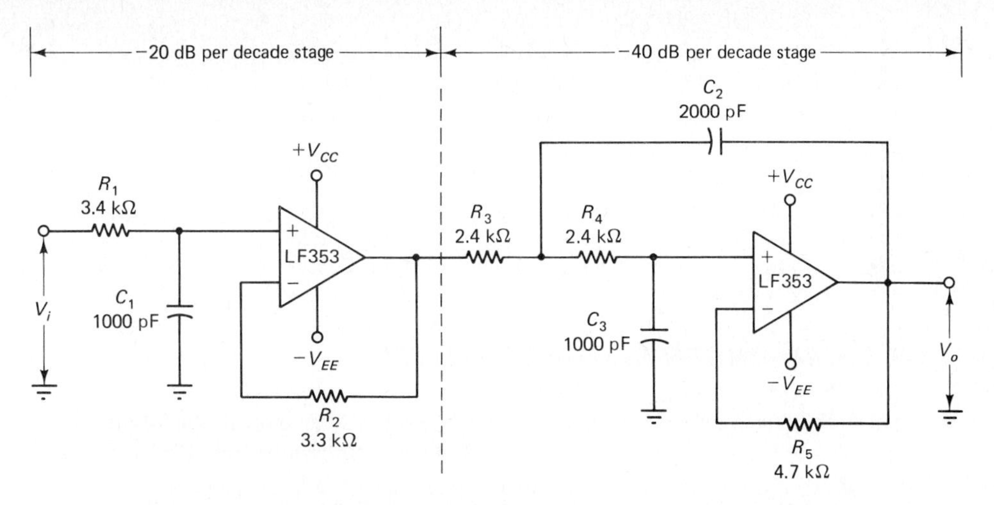

Figure 11-12 Third-order low-pass filter consisting of a −20 dB per decade stage cascaded with a −40 dB per decade stage. For an overall cutoff frequency of f_C, each stage must be designed to have a cutoff frequency $f_C/0.65$. Component values are those calculated in Example 11-6.

the overall circuit gain is down by 6 dB. This cannot be allowed because f_c is defined as the frequency at which the circuit gain is down by 3 dB. So, each stage should be designed to have an attenuation of 1.5 dB at the cutoff frequency. For a low-pass filter, this can be achieved by designing each stage in the usual way, but using a frequency of

$$f = \frac{f_c}{0.65} \tag{11-12}$$

Example 11-6

Design a third-order low-pass filter as in Fig. 11-12 to have a cutoff frequency of 30 kHz. Use BIFET op-amps.

Solution

−20 dB per decade stage (first order)

select $\quad C_1 = 1000 \text{ pF}$

from Eqs. 11-4 and 11-12, $\quad X_{C1} = R_1$ at $f_c/0.65$

giving,

$$R_1 = \frac{0.65}{2\pi f_c C_1} = \frac{0.65}{2\pi \times 30 \text{ kHz} \times 1000 \text{ pF}}$$

$$= 3.4 \text{ k}\Omega \qquad \text{(use 3.4 k}\Omega \pm 1\% \text{ standard value)}$$

$$R_2 \approx R_1 = 3.4 \text{ k}\Omega \qquad \text{(use 3.3 k}\Omega \text{ standard value)}$$

−40 dB per decade stage (second order)

select $\quad C_3 = 1000 \text{ pF}$

from Eqs. 11-5 and 11-11, $\quad X_{C3} = \sqrt{2}\, R_4$ at $f_c/0.65$

so, $R_4 = \dfrac{0.65}{2\pi f_c \sqrt{2}\, C_3} = \dfrac{0.65}{2\pi \times 30\text{ kHz} \times \sqrt{2} \times 1000\text{ pF}}$

$= 2.4\text{ k}\Omega$ (use 2.4 kΩ ±1% standard value)

Eq. 11-6, $C_2 = 2C_3 = 2000\text{ pF}$ (standard value)

$R_3 = R_4 = 2.4\text{ k}\Omega$

$R_5 \approx R_3 + R_4 = 4.8\text{ k}\Omega$ (use 4.7 kΩ standard value)

11-7 THIRD-ORDER HIGH-PASS FILTER

A *third-order* (*−60 dB per decade*) *high-pass filter* circuit consisting of a 20 dB per decade stage and a 40 dB per decade stage is shown in Fig. 11-13. The overall gain of the circuit falls off at a rate of 60 dB per decade at frequencies below the cutoff frequency. As for the 60 dB per decade low-pass circuit, each stage should be designed to have an attenuation of 1.5 dB at the desired cutoff frequency. For a high-pass filter, this is achieved by designing each stage in the usual way, but using a frequency of

$$f = 0.65 f_c \tag{11-13}$$

The cutoff frequency of the op-amps is an important consideration for any high-pass filter. For example, a high-pass filter might be designed to have a low cutoff frequency of 100 kHz and be intended to pass signals up to 1 MHz. If a two-stage circuit is used with each of the op-amps selected to have a high cutoff frequency of 1 MHz, the output of each stage is down by 3 dB (from its mid-frequency level) at 1 MHz. This gives an overall attenuation of 6 dB, and is exactly the same as in the

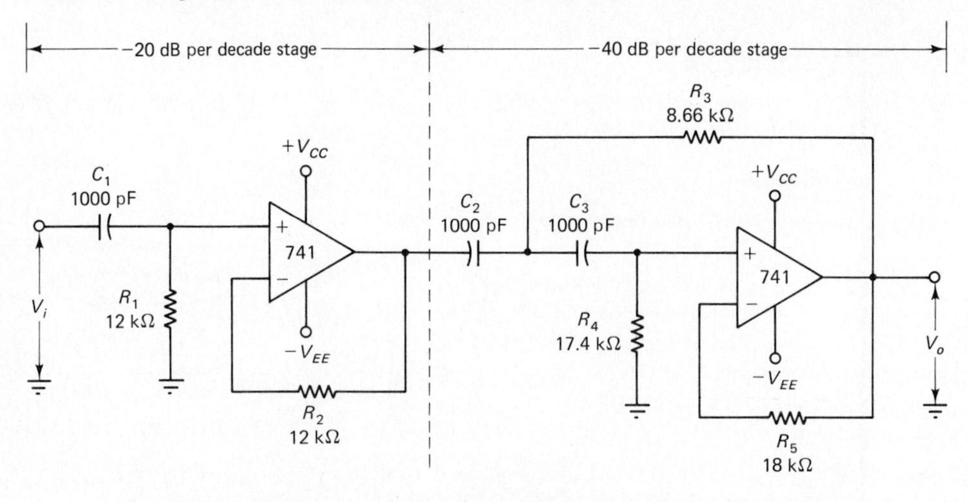

Figure 11-13 Third-order high-pass filter consisting of a −20 dB per decade stage and a −40 dB per decade stage connected in cascade. For an overall cutoff frequency of f_C, each stage must be designed to have a cutoff frequency $0.65 f_C$. The component values shown are determined in Example 11-7.

case of the cutoff frequency for the low-pass two-stage circuit. So, Eq. 11-12 also applies here. For any two stage op-amp circuit (filter circuit or otherwise) to have an overall high cutoff frequency of f_2, the high cutoff frequency of each stage must be $f = f_2/0.65$.

Example 11-7

Design a third-order high-pass filter as in Fig. 11-13 to have a cutoff frequency of 20 kHz. Use 741 op-amps.

Solution

−20 dB per decade stage (first order)

$$R_1 = 120\ \text{k}\Omega \qquad \text{(as in Example 11-2)}$$

from Eqs. 11-4 and 11-13,

$$X_{C1} = R_1 \text{ at } 0.65 f_c$$

giving,

$$C_1 = \frac{1}{2\pi 0.65 f_c R_1} = \frac{1}{2\pi \times 0.65 \times 20\ \text{kHz} \times 120\ \text{k}\Omega}$$

$$= 102\ \text{pF}$$

This is so small that it can be affected by stray capacitance. Redesign, first selecting a suitable capacitance for C_1.

select $\quad C_1 = 1000\ \text{pF}$

now

$$R_1 = \frac{1}{2\pi 0.65 f_c C_1} = \frac{1}{2\pi \times 0.65 \times 20\ \text{kHz} \times 1000\ \text{pF}}$$

$$= 12.2\ \text{k}\Omega \qquad \text{(use 12 k}\Omega\text{ standard value)}$$

$$R_2 = R_1 = 12\ \text{k}\Omega$$

−40 dB per decade stage (second order)

select $\quad C_3 = 1000\ \text{pF}$

from Eqs. 11-8 and 11-13,

$$R_4 = \sqrt{2}\, X_{C3} \text{ at } 0.65 f_c$$

so,

$$R_4 = \frac{\sqrt{2}}{2\pi 0.65 f_c C_3} = \frac{\sqrt{2}}{2\pi \times 0.65 \times 20\ \text{kHz} \times 1000\ \text{pF}}$$

$$= 17.3\ \text{k}\Omega \qquad \text{(use 17.4 k}\Omega\ \pm 1\%\text{ standard value)}$$

Eq. 11-9, $\quad C_2 = C_3 = 1000\ \text{pF}$

Eq. 11-10, $\quad R_3 = R_4/2 = 8.7\ \text{k}\Omega \qquad$ (use 8.66 kΩ $\pm 1\%$ standard value)

$$R_5 \approx R_4 = 17.4\ \text{k}\Omega \qquad \text{(use 18 k}\Omega\text{ standard value)}$$

Example 11-8

Estimate the highest signal frequency that can be passed by the circuit designed in Example 11-7.

Solution

From the 741 data sheet in Appendix 1-1 (enlarged frequency response in Fig. 5-13(a)), the op-amp unity gain cutoff frequency is $f_u \approx 800$ kHz. From Eq. 5-6, the high cutoff frequency of each op-amp is

$$f = \frac{f_u}{A_v}$$

$$\approx 800 \text{ kHz}$$

From Eq 11-12, the cutoff frequency for the two-stage circuit is

$$f_c = 0.65f = 0.65 \times 800 \text{ kHz}$$

$$= 520 \text{ kHz}$$

11-8 BANDPASS FILTERS

Multi-stage Bandpass Filter

A bandpass filter can be constructed simply by connecting low-pass and high-pass filters in cascade (see Fig. 11-14). For example, suppose a low-pass circuit with $f_c = 100$ kHz is cascaded with a high-pass circuit which has $f_c = 10$ kHz. The low-

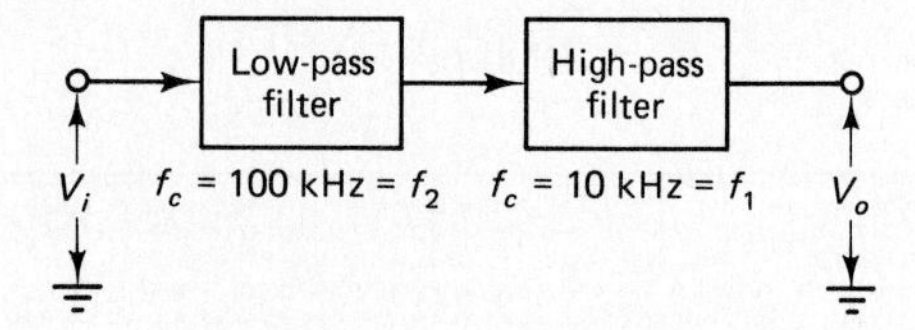

(a) Cascaded low-pass and high-pass filters

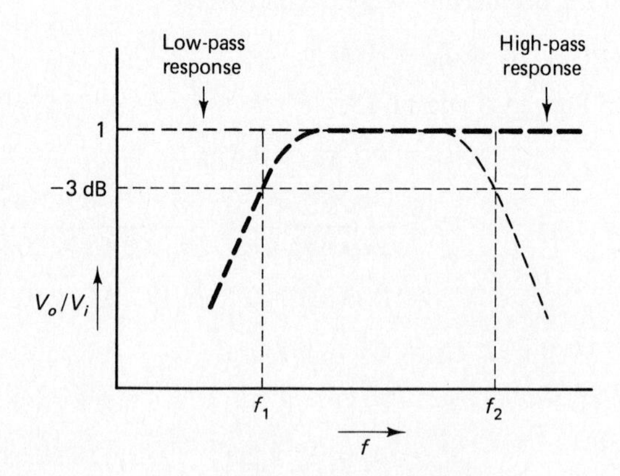

(b) Bandpass frequency response

Figure 11-14 Bandpass filter circuit consisting of low-pass and high-pass stages connected in cascade. The low-pass stage sets the high cutoff frequency f_2 and the high pass stage determines the low cutoff frequency f_1.

pass circuit will pass all frequencies up to 100 kHz, while the high-pass circuit will block all frequencies below 10 kHz. So, the combination gives a filter with a pass band from 10 kHz to 100 kHz.

Single-Stage First-Order Bandpass Filter

A single-stage bandpass filter is shown in Fig. 11-15. If the capacitors were not present, the circuit would look like an inverting amplifier. The capacitors are selected to have X_{C2} large enough to be neglected at low frequencies and X_{C1} small enough to be neglected at high frequencies.

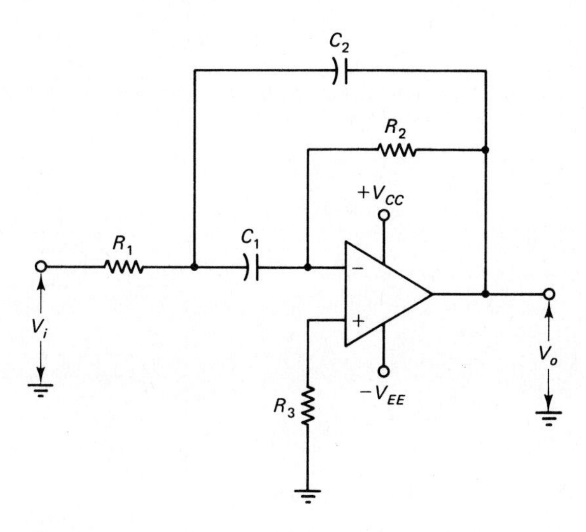

Figure 11-15 Single-stage bandpass filter circuit. For a wide pass band, the cutoff frequencies are set by making $X_{C1} = R_1$ at f_1 and at $X_{C2} = R_2$ at f_2.

At low frequencies, X_{C2} is so large that it can be eliminated from the low frequency equivalent circuit. As shown in Fig. 11-16(a), the circuit is an inverting amplifier with a voltage gain of

$$A_v = \frac{R_2}{Z_1} = \frac{R_2}{\sqrt{(R_1^2 + X_{C1}^2)}} \tag{11-14}$$

At signal frequencies in the pass band of the circuit, X_{C1} becomes much smaller than R_1, So, the circuit gain becomes

$$A_v \approx \frac{R_2}{R_1}$$

Equation 11-14 shows that for the voltage gain to be down by 3 dB (from the mid-frequency gain),

$$X_{C1} = R_1 \text{ at } f_1 \tag{11-15}$$

where f_1 is the low cutoff frequency.

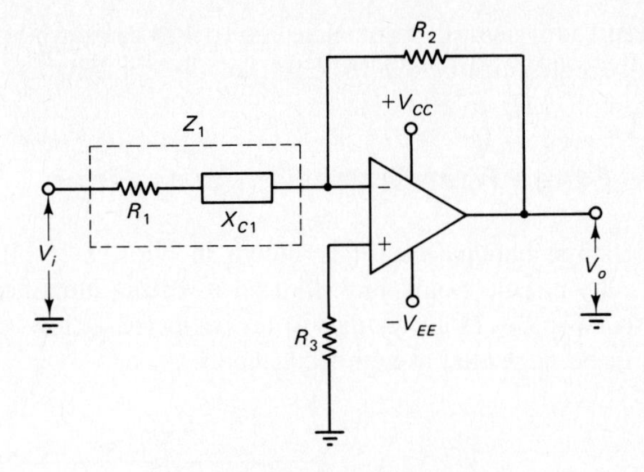

(a) Low frequency equivalent circuit

$$A_v = \frac{R_2}{Z_1}$$

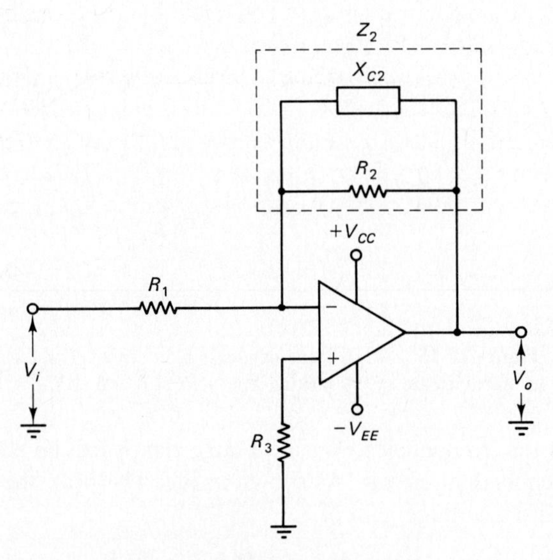

(b) High frequency equivalent circuit

$$A_v = \frac{Z_2}{R_1}$$

Figure 11-16 Low frequency and high frequency equivalent circuits for the single-stage bandpass filter. Capacitor C_2 can be neglected at low frequency and C_1 can be disregarded at high frequency.

At high frequencies, X_{C1} becomes so small compared to R_1 that it can be eliminated from the high-frequency equivalent circuit. But, X_{C2} is no longer large enough to be ignored. So, the high frequency equivalent circuit is as shown in Fig. 11-16(b). Once again, the circuit is an inverting amplifier and its voltage gain is

$$A_v = \frac{X_{C2} \parallel R_2}{R_1} = \frac{1}{R_1\sqrt{[(1/R_2)^2 + (1/X_{C2})^2]}} \quad (11\text{-}16)$$

At frequencies in the pass-band, X_{C2} is much larger than R_1. So, the circuit voltage gain is once again

$$A_v \approx \frac{R_2}{R_1}$$

From Eq. 11-16, the voltage gain is down by 3 dB from its mid-frequency value when

$$X_{C2} = R_2 \text{ at } f_2 \quad (11\text{-}17)$$

where f_2 is the high cutoff frequency.

It is seen that the circuit behaves as an inverting amplifier when the signal frequency is in the pass band, as a high-pass filter for low frequencies, and as a low-pass filter for high frequencies. The low cutoff frequency is determined by Eq. 11-15 and the high cutoff frequency is calculated from Eq. 11-17.

The normal design approach of selecting the resistors first is likely to give unacceptably small capacitance values whether bipolar or BIFET op-amps are used. So, it is usually best to commence by selecting the value of the smallest capacitor (C_2). Then R_2 can be determined from Eq. 11-17 and R_1 can be calculated in relation to R_2 for the desired voltage gain. The value of R_1 can be substituted into Eq. 11-15 to calculate C_1. Since direct current through R_1 is interrupted by C_1, resistor R_3 should be made equal to R_2 for resistance equality at the op-amp input terminals.

Example 11-9

Design a single-stage bandpass filter, as in Fig. 11-15, to have a voltage gain of 1 and a pass band from 300 Hz to 30 kHz.

Solution

Select $\quad C_2 = 1000 \text{ pF}$

From Eq. 11-17, $\quad X_{C2} = R_2 \text{ at } f_2$

or,

$$R_2 = \frac{1}{2\pi f_2 C_2} = \frac{1}{2\pi \times 30 \text{ kHz} \times 1000 \text{ pF}}$$

$$= 5.3 \text{ k}\Omega \qquad \text{(use 5.36 k}\Omega \pm 1\% \text{ standard value)}$$

$$R_3 \approx R_2 = 5.36 \text{ k}\Omega \qquad \text{(use 5.6 k}\Omega \text{ standard value)}$$

For $A_v = 1$ $\quad R_1 = R_2 = 5.36 \text{ k}\Omega$

From Eq. 11-15,

$$C_1 = \frac{1}{2\pi f_1 R_1} = \frac{1}{2\pi \times 300 \text{ Hz} \times 5.36 \text{ k}\Omega}$$

$$= 0.1\ \mu\text{F} \qquad \text{(standard value)}$$

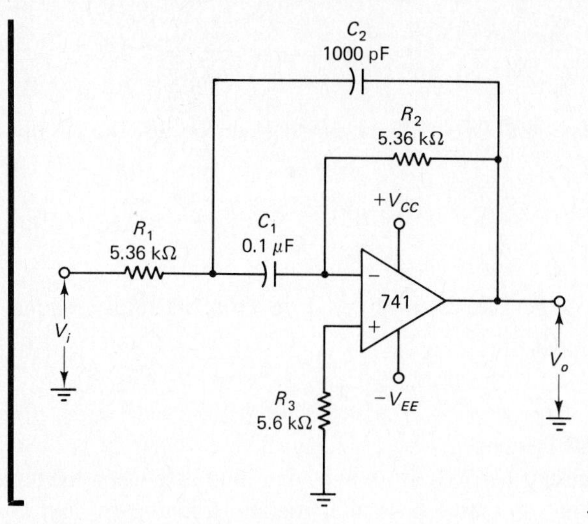

Figure 11-17 Bandpass filter designed in Example 11-9; $f_1 = 300$ Hz and $f_2 = 30$ kHz.

Wide-Band and Narrow-Band Bandpass Filters

The bandpass filter discussed above is essentially a wide-band circuit, and its typical frequency response is illustrated in Fig. 11-18(a). Narrow-band bandpass filters have the kind of frequency response shown in Fig. 11-18(b).

The bandwidth of the filter circuit is,

$$B = f_2 - f_1 \tag{11-18}$$

The circuit *Q factor* is the relationship of the center frequency f_o to the bandwidth.

$$Q = \frac{f_o}{B} \tag{11-19}$$

The Q factor is a *figure of merit* for a filter circuit. It defines the selectivity of the filter in passing the center frequency and rejecting other frequencies. Figure 11-18(b) shows that a filter with a Q of 10 has a much narrower bandwidth than a filter with Q equal to 1. The filter with the higher Q is the most selective of the two. Narrow-band filters are usually classified as those with a Q factor greater than 5, while wide-band circuits have a Q less than 5.

The center frequency of the filter can be determined from,

$$f_o = \sqrt{(f_1 f_2)} \tag{11-20}$$

For the wideband filter designed in Example 11-9,

$$f_o = \sqrt{(f_1 f_2)} = \sqrt{(300 \text{ Hz} \times 30 \text{ kHz})}$$
$$= 3 \text{ kHz}$$

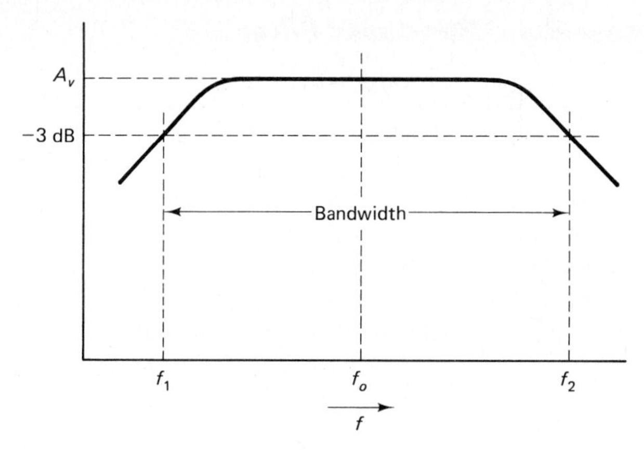

(a) Frequency response of wide-band bandpass filter

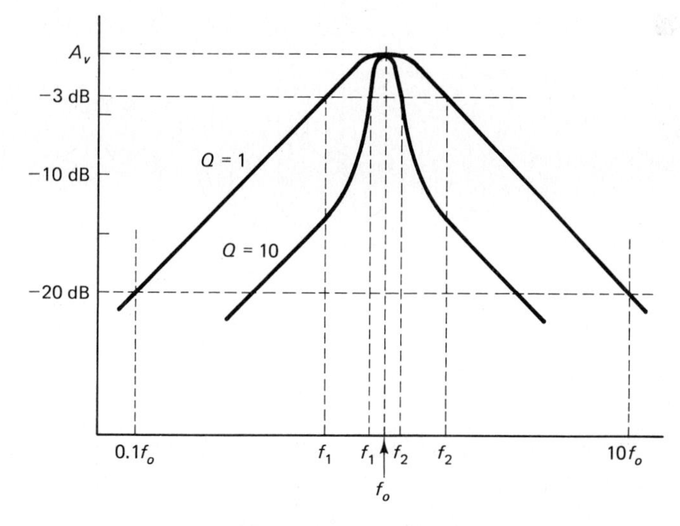

(b) Frequency response of narrow-band bandpass filters

Figure 11-18 Frequency response of wide-band and narrow-band bandpass filters. The circuit Q factor $[Q = f_o/(f_2 - f_1)]$ is very important in narrow-band filters.

$$B = f_2 - f_1 = 30 \text{ kHz} - 300 \text{ Hz}$$

$$= 29.7 \text{ kHz}$$

and,

$$Q = \frac{f_o}{B} = \frac{3 \text{ kHz}}{29.7 \text{ kHz}}$$

$$\approx 0.1$$

Narrow-Band Bandpass Filter

Reconsidering the circuit in Fig. 11-15, it is seen that, as f_1 and f_2 are brought closer together, C_2 will affect the low cutoff frequency, and C_1 will interfere with the high cutoff frequency. To counter this, another resistor (R_4) is included in the circuit, as shown in Fig. 11-19. This additional resistor makes the circuit analysis very complex, but it can be stated that R_4 attenuates the input signal at frequencies at which the impedance of capacitor C_1 is high. When the impedance of C_1 becomes very small compared to R_1, R_4 is in parallel with the op-amp input terminals, and in this position it has very little effect (see Section 5-9).

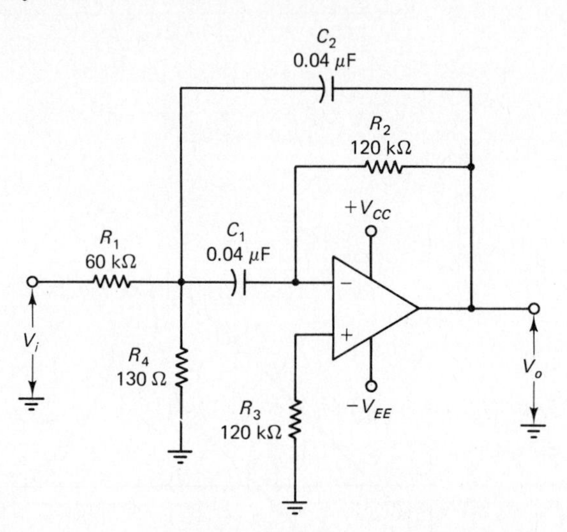

Figure 11-19 Single-stage bandpass filter with additional resistor (R_4) to improve the Q factor. Component values are calculated in Example 11-10; f_o = 1 kHz and B = 66 Hz.

The following equations relate the component values to the circuit performance. With $C_1 = C_2 = C$,

$$R_1 = \frac{R_2}{2} \tag{11-21}$$

$$R_2 = 2\,Q\,X_c \text{ at } f_o \tag{11-22}$$

$$R_4 = \frac{R_1}{2Q^2 - 1} \tag{11-23}$$

Example 11-10

Using a 741 op-amp, design a bandpass filter as in Figure 11-19. The center frequency is to be 1 kHz and the pass band is to be approximately ±33 Hz on each side of 1 kHz.

Solution

$$B = 33\text{ Hz} + 33\text{ Hz} = 66\text{ Hz}$$

$$\text{Eq. 11-19,} \qquad Q = \frac{f_o}{B} = \frac{1 \text{ kHz}}{66 \text{ Hz}}$$

$$\approx 15.2$$

$$R_2 = R_3 = 120 \text{ k}\Omega \qquad \text{(as in Example 11-2)}$$

$$\text{Eq. 11-22,} \qquad R_2 = 2\,Q\,X_c \text{ at } f_o$$

$$\text{which gives} \qquad C = \frac{2Q}{2\pi f_o R_2} = \frac{2 \times 15.2}{2\pi \times 1 \text{ kHz} \times 120 \text{ k}\Omega}$$

$$= 0.04\ \mu\text{F} \qquad \text{(standard value)}$$

$$C_1 = C_2 = C = 0.04\ \mu\text{F}$$

$$\text{Eq. 11-21,} \qquad R_1 = \frac{R_2}{2} = \frac{120 \text{ k}\Omega}{2}$$

$$= 60 \text{ k}\Omega \qquad \text{(use 60.4 k}\Omega \pm 1\% \text{ standard value)}$$

$$\text{Eq. 11-23,} \qquad R_4 = \frac{R_1}{2Q^2 - 1} = \frac{60 \text{ k}\Omega}{(2 \times 15.2^2) - 1}$$

$$= 130\ \Omega \qquad (\pm 1\% \text{ standard value})$$

11-9 STATE-VARIABLE BANDPASS FILTER

The *state-variable* filter circuit shown in Fig. 11-20 is also known as a *biquadratic* or *biquad* filter. This circuit can be designed to function as a low-pass, high-pass, or bandpass filter. As a bandpass circuit, it can have a very high Q factor and a very narrow bandwidth.

If resistor R_2 is ignored for the moment, A_1 together with R_3 and R_4 would appear to function as a noninverting amplifier. Op-amps A_2 and A_3 and their associated components are connected to function as integrating circuits. The input to A_1 is applied to its noninverting terminal, but its output is inverted by A_2. Consequently, A_1 and A_2 together with R_1 and R_2 constitute an inverting amplifier. For dc conditions, feedback via R_2 and R_1 keeps the output of A_2 at ground level when the input is zero. Similarly, feedback via R_3 and R_4 and the $A_2\,A_3$ loop, keeps the outputs of A_1 and A_3 close to ground level.

To understand the ac operation of the filter circuit, first consider the extremes of very low and very high signal frequency.

At very low frequency, capacitors C_1 and C_2 can be expected to have very high impedances compared to resistors R_5 and R_6 respectively. Thus, the stage gains of A_2 and A_3 will be very large. Also, the A_2 and A_3 integrator stages each introduce a 90° phase lag in the output from A_1. So, the voltage fed to the bottom of resistor R_4 will be in antiphase to the output from A_1 and will be amplified by the large stage gains of A_2 and A_3. This is negative feedback and it severely attenuates the output from A_1. Since the stage gain of A_3 is so large, the output of A_2 (which provides an input to

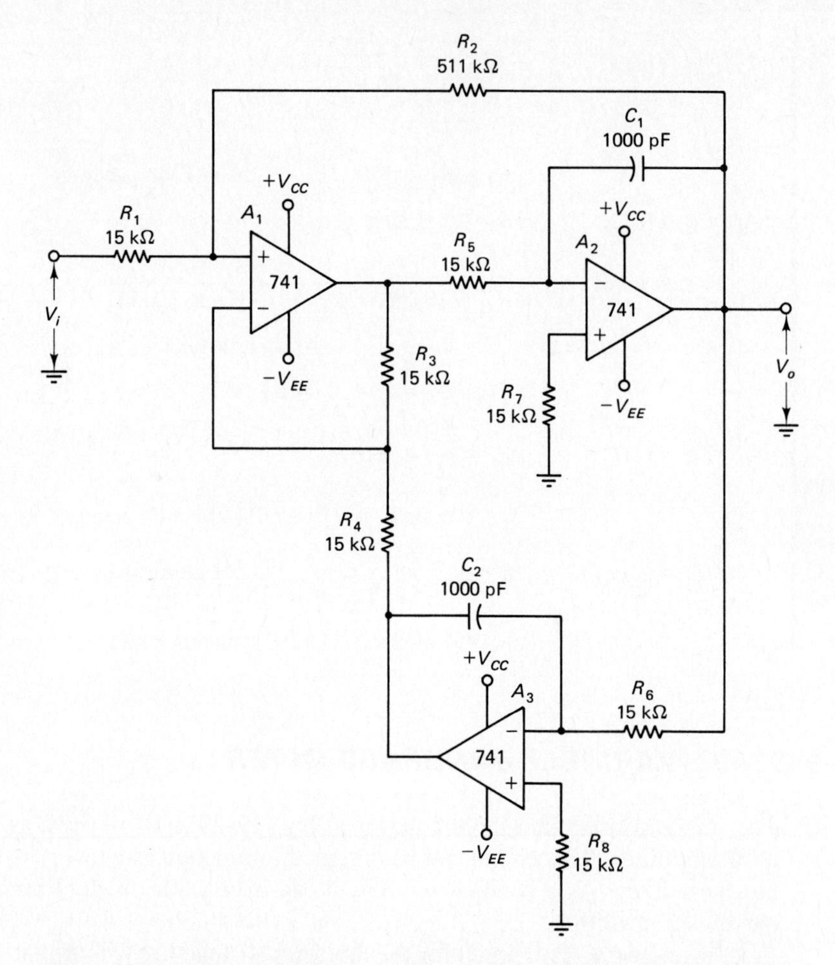

Figure 11-20 State-variable bandpass filter. At low and high frequencies, the A_2 and A_3 stages cause the output to be severely attenuated. At the center frequency, the A_2 and A_3 stages cancel the R_3R_4 feedback network and the voltage gain becomes $-R_2/R_1$. Component values are calculated in Example 11-11.

A_3) must be very small. So the circuit output (from A_2) is much smaller than the input at low frequencies. Since v_o is very much smaller than v_i, any feedback that might have occurred via resistor R_2 is of no consequence.

When the signal frequency is very high, the impedances of C_1 and C_2 become very small compared to resistors R_5 and R_6. Consequently, the A_2 and A_3 stages attenuate the output of A_1. Now, the stage gain of A_1 is $(R_3 + R_4)/R_4$. Usually, R_3 is made equal to R_4 to give an A_1 gain of 2. However, the output of A_1 is severely attenuated by the A_2 stage and once again, v_o is much smaller than v_i. Here again, feedback via resistor R_2 has no significant effect.

The circuit is designed so that at the center frequency of the bandwidth (f_0)

$$X_{C1} = R_5 \text{ and } X_{C2} = R_6$$

Thus, with $C_1 = C_2$ and $R_5 = R_6$,

$$f_o = \frac{1}{2\pi C_1 R_5} \tag{11-24}$$

This means the A_2 and A_3 stages each have a voltage gain of 1 at f_o, as well as each having a 90° phase shift. Now, the voltage fed to the bottom of R_4 is equal to and in antiphase with the output of A_1. Thus, the voltage at the junction of R_3 and R_4 remains constant and there is no negative feedback to restrict the voltage gain of A_1. So, A_1 has a very large gain from its noninverting input terminal to its output. A_2 has a gain of 1 and resistors R_2 and R_1 become effective as a feedback network.

As already explained, the input to A_1 is applied to its noninverting terminal and its output is inverted by A_2. So, A_1 and A_2 together with R_1 and R_2 are an inverting amplifier. The circuit voltage gain at the center of the pass band is then

$$A_v = -\frac{R_2}{R_1}$$

The analysis to determine the relationship between the center frequency and bandwidth of a state-variable filter is very complex. However, the result of that analysis is a very simple equation for Q,

$$Q = \frac{R_1 + R_2}{2R_1} \tag{11-25}$$

which gives

$$R_2 = R_1(2Q - 1) \tag{11-26}$$

Using the appropriate equations, the circuit can be easily designed to meet a given specification.

Circuit design begins with selection of a convenient capacitance value for C_1 and C_2. Then, R_5 and R_6 are calculated from Eq. 11-24. Since R_7 and R_8 are included only to equalize the resistances at the op-amp input terminals, they should be made equal to R_5 and R_6.

Resistors R_3 and R_4 are selected to give an A_1 stage gain of 2 (when R_2 is ineffective). So, R_3 equals R_4. These can also be made equal to the other resistor values. Furthermore, R_1 can be made equal to the common value of the other resistors, leaving R_2 to be determined from Eq. 11-26. Only R_5 and R_6 need to have precise resistance values in relation to C_1 and C_2. The other resistors (except for R_2) can be approximately equal to R_5 and R_6.

The circuit design procedure is seen to consist of only three steps: (1) Select a convenient capacitance value for C_1 and C_2; (2) determine the common resistance value for all resistors except R_2 from Eq. 11-24; (3) calculate R_2 from Eq. 11-26.

If R_5 and R_6 are variable resistors (or partially variable) and are adjusted together to maintain equality, the center frequency of the filter can be adjusted. An alternative to this is to make C_1 and C_2 adjustable.

Example 11-11

Design a state-variable bandpass filter to have $f_1 = 10.3$ kHz and $f_2 = 10.9$ kHz.

Solution

Select $C_1 = C_2 = 1000$ pF

From Eq. 11-20, $f_0 = \sqrt{(f_1 f_2)} = \sqrt{(10.3 \text{ kHz} \times 10.9 \text{ kHz})}$

$= 10.6$ kHz

From Eq. 11-24, $R_5 = R_6 = \dfrac{1}{2\pi C_1 f_o} = \dfrac{1}{2\pi \times 1000 \text{ pF} \times 10.6 \text{ kHz}}$

$= 15.01 \text{ k}\Omega$ (use 15 kΩ standard value)

$R_1 = R_3 = R_4 = R_7 = R_8 \approx R_5 = 15 \text{ k}\Omega$

From Eqs. 11-18 and 11-19, $Q = \dfrac{f_o}{f_2 - f_1} = \dfrac{10.6 \text{ kHz}}{10.9 \text{ kHz} - 10.3 \text{ kHz}}$

$= 17.7$

Eq. 11-26, $R_2 = R_1(2Q - 1) = 15 \text{ k}\Omega[(2 \times 17.7) - 1]$

$= 515 \text{ k}\Omega$ (use 511 kΩ ±1%)

11-10 BANDSTOP FILTER

The function of a *bandstop filter,* also known as a *band-reject filter,* is to block a band of signal frequencies. This is the inverse of the bandpass filter function. A bandstop filter with a very narrow stop band is sometimes known as a *notch* filter.

Low-Pass and High-Pass Filters as Bandstop Filter

Figure 11-21(a) shows how a bandstop filter can be constructed by parallel-connecting a low-pass filter and a high-pass filter. The circuit inputs can be connected in

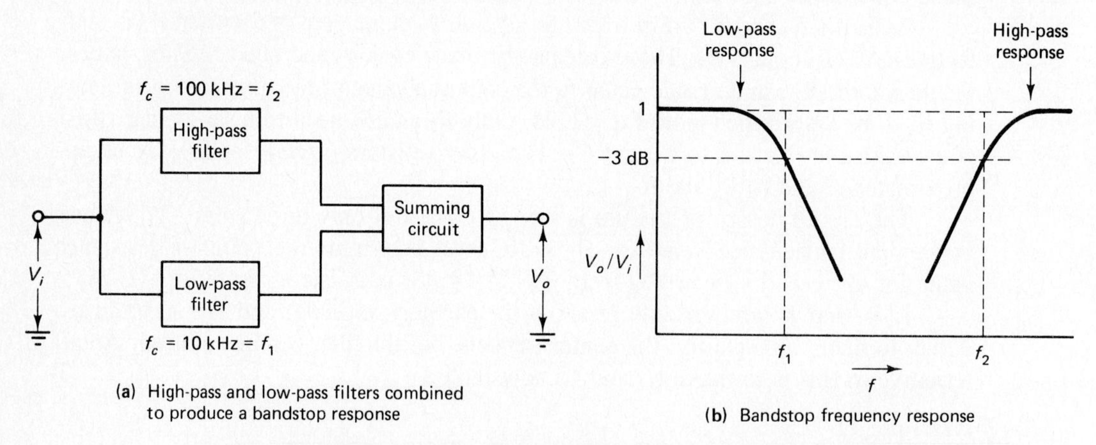

Figure 11-21 Bandstop filter consisting of parallel-connected low-pass and high-pass circuits with a summing circuit to combine their outputs. The high-pass stage determines f_2, and the low-pass stage sets f_1.

parallel without any problem, but the outputs must be applied to a summing circuit, as illustrated, to avoid one output overloading the other. Suppose that the low-pass circuit has a cutoff frequency of 10 kHz and that the cutoff frequency of the high-pass circuit is 100 kHz. The low-pass circuit will block signal frequencies above 10 kHz, while the high-pass circuit will block frequencies below 100 kHz. The result is a stop band of 10 kHz to 100 kHz, as illustrated in Fig. 11-21(b).

Note that the frequency limits of the stop band are those frequencies at which the signal is −3 dB from its normal output level. The low-pass and high-pass filters can be first-order, second-order, or higher, for any desired fall-off rate of the band-stop frequency response.

Bandpass and Summing Circuit as Bandstop

Another method of creating a bandstop filter is to sum the output of a bandpass filter with its own input signal, as illustrated in Fig. 11-22(a). In this case, the bandpass filter must have a voltage gain of −1, that is, its output must be the inverse of the input signal. Alternatively, the signal might be amplified to the same level as the bandpass output and inverted if necesary.

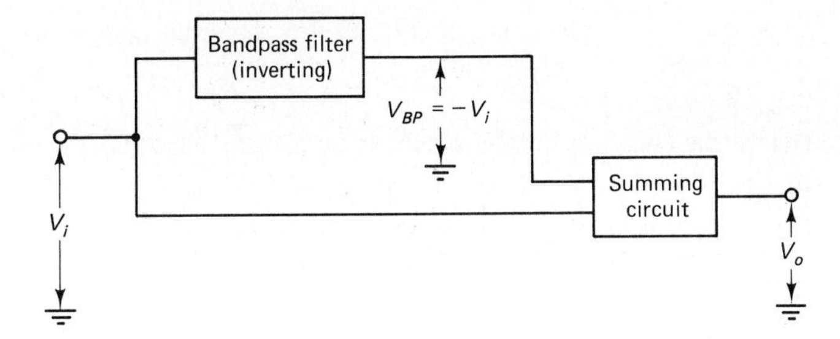

(a) Bandpass filter and summing circuit as bandstop filter

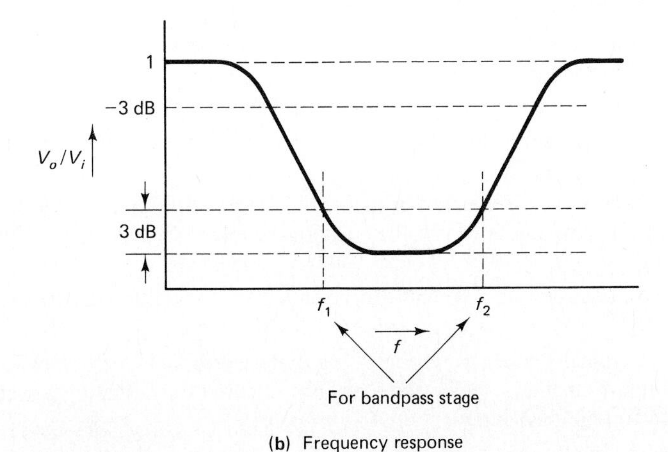

(b) Frequency response

Figure 11-22 Bandstop filter made up of a bandpass stage and a summing circuit. The output is 3 dB above the attenuated level at the cutoff frequencies of the band-pass circuit.

During the passband of the bandpass circuit, its output (v_{BP}) equals $-v_i$. Thus, the two inputs to the summing circuit cancel and the output voltage is zero, or is attenuated to some low level. Above and below the passband of the bandpass circuit, v_{BP} is very much smaller than v_i. So, the summing circuit output equals v_i and the combination has the bandstop frequency response shown in Fig. 11-22(b).

The frequency response of the bandpass circuit has v_{BP} equal to $0.707v_i$ at f_1 and f_2. This gives a summing circuit output at these frequencies of,

$$v_o = v_i - 0.707v_i$$

$$= 0.293v_i$$

This is a 3 dB increase from the attenuated level, as illustrated in Fig. 11-22(b). To produce a bandstop circuit which has the signal attenuated by 3 dB at the cutoff frequencies, as in Fig. 11-21(b), the bandpass circuit must be designed to have an output of $0.293v_i$ at f_1 and f_2.

REVIEW QUESTIONS

11-1. Draw an all-pass phase lag circuit. Sketch the input and output waveforms and the typical frequency response, and explain the circuit operation. Show how the phase lag may be made adjustable.

11-2. Derive the equation for phase shift in an all-pass phase lag circuit.

11-3. Briefly discuss the procedure for designing an all-pass phase lag circuit.

11-4. Draw an all-pass phase lead circuit. Sketch the input and output waveforms and the typical frequency response and explain the circuit operation.

11-5. Derive the equation for phase shift in an all-pass phase lead circuit.

11-6. Sketch the circuits of first-order low-pass and first-order high-pass active filters. Also, sketch the typical frequency response for each circuit, and briefly explain the operation of each filter.

11-7. Write the equation for the voltage gain of a first-order low-pass active filter, and briefly discuss the circuit design procedure.

11-8. Discuss the circuit design procedure for a first-order high-pass active filter.

11-9. Draw the circuit diagram of a second-order active low-pass filter and explain its operation.

11-10. Sketch the typical frequency responses of Butterworth and Chebyshev second-order active low-pass filters. Write the equations involved in the design of the Butterworth circuit.

11-11. Sketch the circuit of a second-order active high-pass filter. Briefly explain its operation.

11-12. Sketch typical frequency responses for Butterworth and Chebyshev second-order active high-pass filters. Write the equations required for designing a second-order Butterworth high-pass filter.

11-13. Draw the circuit of a third-order active low-pass filter. Briefly explain the circuit operation and discuss the design procedure.

11-14. Draw the circuit of a third-order active high-pass filter and briefly explain its operation. Discuss the design procedure for this circuit.

11-15. Show how a bandpass filter circuit can be constructed by the use of a low-pass filter and a high-pass filter. Sketch the expected frequency response and explain the bandpass filter operation.

11-16. Sketch the circuit of a single-stage bandpass filter. Explain the low-pass and high-pass operation of the circuit and briefly discuss the design procedure.

11-17. Discuss the differences between wide-band and narrow-band bandpass filters. Sketch typical frequency responses for each. Write the equations relating Q, B, f_1, and f_2.

11-18. Show how the circuit of a single-stage wide-band bandpass filter should be modified for narrow-band operation. Briefly explain.

11-19. Draw the circuit of a state-variable bandpass filter. Explain the operation of the circuit and briefly discuss the design procedure.

11-20. Show how a bandstop filter circuit can be constructed by the use of low-pass and high-pass filters. Sketch the expected frequency response, and briefly explain.

11-21. Show how a bandstop filter can be constructed by the use of a bandpass filter and a summing circuit. Sketch the expected frequency response, and discuss the relationship between bandpass and bandstop 3 dB frequencies.

PROBLEMS

11-1. Using a BIFET op-amp, design an all-pass circuit to introduce a phase lag adjustable from 70° to 100° into a 3 V, 3.3 kHz signal.

11-2. Design an all-pass phase lead circuit to give a phase lead, adjustable from 90° to 130°, when the input is a 2 V, 1.5 kHz signal. Use a bipolar op-amp.

11-3. A first-order active low-pass filter circuit with a cutoff frequency of 3 kHz is to be designed. First, design the circuit to use a bipolar op-amp. Then, redesign it to use a BIFET op-amp.

11-4. Modify each of the circuits designed for Problem 11-3 to make the cutoff frequency adjustable from 2 kHz to 4 kHz.

11-5. Design a second-order low-pass filter circuit to have a cutoff frequency of 5 kHz. Use a 741 op-amp.

11-6. A second-order low-pass filter has the following components: R_1 = 56 kΩ, R_2 = 56 kΩ, R_3 = 120 kΩ, C_1 = 600 pF, C_2 = 1200 pF. Calculate the cutoff frequency.

11-7. Using a BIFET op-amp, design a first-order active high-pass filter to have a cutoff frequency of 15 kHz. If the op-amp has a unity gain frequency of 1.2 MHz, estimate the highest signal frequency that will be passed.

11-8. A first-order high-pass filter using a 741 op-amp has R_1 = 56 kΩ, and C_1 = 600 pF. Calculate the circuit cutoff frequency and determine the highest signal frequency that can be passed.

11-9. Design a second-order high-pass filter circuit to have a cutoff frequency of 7 kHz. Use an LF353 op-amp and estimate the highest signal frequency that can be passed.

11-10. A second-order high-pass filter has the following components: R_1 = 28 kΩ, R_2 = 56 kΩ, C_1 = C_2 = 1000 pF. Calculate the circuit cutoff frequency.

11-11. Using 741 op-amps, design a third-order active low-pass filter to have a cutoff frequency of 18 kHz.

11-12. Using LM108 op-amps, design a third-order active high-pass filter circuit with a cutoff frequency of 8 kHz.

11-13. Estimate the highest signal frequency that will be passed by the filter designed for Problem 11-12.

11-14. A bandpass filter using low-pass and high-pass circuits as in Fig. 11-14, is to have a bandwidth from 400 Hz to 30 kHz. Using 741 op-amps, design suitable second-order stages.

11-15. Design a single-stage bandpass filter (Fig. 11-15) to have cutoff frequencies of 1 kHz and 50 kHz, and a voltage gain of 1.

11-16. A single-stage bandpass filter circuit as in Fig. 11-15, has the following components: $R_1 = R_2 = R_3 = 3.9\ \text{k}\Omega$, $C_1 = 0.12\ \mu\text{F}$, $C_2 = 600$ pF. Calculate its upper and lower cutoff frequencies.

11-17. A single-stage bandpass filter circuit, as in Fig. 11-19, is to be designed using a 715 op-amp. The center frequency is to be 3.3 kHz with the pass band approximately ± 50 Hz on each side. Determine suitable component values.

11-18. Redesign the bandpass filter in Example 11-10 to use a BIFET op-amp.

11-19. A state-variable bandpass filter has the following components: $R_1 = 12\ \text{k}\Omega$, $R_2 = 560\ \text{k}\Omega$, $R_3 = R_4 = 27\ \text{k}\Omega$, $R_5 = R_6 = R_7 = R_8 = 33\ \text{k}\Omega$, $C_1 = C_2 = 1200$ pF. Calculate the width of the pass band.

11-20. A bandpass filter is to be designed to pass signal frequencies ranging from 2 kHz to 2.5 kHz. Design a suitable state-variable circuit.

11-21. Design a state-variable bandpass filter to meet the specification in Problem 11-17.

11-22. The bandstop filter in Fig. 11-21 is to have $f_1 = 10$ kHz and $f_2 = 100$ kHz. The input signal amplitude is 1 V. Design the complete circuit, using first-order high-pass and low-pass filters.

11-23. The bandpass filter circuit designed in Example 11-9 is used in the bandstop circuit in Fig. 11-22. Determine the signal frequencies at which the output voltage is 3 dB down from its low-frequency and high-frequency level.

COMPUTER PROBLEMS

11-24. Write a computer program to solve the type of all-pass phase lag circuit design problem presented in Example 11-1.

11-25. Write a computer program to design a third-order low-pass filter circuit as in Fig. 11-12. Given: bipolar op-amps and cutoff frequency.

11-26. Write a computer program to determine the lower cutoff frequency of the third-order high-pass filter circuit in Fig. 11-13. Given all component values.

11-27. Write a computer program to design the type of bandpass filter circuit in Fig. 11-19. Given: cutoff frequency and bandwidth.

LABORATORY EXERCISES

11-1 Low-Pass Filter

1. Construct the second-order low-pass filter circuit shown in Fig. 11-7, as designed in Example 11-3.

2. Apply a ±1 V, 100 Hz sine wave input and monitor the input and output waveforms on an oscilloscope.
3. Increase the signal frequency until the output falls to approximately ±0.707 V. Note the frequency and compare it to the designed cutoff frequency.
4. Double the signal frequency, measure the level of the output and calculate the rate of output fall-off.

11-2 High-Pass Filter

1. Construct the second-order high-pass filter circuit shown in Fig. 11-11, as designed in Example 11-5. Include correct compensation for the 715 op-amp. (A 741 could be used with this circuit.)
2. Apply a ±1 V, 20 kHz sine wave input, and monitor the input and output waveforms on an oscilloscope.
3. Reduce the signal frequency until the output falls to approximately ±0.707 V. Note the frequency and compare it to the designed cutoff frequency.
4. Half the signal frequency, measure the level of the output, and calculate the rate of output fall-off.
5. Increase the signal frequency to discover the high cutoff frequency for the circuit. Compare it to the estimated high cutoff frequency.

11-3 Bandpass Filter

1. Construct the bandpass filter circuit shown in Fig. 11-17, as designed in Example 11-9.
2. Apply a ±1 V, 3 kHz sine wave input, and monitor the input and output waveforms on an oscilloscope.
3. Reduce the signal frequency until the output falls to approximately ±0.707 V. Note the frequency and compare it to the designed lower cutoff frequency.
4. Half the signal frequency, measure the level of the output, and calculate the rate of output fall-off.
5. Increase the signal frequency through 3 kHz until the output falls to approximately ±0.707 V. Note the frequency and compare it to the designed higher cutoff frequency.
6. Double the signal frequency, measure the level of the output, and calculate the rate of output fall-off.

11-4 State-Variable Bandpass Filter

1. Construct the state-variable bandpass filter circuit shown in Fig. 11-20, as designed in Example 11-11.
2. Apply a 10 kHz sine wave input, and monitor the input and output waveforms on an oscilloscope.
3. Carefully adjust the signal frequency to determine the filter center frequency (f_0). Adjust the signal amplitude to give a maximum output of ±1 V. Compare the measured and designed values of f_0.
4. Investigate the circuit bandwidth, and compare it to the designed bandwidth.

CHAPTER 12

DC Voltage Regulators

Objectives

- Explain the action of a dc voltage regulator.
- Define and write equations for *line regulation, load regulation,* and *ripple rejection,* for a dc voltage regulator.
- Sketch the following op-amp dc voltage regulator circuits and explain the operation of each: *voltage follower regulator, adjustable output regulator, precision voltage regulator, high-current voltage regulator, current-limiting circuit, foldback current-limiting circuit, plus-minus supply, tracking plus-minus supply.*
- Design each of the above circuits to meet a given specification.
- Analyze each of the above circuits to determine its performance.
- Sketch the basic circuit of a 723 integrated circuit voltage regulator and explain its operation.
- Show how a 723 IC regulator can be employed as a *positive voltage regulator* and as a *negative voltage regulator.*
- Perform all design calculations for the use of a 723 IC regulator.
- Sketch a regulator circuit using an LM217 IC regulator, explain its operation, and determine the required external component values and supply voltage.
- Sketch a regulator circuit using an LM340 IC regulator; discuss its performance.

INTRODUCTION

The dc supply voltage required by electronic circuits is normally derived by transforming and rectifying the standard domestic or industrial ac supply. The *raw dc* voltage produced in this way is not sufficiently stable for most purposes, and it usually contains an unacceptably large ripple waveform. DC voltage regulator circuits are employed to stabilize the voltage and to attenuate the ripple.

All voltage regulators employ a Zener diode as a stable reference voltage source. A Zener diode with a series resistor can be used as a simple low-current voltage regulator. The addition of a transistor allows a larger load current to be supplied. A vast improvement in the regulator performance results when an error amplifier is added to detect and amplify the difference between the regulator output and the reference voltage. IC operational amplifiers make ideal error amplifiers.

Complete dc voltage regulator circuits are available as integrated circuits. Some of these can be widely applied as different types of voltage regulators. Others are simply connected to a suitable supply to produce a fixed output voltage.

12-1 VOLTAGE REGULATOR BASICS

Regulator Action

A dc voltage regulator accepts an unregulated supply voltage and produces a stable dc output voltage. This is illustrated in Fig. 12-1(a), where the regulator supply voltage is identified V_s and the output is V_o. The unregulated supply is usually derived by transforming, rectifying, and smoothing an ac supply voltage, or *line voltage*. The regulator normally produces a reasonably constant output voltage regardless of variations in the line voltage, and it normally supplies widely varying load currents without a significant change in the output voltage.

As shown in Fig. 12-1(b), the unregulated supply has a substantial ripple voltage (V_{rs}) superimposed upon its average dc level. The peak-to-peak amplitude of the ripple voltage depends on the level of load current, becoming greatest at the highest load current, and smallest when the load current falls to zero. With an average dc level of 18 V, the ripple amplitude might typically be 2 V peak-to-peak. There is usually some output voltage ripple (V_{ro}), but this is normally very much smaller than the supply voltage ripple. Figure 12-1(b) also shows that the output voltage (V_o) is lower than the minimum level of the supply voltage. The maximum dc output voltage that can be provided by the regulator is normally at least 3 V lower than $V_{s(\text{min})}$. This is explained in Section 12.2.

Regulator Performance

Ideally, the output voltage of a regulator should remain perfectly constant regardless of variations in the supply voltage and load current. There should also be no ac ripple voltage at the output. Practical voltage regulators do have some output ripple, and the output voltage is affected to some extent by variations in load current and

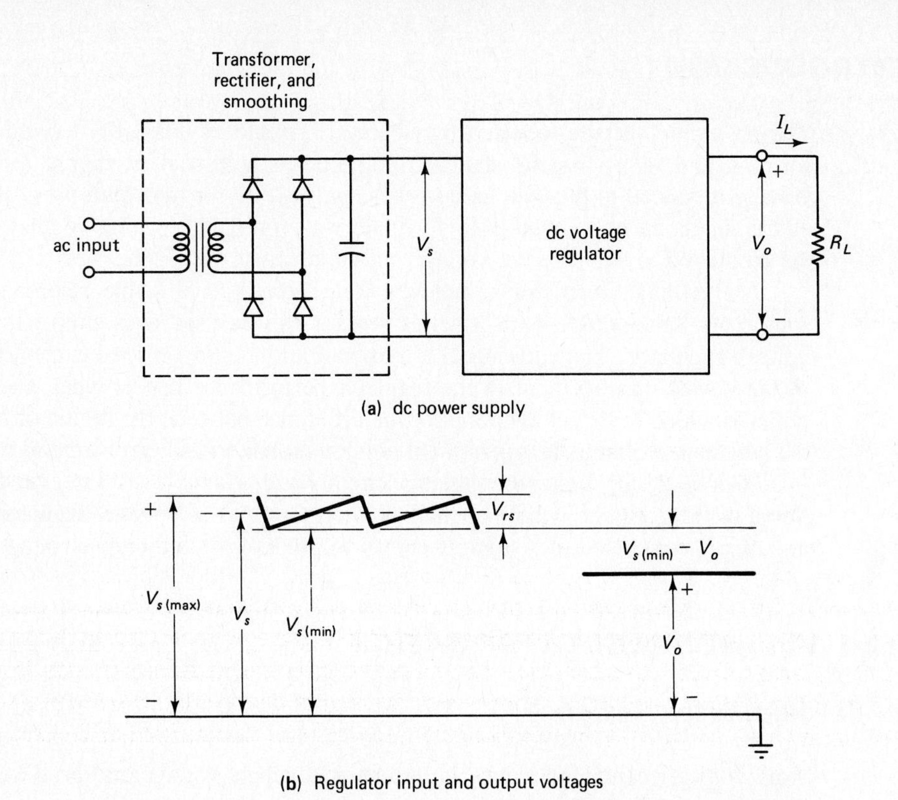

Figure 12-1 A dc voltage regulator is used to stabilize the supply voltage derived from a rectifier and smoothing circuit. The high ripple voltage present in the supply is also attenuated by the regulator.

line voltage. The performance of a voltage regulator is usually defined in terms of *line regulation, load regulation,* and *ripple rejection.*

The line regulation defines the variation in output voltage (ΔV_o) that occurs when the supply voltage (V_s) increases or decreases by a specified amount, usually 10%. The output voltage change is expressed as a percentage of the normal dc output voltage (V_o). Thus,

$$\text{Line regulation} = \frac{(\Delta V_o \text{ for } \pm 10\%\ V_s \text{ change}) \times 100\%}{V_o} \qquad (12\text{-}1)$$

The load regulation defines the regulator performance in relation to load current changes. When the load current changes from zero to full load the output voltage changes by an amount (ΔV_o). The load regulation is ΔV_o expressed as a percentage of the normal output voltage.

$$\text{Load regulation} = \frac{(\Delta V_o \text{ for } \Delta I_L = I_{L(\max)}) \times 100\%}{V_o} \qquad (12\text{-}2)$$

The ripple rejection is a measure of how much the voltage regulator attenuates the supply voltage ripple. It is usually expressed in decibels. With a supply ripple voltage of V_{rs}, and an output ripple of V_{ro},

$$\text{ripple rejection} = 20 \log \left[\frac{V_{rs}}{V_{ro}}\right] \text{ dB} \tag{12-3}$$

12-2 VOLTAGE FOLLOWER REGULATOR

Circuit Operation

A *voltage follower dc voltage regulator* is illustrated in Fig. 12-2. This is similar to the circuit in Fig. 6-1(b), except that the op-amp supply terminals are connected to the regulator supply voltage. Thus, the op-amp supply is $+V_s$ and ground. With the op-amp inverting input terminal connected to the output, V_o will remain close to the voltage at the noinverting input terminal; that is, V_o equals the Zener diode voltage V_z. The function of *series-pass transistor* Q_1 is to provide a load current larger than the maximum supplied by the op-amp alone. With a large load current, the op-amp output has only to supply the transistor base current. If the load current is less than 25 mA, Q_1 can be omitted and the op-amp can be connected directly as a voltage follower. C_1 is a large value capacitor, usually 50 μF to 100 μF, which is connected at the output to help supply fast demands on the regulator (transients), and to suppress any tendency of the regulator to oscillate.

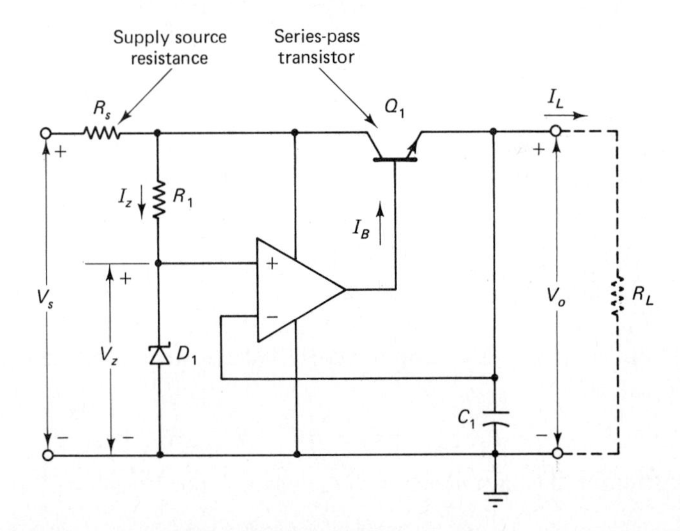

Figure 12-2 In a voltage follower regulator the op-amp is connected to function as a voltage follower with a stable input provided by a Zener diode. The output voltage is held constant at the Zener diode voltage, and the series-pass transistor supplies a larger load current than the op-amp can handle.

Design

Design procedure for the circuit in Fig. 12-2 is similar to that discussed in Section 6-1, except that, with the op-amp negative supply terminal grounded, the op-amp input terminals should be biased at approximately half the supply voltage. This is discussed in Section 4-8. The 741 op-amp, or any other internally compensated op-amp, is normally quite suitable for use in dc voltage regulators.

The series-pass transistor may have to be fitted with a *heat sink** to avoid damaging the device by overheating. There must typically be a minimum V_{CE} of 3 V across the series-pass transistor to keep it operational. This means that the supply voltage must always be at least 3 V greater than the regulated output. With less than 3 V, substantial ripple may be passed to the output, and the line and load regulations may be impaired.

Performance

When the supply voltage changes, the voltage drop across R_1 changes by the same amount (see Fig. 12-2). Consequently, the Zener diode current changes. This causes a slight change in V_z, and, because V_o equals V_z, the output voltage also changes. To determine the change in output voltage by a given change in the supply level, the Zener current variation is first calculated,

$$\Delta I_z = \frac{\Delta V_s}{R_1}$$

Using the dynamic impedance specified for the Zener diode (Z_z), ΔV_o is determined as,

$$\Delta V_o = \Delta V_z = \Delta I_z Z_z$$

or,

$$\Delta V_o = \frac{\Delta V_s Z_z}{R_1} \tag{12-4}$$

If ΔV_o is calculated for a 10% change in V_s, it can be substituted into Equation 12-1 to determine the line regulation of the regulator.

To calculate the load regulation, the supply source resistance R_s must be known (see Fig. 12-2). The value of R_s depends on the components of the rectifying and smoothing circuit. The drop in supply voltage when the load current goes from zero to $I_{L(\max)}$ is,

$$\Delta V_s = I_{L(\max)} R_s$$

Substituting this as a supply voltage change into Equation 12-4,

$$\Delta V_o = \frac{I_{L(\max)} R_s Z_z}{R_1} \tag{12-5}$$

*David A. Bell, *Electronic Devices and Circuits,* 3rd ed. (Englewood Cliffs NJ: Prentice-Hall, Inc., 1986) pp. 196–202.

This output voltage change, due to the load current change, can now be substituted into Equation 12-2 to calculate the circuit load regulation.

The supply ripple voltage (V_{rs}) can be treated like a supply voltage change to determine the output ripple voltage. Substituting V_{rs} for ΔV_s and V_{ro} for ΔV_o in Equation 12-4, the output ripple voltage is,

$$V_{ro} = \frac{V_{rs} Z_z}{R_1} \qquad (12\text{-}6)$$

The V_{rs} and V_{ro} values can be substituted into Equation 12-3 to determine the ripple rejection in decibels.

From Equations 12-4, 12-5, and 12-6 it would seem that the operational amplifier has no effect on the regulator performance. However, if the op-amp were not present, the transistor base-emitter voltage (V_{BE}), and the changes in V_{BE}, would have to be taken into account.

Example 12-1

The dc voltage source designed in Example 6-1 has $V_s = V_{cc} = 12$ V, $V_o = 6.3$ V, $R_1 = 270\ \Omega$. D_1 is a 1N753 Zener diode and $I_{L(\max)} = 42$ mA. If the supply source resistance is 25 Ω, determine the line regulation, load regulation, and ripple rejection for the circuit.

Solution

From the 1N753 data sheet, $Z_z = 7\ \Omega$

A 10% change in V_s is, $\Delta V_s = 10\%$ of 12 V

$= 1.2$ V

Eq. 12-4, $$\Delta V_o = \frac{\Delta V_s Z_z}{R_1} = \frac{1.2\text{ V} \times 7\ \Omega}{270\ \Omega}$$

$= 31$ mV

Eq. 12-1, $$\text{Line regulation} = \frac{(\Delta V_o \text{ for } \pm 10\%\ V_s \text{ change}) \times 100\%}{V_o}$$

$$= \frac{31\text{ mV} \times 100\%}{6.3\text{ V}}$$

$\approx 0.5\%$

Eq. 12-5, $$\Delta V_o = \frac{I_{L(\max)} R_s Z_z}{R_1} = \frac{42\text{ mA} \times 25\ \Omega \times 7\ \Omega}{270\ \Omega}$$

$= 27$ mV

Eq. 12-2, $$\text{Load regulation} = \frac{(\Delta V_o \text{ for } \Delta I_L = I_{L(\max)}) \times 100\%}{V_0}$$

$$= \frac{27\text{ mV} \times 100\%}{6.3\text{ V}}$$

$\approx 0.4\%$

Eq. 12-6,
$$V_{ro} = \frac{V_{rs} Z_z}{R_1} = \frac{V_{rs} \times 7\ \Omega}{270\ \Omega}$$
$$= 25.9 \times 10^{-3} V_{rs}$$

Eq. 12-3,
$$\text{ripple rejection} = 20 \log \left[\frac{V_{rs}}{V_{ro}}\right] \text{dB} = 20 \log \left[\frac{V_{rs}}{25.9 \times 10^{-3} V_{rs}}\right]$$
$$= 31.7 \text{ dB}$$

12-3 ADJUSTABLE OUTPUT REGULATOR

Circuit Operation

Figure 12-3(a) shows how the simple voltage follower regulator can be modified to produce an output voltage which is greater than the zener diode voltage. In this case, the voltage across resistor R_3 is always going to equal the Zener diode voltage. If V_{R3} were to become greater than V_z, the op-amp output would fall, thus reducing V_o until V_{R3} again equals V_z. If V_{R3} drops below the level of V_z, the op-amp output moves in a positive direction until V_{R3} equals V_z once again. Because there is a precise relationship between V_o and V_{R3} defined by R_2 and R_3, the output voltage is maintained constant by keeping V_{R3} equal to V_z. When used in a voltage regulator, the op-amp is referred to as an *error amplifier* because it amplifies any error in the output to keep the output stabilized at the desired level.

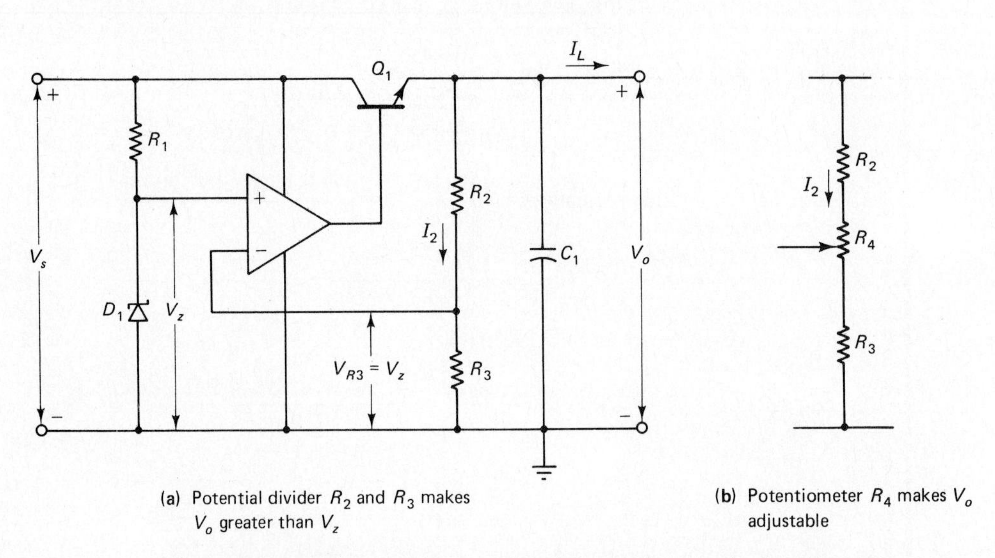

(a) Potential divider R_2 and R_3 makes V_o greater than V_z

(b) Potentiometer R_4 makes V_o adjustable

Figure 12-3 Including a potential divider (R_2 and R_3) in the simple voltage follower regulator circuit permits the output voltage to be larger than the Zener reference voltage; $V_o = V_z(R_2 + R_3)/R_3$. The use of a potentiometer (R_4) facilities adjustment of the output voltage.

$$V_{R3} = V_z$$

and
$$V_o = \frac{V_z(R_2 + R_3)}{R_3} \tag{12-7}$$

The output voltage can be made adjustable by including potentiometer R_4 between R_2 and R_3, as shown in Fig. 12-3(b).

The design procedure for this circuit is the same as already discussed, with the addition of the potential divider design (R_2, R_3, and R_4).

Performance

Analysis of the circuit in Fig. 12-3 must take account of the potential divider. Just as V_z is multiplied by $(R_2 + R_3)/R_3$ to determine V_o, so too any variations in V_z must be multiplied by the same factor to determine the output voltage changes. Therefore, modifying Eq. 12-4,

$$\Delta V_o = \frac{\Delta V_s\, Z_z}{R_1} \times \frac{R_2 + R_3}{R_3} \tag{12-8}$$

All supply voltage changes, whether due to line voltage variations, load current changes, or ripple voltages, can be substituted into Eq. 12-8 to determine the effect on the output.

Example 12-2

Using a 741 op-amp, design a voltage regulator circuit as in Fig. 12-3(a) to produce an output of 12 V with a maximum load current of 50 mA.

Solution

$$V_{s(\min)} = V_o + 3\text{ V} = 12\text{ V} + 3\text{ V}$$
$$= 15\text{ V}$$

allowing $V_{rs} = 2$ V peak-to-peak,

$$V_s = V_{s(\min)} + \frac{V_{rs}}{2} = 15\text{ V} + 1\text{ V}$$
$$= 16\text{ V}$$

Supply voltage, $V_s = 16$ V with a 2 V (max) ripple superimposed

Let $V_z \approx \dfrac{V_s}{2} = \dfrac{16\text{ V}}{2}$

≈ 8 V (use a 1N756 Zener diode which has $V_z = 8.2$ V)

$I_z \approx 20$ mA

$$R_1 = \frac{V_s - V_z}{I_z} = \frac{16\text{ V} - 8.2\text{ V}}{20\text{ mA}}$$

$= 390\ \Omega$ (standard value)

$$I_2 \gg I_{B(max)}$$

Let $I_2 = 50\ \mu A$

$$R_2 = \frac{V_o - V_z}{I_2} = \frac{12\ V - 8.2V}{50\ \mu A}$$

$$= 76\ k\Omega \qquad \text{(use 68 k}\Omega\text{ standard value and recalculate } I_2\text{)}$$

$$I_2 = \frac{V_o - V_z}{R_2} = \frac{12\ V - 8.2\ V}{68\ k\Omega}$$

$$= 55.9\ \mu A$$

$$R_3 = \frac{V_z}{I_2} = \frac{8.2\ V}{55.9\ \mu A}$$

$$= 147\ k\Omega \qquad \text{(use 150 k}\Omega\text{ standard value)}$$

Select $C_1 = 50\ \mu F$

Q_1 specification,

$$V_{CE(max)} = V_{s(max)} = V_s + V_{rs}/2$$

$$= 16\ V + 2\ V/2 = 17\ V$$

$$I_E = I_L = 50\ mA$$

$$P = V_{CE} \times I_L = (V_s - V_o) \times I_L$$

$$= (16\ V - 12\ V) \times 50\ mA$$

$$= 200\ mW$$

A 2N718 is a suitable device

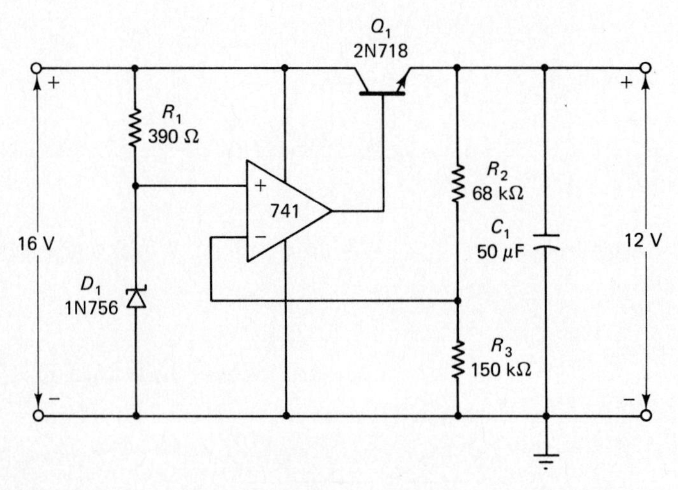

Figure 12-4 Regulator circuit designed in Example 12-2. $V_o = 12$ V, $I_{L(max)} =$ 50 mA.

Example 12-3

Analyze the dc voltage regulator designed in Example 12-2 to determine the line regulation, load regulation, and ripple rejection. Assume that the supply source resistance is 10 Ω.

Solution

From the 1N756 data sheet, $Z_z = 8\ \Omega$

A 10% change in V_s is $\Delta V_s = 10\%\text{ of }16\text{ V}$

$$= 1.6\text{ V}$$

Eq. 12-8,
$$\Delta V_o = \frac{\Delta V_s\, Z_z\,(R_2 + R_3)}{R_1 R_3} = \frac{1.6\text{ V} \times 8\ \Omega\,(68\text{ k}\Omega + 150\text{ k}\Omega)}{390\ \Omega \times 150\text{ k}\Omega}$$

$$\approx 48\text{ mV}$$

Eq. 12-1,
$$\text{Line regulation} = \frac{(\Delta V_o\text{ for }\pm 10\%\ V_s\text{ change}) \times 100\%}{V_o}$$

$$= \frac{48\text{ mV} \times 100\%}{12\text{ V}}$$

$$= 0.4\%$$

For an I_L change of 50 mA,

$$\Delta V_s = I_L R_s = 50\text{ mA} \times 10\ \Omega$$

$$= 0.5\text{ V}$$

Eq. 12-8,
$$\Delta V_o = \frac{\Delta V_s\, Z_z\,(R_2 + R_3)}{R_1 R_3} = \frac{0.5\text{ V} \times 8\ \Omega\,(68\text{ k}\Omega + 150\text{ k}\Omega)}{390\ \Omega \times 150\text{ k}\Omega}$$

$$\approx 15\text{ mV}$$

Eq. 12-2,
$$\text{Load regulation} = \frac{(\Delta V_o\text{ for }\Delta I_L = I_{L(\max)}) \times 100\%}{V_o}$$

$$= \frac{15\text{ mV} \times 100\%}{12\text{ V}}$$

$$= 0.125\%$$

Substituting for V_{rs} into Eq. 12-8,

$$V_{ro} = \frac{V_{rs} Z_z\,(R_2 + R_3)}{R_1 R_3} = \frac{V_{rs} \times 8\ \Omega\,(68\text{ k}\Omega + 150\text{ k}\Omega)}{390\ \Omega \times 150\text{ k}\Omega}$$

$$= 28.9 \times 10^{-3} V_{rs}$$

Eq. 12-3,
$$\text{ripple rejection} = 20\log\left[\frac{V_{rs}}{V_{ro}}\right]\text{dB} = 20\log\left[\frac{V_{rs}}{28.9 \times 10^{-3} V_{rs}}\right]$$

$$\approx 31\text{ dB}$$

12-4 PRECISION VOLTAGE REGULATOR

The regulator circuit in Fig. 12-3 has the disadvantage that its Zener diode circuit is connected to the supply voltage. Thus, as already found in the circuit analysis, each change in supply voltage causes a small change in the Zener voltage which results in an output voltage change. The regulator shown in Fig. 12-5 has the Zener diode circuit connected to the (stabilized) output voltage, otherwise the circuit is identical to that in Fig. 12-3. Now the supply voltage variations have virtually no effect on the Zener diode voltage, so the output voltage is a more stable quantity.

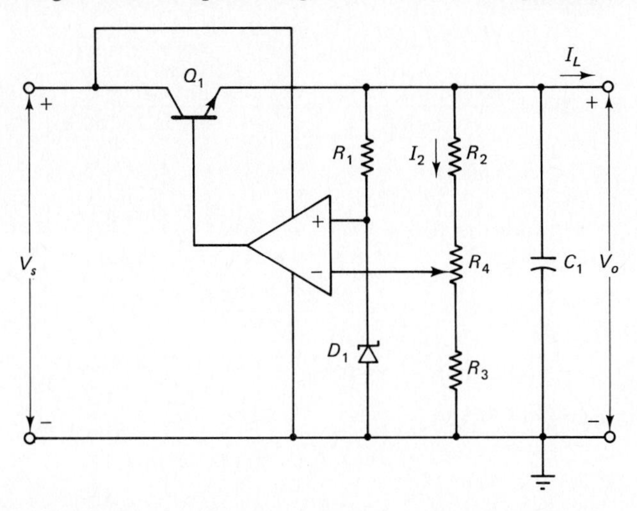

Figure 12-5 Precision voltage regulator with the Zener diode circuit connected to the regulator output. Since the Zener voltage is now unaffected by changes in the supply, the output voltage is much more stable.

The op-amp output (connected to Q_1 base) in Fig. 12-5 must be more positive than the regulated output voltage. This would not be possible if the op-amp positive supply terminal were connected to the regulated output. Therefore, the positive supply terminal of the op-amp remains connected to the positive (unregulated) supply voltage. When the supply voltage varies, it tends to change the op-amp output voltage. But this is corrected by a small change in the regulated output, which is attenuated by the potential divider and amplified by the open circuit gain (M) of the op-amp. For a supply voltage change of ΔV_s, the output changes by

$$\Delta V_o = \frac{\Delta V_s}{M} \times \frac{R_2 + R_3 + R_4}{R_3} \tag{12-9}$$

All supply voltage changes, whether due to line, load, or ripple effects, are attenuated as in Eq. 12-9.

The design procedure for the circuit in Fig. 12-5 is the same as already discussed, except the voltage across R_1 is now $(V_o - V_z)$ instead of $(V_s - V_z)$.

When a large output current is demanded from a voltage regulator, the base current to Q_1 may be too large to be supplied by the op-amp. For example, suppose that the output current is 600 mA and that Q_1 has a current gain factor (h_{FE}) of 20. The transistor base current to be supplied by the op-amp would be

$$I_B = \frac{I_L}{h_{FE}} = \frac{600\text{ mA}}{20}$$

$$= 30\text{ mA}$$

To further reduce this current so that it is not too large for the op-amp, another transistor (Q_2) is connected at the base of Q_1. So Q_1 and Q_2 form a *Darlington circuit,* as illustrated in Fig. 12-6. Now the op-amp output current is

$$I_{B2} = \frac{I_L}{h_{FE1}\, h_{FE2}} \tag{12-10}$$

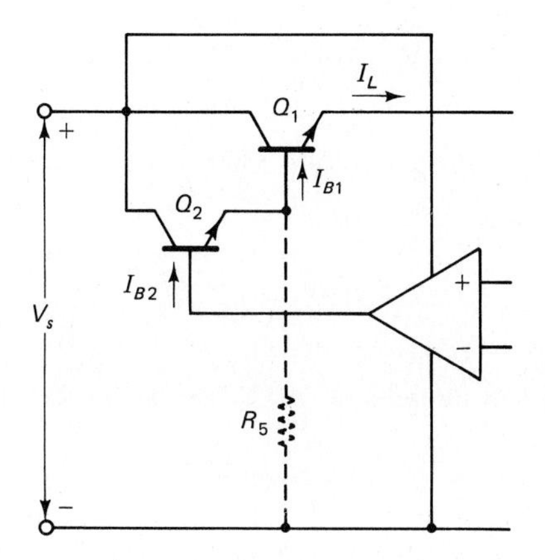

Figure 12-6 An additional transistor (Q_2) is usually Darlington-connected with the series-pass transistor when the regulator has to supply a large output current. This reduces the Q_1 base current to a level (I_{B2}) that can be supplied by the op-amp.

When the load current is zero, the base current of Q_1 can be very low. This is also the emitter current for Q_2 and it may not be large enough to keep Q_2 operating satisfactorily. In this case, resistor R_5 should be included at the emitter terminal of Q_2, as illustrated. The resistance of R_5 is determined as,

$$R_5 = \frac{V_o + V_{BE1}}{I_{E2(\min)}} \tag{12-11}$$

Example 12-4

Design a dc voltage regulator to have an output voltage adjustable from 10 V to 15 V, and a maximum output current of 400 mA. Use the circuit in Fig. 12-5, with the modification shown in Fig. 12-6.

Solution

Supply voltage

$$V_{s(\min)} = V_{o(\max)} + 3\text{ V} = 15\text{ V} + 3\text{ V}$$

$$= 18\text{ V}$$

Allowing a supply voltage ripple of $V_{rs} = 3$ V peak-to-peak,

$$V_s = V_{s(\min)} + \frac{V_{rs}}{2} = 18\text{ V} + 1.5\text{ V}$$

$$= 19.5\text{ V}$$

$$V_s = 19.5\text{ V with a 3 V (max) ripple superimposed}$$

Zener circuit

Let $\quad V_z \approx \frac{V_o}{2} = \frac{15\text{ V}}{2}$

≈ 7.5 V (use a 1N755 Zener diode which has $V_z = 7.5$ V)

$$I_z \approx 20\text{ mA}$$

$$R_1 = \frac{V_o - V_z}{I_z} = \frac{15\text{ V} - 7.5\text{ V}}{20\text{ mA}}$$

$= 375\ \Omega$ (use 330 Ω standard value)

Potential divider

$$I_2 \gg I_{B(\max)}$$

Let $\quad I_2 = 50\ \mu\text{A}$

When V_o is at its minimum of 10 V, the moving contact is at the top of R_4,

$$R_2 = \frac{V_{o(\min)} - V_z}{I_2} = \frac{10\text{ V} - 7.5\text{ V}}{50\ \mu\text{A}}$$

$= 50$ kΩ (use 47 kΩ standard value and recalculate I_2)

$$I_2 = \frac{10\text{ V} - 7.5\text{ V}}{47\text{ k}\Omega} = 53.2\ \mu\text{A}$$

$$R_3 + R_4 = \frac{V_z}{I_2} = \frac{7.5\text{ V}}{53.2\ \mu\text{A}}$$

$$= 141\text{ k}\Omega$$

When V_o is at its maximum of 15 V, the moving contact is at the bottom of R_4.

$$I_2 = \frac{V_o}{R_2 + R_3 + R_4} = \frac{15\text{ V}}{47\text{ k}\Omega + 141\text{ k}\Omega}$$

$$= 79.8\ \mu\text{A}$$

$$R_3 = \frac{V_z}{I_2} = \frac{7.5\text{ V}}{79.8\ \mu\text{A}}$$

≈ 94 kΩ (use 100 kΩ)

$$R_4 = (R_3 + R_4) - R_3 = 141\text{ k}\Omega - 100\text{ k}\Omega$$

$= 41$ kΩ (use 50 kΩ standard value potentiometer)

Capacitor

$$C_1 = 100\ \mu\text{F}$$

***Q_1* specification,**

$$V_{CE(\max)} = V_{s(\max)} = V_s + V_{rs}/2$$
$$= 19.5\text{ V} + 3\text{ V}/2$$
$$= 21\text{ V}$$
$$I_E = I_L = 400\text{ mA}$$
$$P = V_{CE}\, I_L = (V_s - V_o) \times I_L$$
$$= (19.5\text{ V} - 10\text{ V}) \times 400\text{ mA}$$
$$= 3.8\text{ W}$$

A 2N3055 is a suitable device

***Q_2* specification**

$$V_{CE(\max)} = V_{s(\max)} = V_s + V_{rs}/2$$
$$= 19.5\text{ V} + 3\text{ V}/2$$
$$= 21\text{ V}$$
$$I_E = I_L/h_{FE1} = \frac{400\text{ mA}}{20} \text{ (using } h_{FE1} = 20 \text{ typically for } Q_1\text{)}$$
$$= 20\text{ mA}$$
$$P = V_{CE} I_L \simeq (V_s - V_o) \times I_L$$
$$= (19.5\text{ V} - 10\text{ V}) \times 20\text{ mA}$$
$$= 190\text{ mW}$$

A 2N3904 is a suitable device

***R_5* calculation**

Let $I_{E2(\min)} = 0.5\text{ mA}$

Eq. 12-11, $$R_5 = \frac{V_{o(\min)} + V_{BE1}}{I_{E2(\min)}} = \frac{10\text{ V} + 0.7\text{ V}}{0.5\text{ mA}}$$
$$= 21.4\text{ k}\Omega \qquad \text{(use 18 k}\Omega\text{ standard value)}$$

Operational amplifier
Because I_2 was selected for a bipolar op-amp, either a bipolar or a BIFET op-amp can be used.

Supply voltage, $V_s = +19.5\text{ V}$

$$\text{Input voltage range} = \frac{V_s}{2} - V_z = \frac{19.5\text{ V}}{2} - 7.5\text{ V}$$
$$\approx 2.3\text{ V}$$

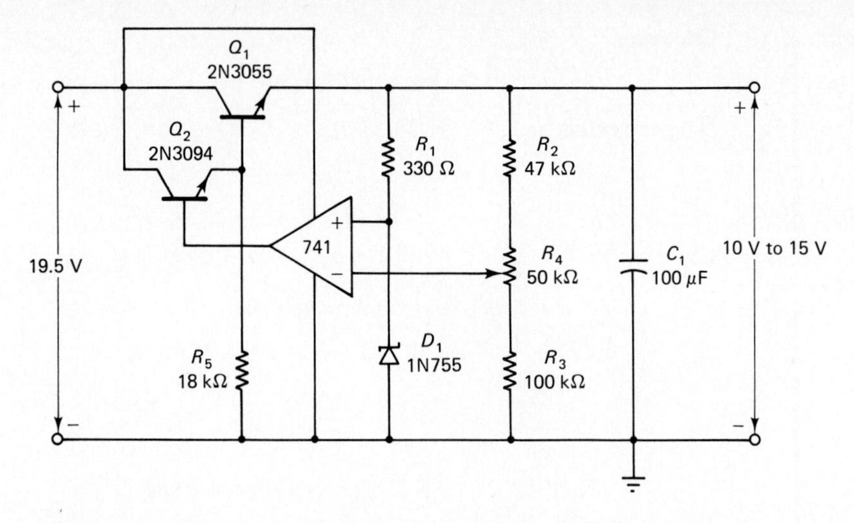

Figure 12-7 Regulator circuit designed in Example 12-4; V_o = 10 V to 15 V, $I_{L(max)}$ = 400 mA.

12-5 OUTPUT CURRENT LIMITING

Short-Circuit Protection

Power supplies used in laboratories are subject to overloads and short-circuits. *Short-circuit protection,* or *current-limiting circuits,* are necessary in such equipment to prevent destruction of components when an overload occurs. Transistor Q_3 and resistors R_6 and R_7 in Fig. 12-8(a) constitute a current-limiting circuit. When the load current I_L flowing through resistor R_6 is at a low or normal level, the voltage drop V_{R6} is not large enough to forward bias the base-emitter junction of Q_3. In this case, Q_3 has no effect on the regulator performance. When the current reaches a predetermined limit, V_{R6} biases Q_3 *on.* Collector current I_{C3} flows through resistor R_7, producing a voltage drop which drives the regulator output voltage down to near zero. The output terminal of the op-amp goes up to its positive saturation level, as it attempts to correct for the falling output voltage on overload. So, resistor R_7 must be included to allow I_{C3} to drop the voltage at the base of the series transistor to near ground level. The resistor of R_7 is calculated as

$$R_7 \approx \frac{V_s}{I_{C3}} \tag{12-12}$$

Resistor R_7 must not be so large that it has an excessive voltage drop across it when the regulator is supposed to be operating normally.

The voltage/current characteristic of the regulator with short-circuit protection is shown in Fig. 12-8(b). It is seen that output voltage V_o remains constant with load current increase up to $I_{L(max)}$. Beyond $I_{L(max)}$, V_o drops to zero and a short-circuit current (I_{SC}) flows. This is slightly greater than $I_{L(max)}$. Under these conditions, series-pass transistor Q_1 is carrying all of the short-circuit current and has virtually all of

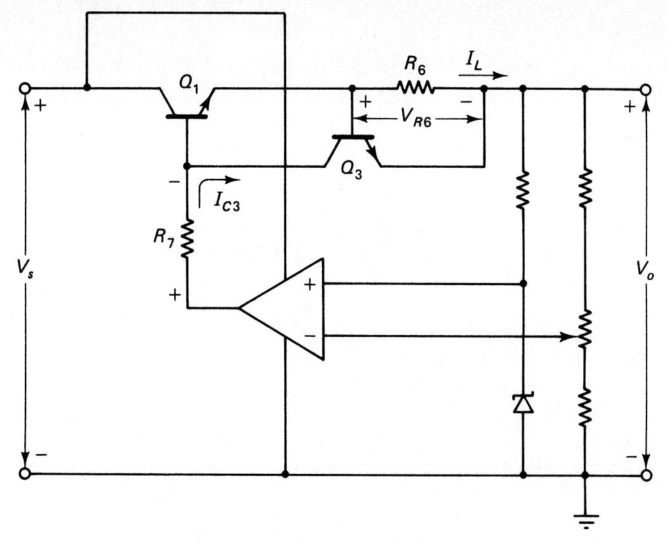

(a) Voltage regulator with a current limiting circuit

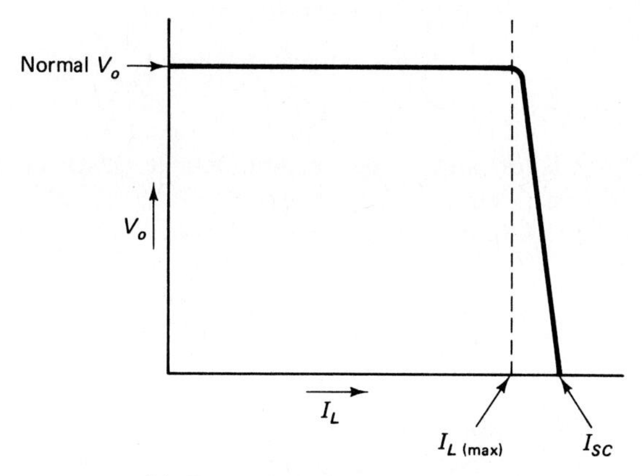

(b) Characteristic of current limiting circuit

Figure 12-8 Current limiting circuit (Q_3, R_6 and R_7). Q_3 remains off until I_L becomes large enough for V_{R6} to forward bias its base-emitter junction. Then, I_{C3} produces a voltage drop across R_7 that reduces the output voltage to near zero.

supply voltage V_s developed across its terminals. The power dissipation in Q_1 is

$$P_1 = V_s \times I_{sc} \tag{12-13}$$

Q_1 must be selected to survive this power dissipation.

Design of the current-limiting circuit in Fig. 12-8 is quite simple. Assuming that Q_3 is a silicon transistor, it should begin to conduct when $V_{R6} \approx 0.5$ V. Therefore, R_6 is calculated as

$$R_6 \approx \frac{0.5 \text{ V}}{I_{L(\max)}} \tag{12-14}$$

Example 12-5

The voltage regulator designed in Example 12-2 is to have the short-circuit output current limited to 60 mA. Design a suitable current limiting circuit, as in Fig. 12-8.

Solution

Eq. 12-14,
$$R_6 \approx \frac{0.5\text{ V}}{I_{sc}} = \frac{0.5\text{ V}}{60\text{ mA}}$$
$$\approx 8.3\ \Omega \qquad \text{(use } 8.25\ \Omega \pm 1\% \text{ standard value)}$$

Let
$$I_{C3} = 5\text{ mA}$$

Eq. 12-12,
$$R_7 \approx \frac{V_s}{I_{C3}} = \frac{16\text{ V}}{5\text{ mA}}$$
$$\approx 3.2\text{ k}\Omega \qquad \text{(use } 3.3\text{ k}\Omega \text{ standard value)}$$

Using $h_{FE1} = 50$ typically, under normal operating conditions
$$I_{B1(\max)} = \frac{I_{L(\max)}}{h_{FE1}} = \frac{50\text{ mA}}{50}$$
$$= 1\text{ mA}$$

This gives,
$$V_{R7} = 1\text{ mA} \times 3.3\text{ k}\Omega$$
$$= 3.3\text{ V}$$

This voltage drop is too large for the circuit to operate satisfactorily. To overcome the problem an additional transistor must be included in the circuit to give the Darlington connection in Fig. 12-6. Now, again using the typical value of 50 for h_{FE}, Eq. 12-10 gives
$$I_{B2(\max)} = \frac{I_{L(\max)}}{h_{FE1}\,h_{FE2}} = \frac{50\text{ mA}}{50 \times 50}$$
$$= 20\ \mu\text{A}$$

and, under normal operating conditions
$$V_{R7} = 20\ \mu\text{A} \times 3.3\text{ k}\Omega$$
$$= 66\text{ mV}$$

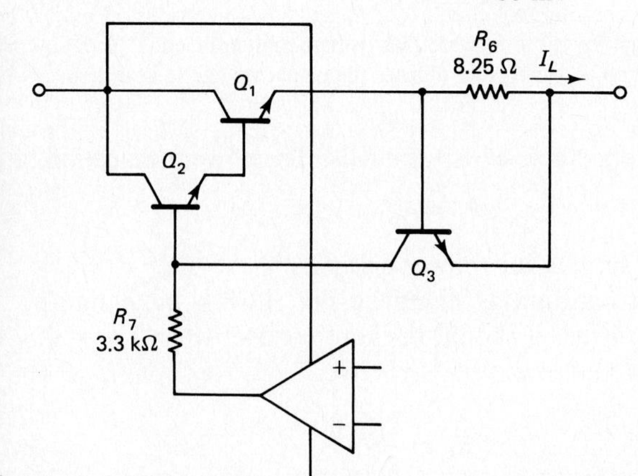

Figure 12-9 Current limiting circuit designed in Example 12-5; $I_{SC} = 60$ mA.

Foldback Current Limiting

A problem with the simple short-circuit protection method discussed above is that a large amount of power is dissipated in the series-pass transistor while the regulator remains short-circuited. The *foldback current-limiting circuit* in Fig. 12-10(a) minimizes this power dissipation. The V_o/I_L graph in Fig. 12-10(b) shows that the regulator output voltage remains constant until $I_{L(\max)}$ is approached. Then the current reduces (or folds back) to a lower short-circuit current level I_{SC} to produce a lower power dissipation in transistor Q_1.

To understand how the circuit in Fig. 12-10 operates, first note that when the output is shorted V_o equals zero. Consequently, the voltage drop across resistor R_8 is

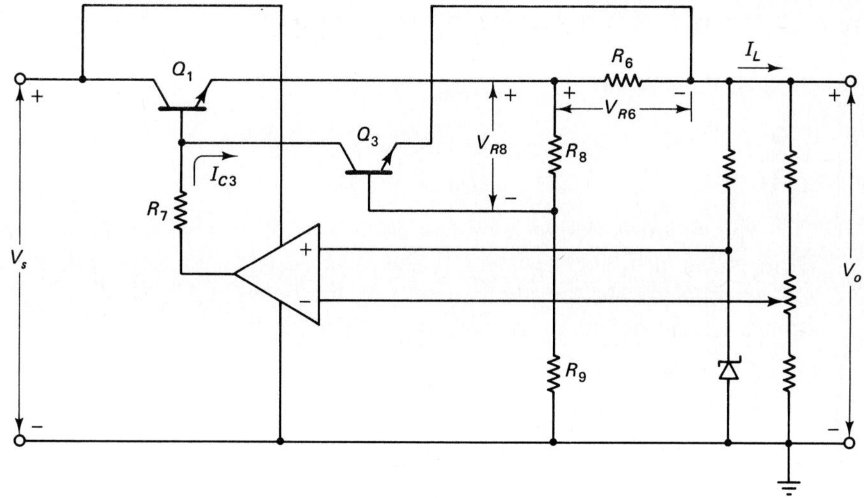

(a) Voltage regulator with a foldback current limiting circuit

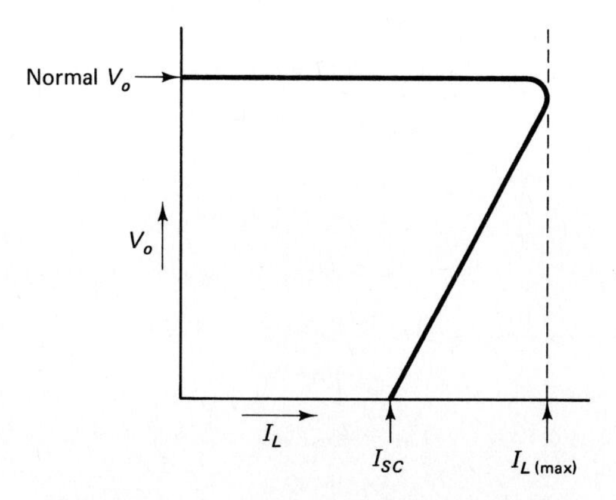

(b) Characteristics of foldback current limiting circuit

Figure 12-10 Foldback current limiting circuit. When the output current exceeds the maximum level, the current folds back to a lower short-circuit current. If $V_{R8} = V_{R6}$, $I_{SC} = I_{L(\max)}/2$.

negligibly small. The voltage across R_6 (at short circuit) is ($I_{SC} \times R_6$) and, as in the simple short-circuit protection circuit, this is designed to just keep transistor Q_3 biased *on*. When the regulator is operating normally with I_L less than $I_{L(max)}$, the voltage drop across R_8 is

$$V_{R8} \approx \frac{V_o \times R_8}{R_8 + R_9}$$

To forward bias the base-emitter junction of Q_3, the voltage drop across R_6 must become

$$I_{L(max)} R_6 = V_{BE} + V_{R8} \tag{12-15}$$

Using $V_{BE} \approx 0.5$ V and making $V_{R8} \approx 0.5$ V gives

$$I_{L(max)} R_6 = 0.5 \text{ V} + 0.5 \text{ V}$$

Since $I_{SC} R_6 \approx 0.5$ V, $\quad I_{L(max)} \approx 2\ I_{SC}$

If $V_{R8} \approx 1$ V, $\quad I_{L(max)} R_6 = 0.5 \text{ V} + 1 \text{ V}$

or, $\quad I_{L(max)} \approx 3\ I_{SC}$

The circuit is designed by first calculating R_6 to give the desired level of I_{SC}. Then, potential divider resistors R_8 and R_9 are determined to provide the necessary voltage drop V_{R8} for the required relationship between I_{SC} and $I_{L(max)}$. The current through R_8 and R_9 has to be much larger than the maximum base current in Q_3.

Example 12-6

Design a foldback current-limiting circuit for the voltage regulator designed in Example 12-4. The maximum output current is to be 400 mA, and when limited it is to fold back to 200 mA.

Solution

$$I_{SC} = 200 \text{ mA}$$

and, $\quad I_{L(max)} = 400 \text{ mA} = 2I_{SC}$

let $\quad V_{R6} \approx 0.5$ V at short circuit

$$R_6 \approx \frac{V_{R6}}{I_{SC}} = \frac{0.5 \text{ V}}{200 \text{ mA}}$$

$$\approx 2.5\ \Omega \qquad \text{(use 2.7 } \Omega \text{ standard value)}$$

at $I_{L(max)}$, $\quad V_{R6} = I_{L(max)} R_6 = 400 \text{ mA} \times 2.7\ \Omega$

$$\approx 1 \text{ V}$$

From Eq. 12-15, $\quad V_{R8} = I_{L(max)} R_6 - 0.5 \text{ V} = 1 \text{ V} - 0.5 \text{ V}$

$$= 0.5 \text{ V}$$

$$I_{R8} \gg I_{B3}$$

let $$I_{C3} = 5 \text{ mA}$$

$$I_{B3} = \frac{I_{C3}}{h_{FE3}} \approx \frac{5 \text{ mA}}{50}$$

$$= 100\ \mu\text{A}$$

let $$I_{R8} = 1 \text{ mA}$$

$$R_8 = \frac{V_{R8}}{I_{R8}} = \frac{0.5 \text{ V}}{1 \text{ mA}}$$

$$= 500\ \Omega \qquad \text{(use 470 } \Omega \text{ standard value)}$$

Using the average level of V_o as 12.5 V,

$$R_9 \approx \frac{V_o - V_{R8}}{I_{R8}} = \frac{12.5 \text{ V} - 0.5 \text{ V}}{1 \text{ mA}}$$

$$= 12 \text{ k}\Omega \qquad \text{(standard value)}$$

Eq. 12-12, $$R_7 \approx \frac{V_s}{I_{C3}} = \frac{19.5 \text{ V}}{5 \text{ mA}}$$

$$\approx 3.9 \text{ k}\Omega \qquad \text{(standard value)}$$

Using $h_{FE1} = 20$ and $h_{FE2} = 50$ typically, under normal operating conditions,

$$I_{B2(\max)} = \frac{I_{L(\max)}}{h_{FE1}\, h_{FE2}} \simeq \frac{400 \text{ mA}}{20 \times 50}$$

$$= 400\ \mu\text{A}$$

so, $$V_{R7} = 400\ \mu\text{A} \times 3.9 \text{ k}\Omega$$

$$\approx 1.6 \text{ V (normally)}$$

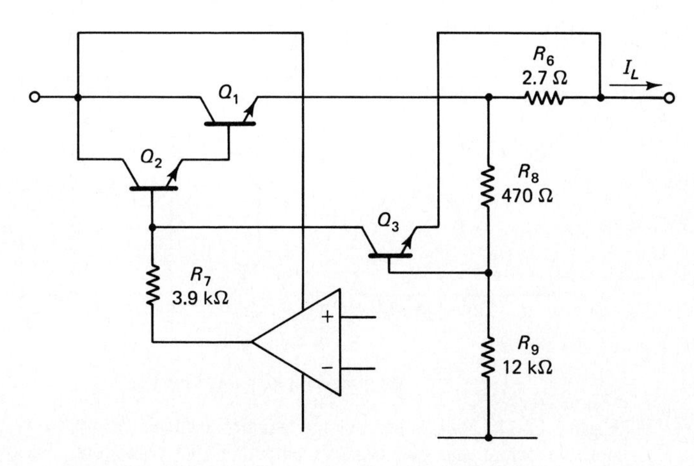

Figure 12-11 Foldback current limiting circuit designed in Example 12-6; $I_{L(\max)} = 400$ mA, and $I_{SC} = 200$ mA.

12-6 PLUS-MINUS SUPPLIES

The negative output terminal of a voltage regulator is usually grounded to give an output voltage which is positive with respect to ground. So long as there are no grounded terminals in the unregulated supply, the positive output of the regulator can be grounded to produce an output which is negative with respect to ground. Us-

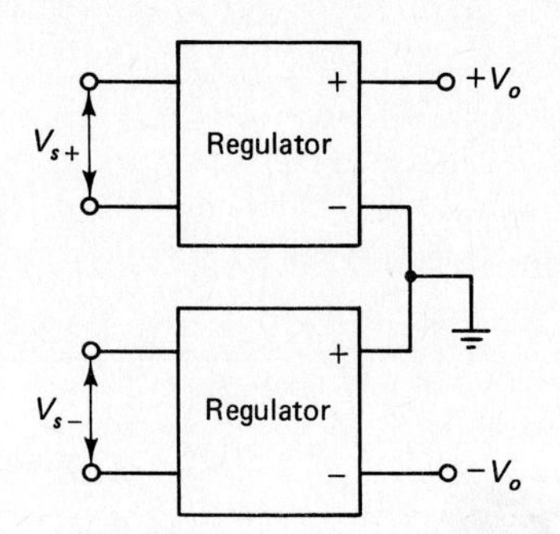

(a) Two similar regulators connected as a plus-minus supply

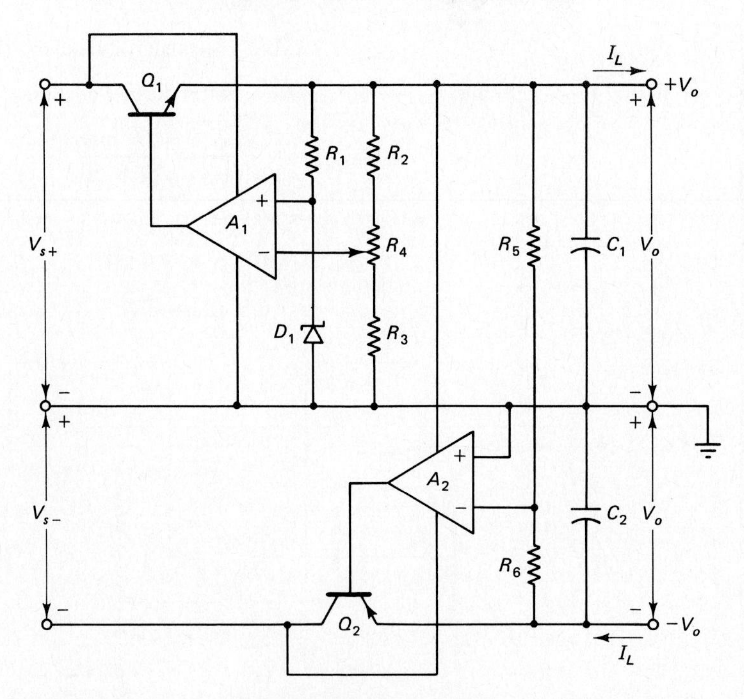

(b) Tracking plus-minus supplies

Figure 12-12 Plus-minus voltage supplies are usually created by grounding one positive output and one negative output of two regulators. Tracking plus-minus voltages have a negative voltage regulator which uses the positive regulator output as a reference voltage source.

ing two dc voltage regulators, one positive and one negative terminal can be connected together and grounded, as illustrated in Fig. 12-12(a). This produces the type of plus-minus voltage often used as a supply for operational amplifiers. As already stated, this arrangement is possible only when there are no ground connections in the unregulated supply. Also, each regulator must have its own independant supply. There must be no common connections in the unregulated dc supplies.

Figure 12-12(b) shows another way of producing a plus-minus regulated dc voltage. The top half of the circuit (Q_1, A_1 and their associated components) is a precision dc voltage regulator, as discussed in Section 12-4. Components A_2, Q_2, R_5, R_6, and C_2 constitute a negative voltage regulator. Instead of a Zener diode, the negative regulator uses the regulated positive output as a reference voltage. Potential divider R_5 and R_6 is connected between the positive and negative output terminals. So, assuming that the positive output remains constant, any change in the negative output will be applied to A_2, where it will be amplified and inverted to correct the change. Note that Q_2 is a *pnp* transistor.

Resistors R_5 and R_6 in Fig. 12-12(b) are normally equal, so the output voltage measured between the positive and negative terminals is exactly twice the positive voltage. This gives a negative output which equals the positive output. If the positive output is +12 V, the negative output will be −12 V, to give a ±12 V regulated supply. When the positive output is adjusted to +15 V, the negative output will follow it to −15 V. This type of negative voltage regulator is sometimes termed a *tracking regulator,* because it follows any changes in the positive voltage output.

To design the plus-minus regulator circuit in Fig. 12-12(b), the positive voltage regulator is first designed in the usual way. Then the common value of R_5 and R_6 is calculated as V_o/I_{R5}, where the potential divider current is much larger than the op-amp input bias current. Q_2 is the *pnp* complement of *npn* transistor Q_1, and V_{s-} should equal V_{s+}.

12-7 INTEGRATED CIRCUIT VOLTAGE REGULATORS

The 723 IC Regulator

The basic circuit of a 723 integrated circuit dc voltage regulator in a dual-in-line package is illustrated in Fig. 12-13. This IC has a voltage reference source (D_1), an error amplifier (A_1), a series-pass transistor (Q_1), and a current limiting transistor (Q_2) all contained in one small package. An additional Zener diode (D_2) is included with its output at terminal 9 identified as V_z. This is used solely for voltage dropping purposes in some applications. The IC can be connected to function as a positive or negative voltage regulator with an output voltage ranging from 2 V to 37 V and output current levels up to 150 mA. An external transistor may be Darlington-connected to Q_1 to handle large load currents. The maximum supply voltage is 40 V, and the line and load regulation are each specified as 0.01%. A partial specification for the 723 regulator is given in Appendix 1-6.

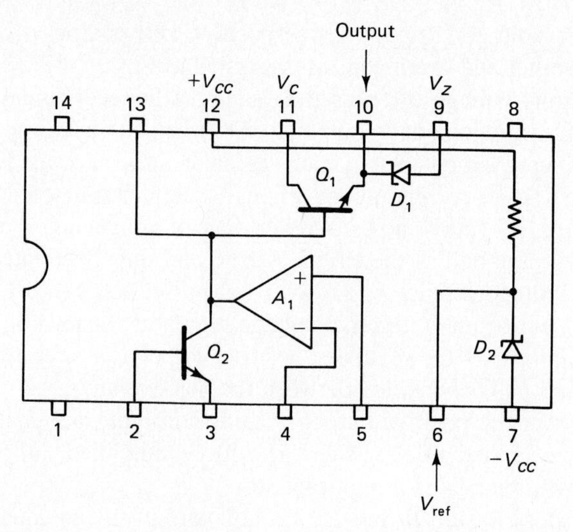

Figure 12-13 The 723 integrated circuit voltage regulator contains a reference voltage source (D_1), an error amplifier (A_1), a series-pass transistor (Q_1), and a current limiting transistor (Q_2).

Figure 12-14(a) shows a 723 connected to function as a positive voltage regulator. The complete arrangement (including the internal circuitry) is similar to the op-amp regulator circuit in Fig. 12-4. One difference between the two circuits is the 100 pF capacitor C_1 connected between the error amplifier output and its inverting input terminal. This is used instead of a large capacitor at the output terminals to prevent the regulator from oscillating. (Sometimes both capacitors may be required). By appropriate selection of resistors R_1 and R_2 in Fig. 12-14(a), the regulator output can be set to any voltage between approximately 7 V (the reference voltage) and 37 V. A potentiometer may be included between R_1 and R_2 to make the output voltage adjustable. The broken lines show connections for simple (non-foldback) current limiting.

It is important to note that, as for other regulator circuits, the supply voltage at the lowest point on the ripple waveform should be at least 3 V greater than the regulator output. Otherwise a high amplitude output ripple may occur.

A negative voltage regulator using the 723 is shown in Fig. 12-14(b). In this case, an external *pnp* transistor (Q_3) must be used and its base is connected to terminal 9, (to the anode of the internal voltage-dropping Zener diode). Resistors R_3 and R_4 are made equal, so the reference voltage is potentially divided to approximately 3.5 V. Then,

$$V_o \approx \frac{3.5\ \text{V}(R_1 + R_2)}{R_1} \tag{12-16}$$

To understand the operation of this circuit, the complete internal and external circuitry must be drawn.

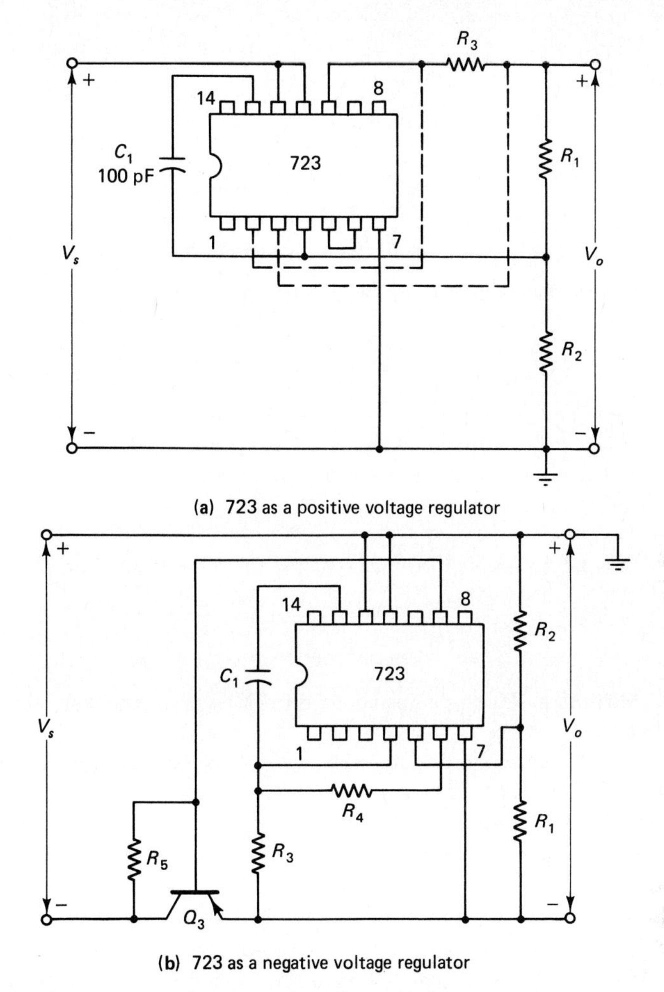

(a) 723 as a positive voltage regulator

(b) 723 as a negative voltage regulator

Figure 12-14 The 723 can be connected to function as a positive voltage regulator or as a negative voltage regulator. The output voltage can range from 2 V to 37 V and the output current (without an external series-pass transistor) can be as high as 150 mA.

Example 12-7

The positive voltage regulator circuit in Fig. 12-14(a) is to have an output of 18 V. Determine suitable resistance values for R_1 and R_2, select an appropriate input voltage, and determine the maximum load current that may be supplied.

Solution

$$I_2 \gg \text{(error amplifier input bias current)}$$

let

$$I_2 = 1 \text{ mA}$$

$$V_{R2} = V_{\text{ref}} = 7.15 \text{ V}$$

$$R_2 = \frac{V_{\text{ref}}}{I_2} = \frac{7.15 \text{ V}}{1 \text{ mA}}$$

$$= 7.15\ \text{k}\Omega \qquad \text{(use 6.8 k}\Omega\text{ and recalculate } I_2\text{)}$$

$$I_2 = \frac{7.15\ \text{V}}{6.8\ \text{k}\Omega} = 1.05\ \text{mA}$$

$$R_1 = \frac{V_o - V_{R2}}{I_2} = \frac{18\ \text{V} - 7.15\ \text{V}}{1.05\ \text{mA}}$$

$$\approx 10\ \text{k}\Omega \qquad \text{(standard value)}$$

For satisfactory operation of the series-pass transistor,

let $\qquad V_s - V_o = 5\ \text{V} \qquad$ (the specification lists 3 V min)

$$V_s = V_o + 5\ \text{V} = 18\ \text{V} + 5\ \text{V}$$

$$= 23\ \text{V}$$

The internal circuit current is approximately,

$$I_{(\text{standby})} + I_{(\text{ref})} \approx 25\ \text{mA}$$

Internal power dissipation, excluding the series-pass transistor,

$$P_i = V_s \times (I_{(\text{standby})} + I_{(\text{ref})}) = 23\ \text{V} \times 25\ \text{mA}$$

$$= 575\ \text{mW}$$

Maximum Power dissipated in the series-pass transistor (for an air temperature of 25°C),

$$P = (\text{specified } P_{D(\max)}) - P_i = 1000\ \text{mW} - 575\ \text{mW}$$

$$= 425\ \text{mW}$$

Maximum load current,

$$I_{L(\max)} = \frac{P}{V_s - V_o} = \frac{425\ \text{mW}}{5\ \text{V}}$$

$$= 85\ \text{mA}$$

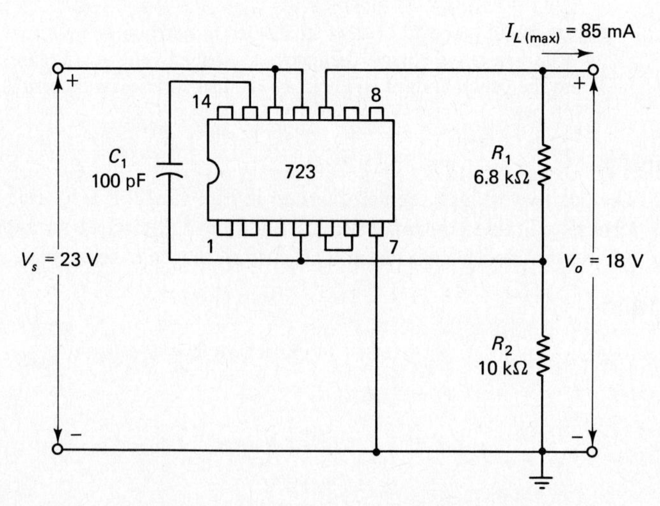

Figure 12-15 723 positive voltage regulator designed in Example 12-7.

LM217 and LM237 IC Regulators

The LM217 and LM237 integrated circuit voltage regulators are three-terminal devices which are very easy to use. The 217, shown in Fig. 12-16(a), is a positive voltage regulator. The 237 is a negative voltage regulator with similar performance to the 217. In each case, input and output terminals are provided for supply and regulated output voltage, and a third adjust terminal (*ADJ*) is included to facilitate output voltage selection. The terminal connections for the TO-3 metal can package are shown in Fig. 12-16(b).

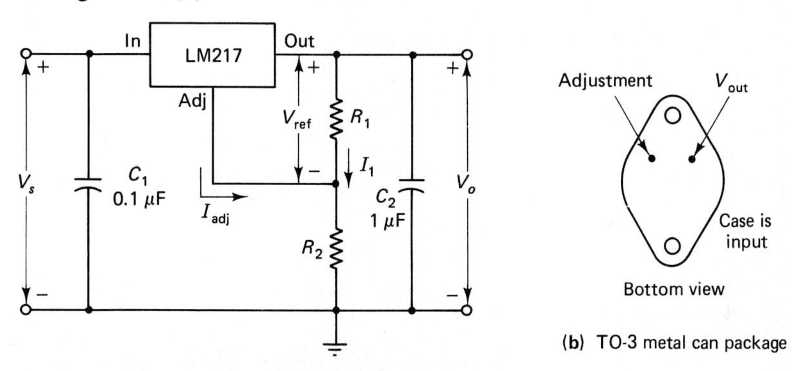

Figure 12-16 LM217 integrated circuit positive voltage regulator and its terminal connections. The output voltage is $V_o = V_{ref}(R_1 + R_2)/R_1$.

The internal reference voltage is typically 1.25 V, and it appears across the *ADJ* and *out* terminals, giving a regulator output voltage of

$$V_o = \frac{V_{ref}(R_1 + R_2)}{R_1} \tag{12-17}$$

To determine suitable resistance values for R_1 and R_2 for a desired output voltage, I_1 is first selected to be much larger than the current that flows in the adjustment terminal of the device. This is a maximum of 100 μA, according to the device data sheet. Then, the resistors are calculated using Eq. 12-17.

An appropriate supply voltage should be selected each time a regulator is designed. Once again, to avoid a high amplitude output ripple, the supply voltage at the lowest point on the ripple waveform should be at least 3 V greater than the regulated output voltage. The total power dissipation in the device should be calculated to ensure that the IC will operate satisfactorily. The regulator specification lists 2000 mW as the maximum power dissipation at a free air temperature of 25°C. This must be derated at 62.5 mW/°C for higher temperature. When a heat sink is used with the LM217 or LM237, a maximum power dissipation of 20 W is possible up to a case temperature of 70°C. Beyond this, the derating factor is 250 mW/°C.

Note the input and output capacitors included in the regulator circuit in Fig. 12-16. Capacitor C_1 helps eliminate oscillation tendencies that might occur with long connecting leads between the filter and regulator circuit. This is necessary only when the regulator is not connected close to its rectifier and filter power supply circuit. Capacitor C_2 improves the transient response of the regulator. C_1 should typically be 0.1 μF and C_2 should be a minimum of 1 μF, but might be as high as 1000 μF if necessary for transient rejection.

Example 12-8

Calculate the resistances of R_1 and R_2 for the LM217 voltage regulator in Fig. 12-16, to produce an output voltage of 9 V.

Solution

$$I_1 \gg (I_{ADJ} = 100\ \mu\text{A})$$

let

$$I_1 = 5\ \text{mA}$$

$$R_1 = \frac{V_{\text{ref}}}{I_1} = \frac{1.25\ \text{V}}{5\ \text{mA}}$$

$$= 250\ \Omega \qquad \text{(use 270 }\Omega\text{ and recalculate } I_1)$$

$$I_1 = \frac{V_{\text{ref}}}{R_1} = \frac{1.25\ \text{V}}{270\ \Omega}$$

$$\approx 4.6\ \text{mA}$$

$$R_2 = \frac{V_o - V_{R1}}{I_1} = \frac{9\ \text{V} - 1.25\ \text{V}}{4.6\ \text{mA}}$$

$$= 1.7\ \text{k}\Omega \qquad \text{(use a 1.5 k}\Omega\text{ and a 220 }\Omega\text{ in series)}$$

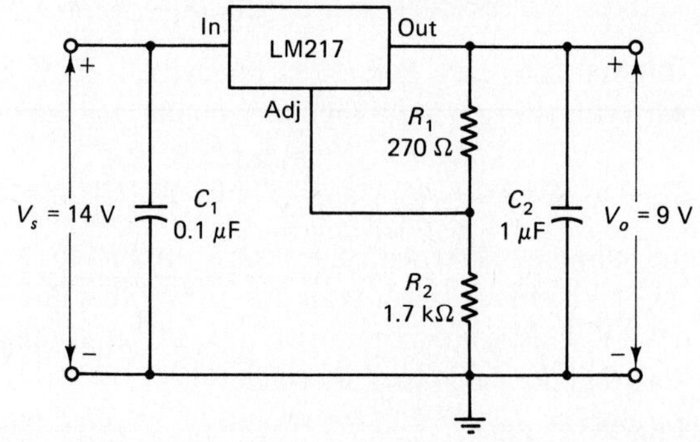

Figure 12-17 LM217 regulator circuit designed in Example 12-8.

LM340 Three-Terminal Regulators

The LM340 are three-terminal positive voltage regulators with fixed output voltages ranging from a low of 5 V to high of 23 V. The regulator is selected for the desired output voltage and then simply connected to a power supply filter circuit, as shown in Fig. 12-18(a). Capacitor C_1 is required only when the regulator is not located close to unregulated supply. It should typically be 0.22 μF. C_2 is not really necessary, but it improves the transient response of the circuit. A 0.1 μF ceramic disc ca-

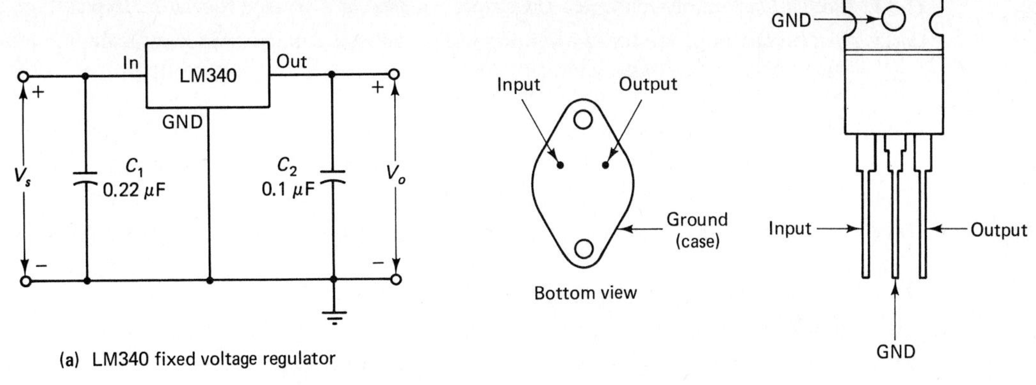

Figure 12-18 The LM340 integrated circuit voltage regulator is available in a range of fixed output voltages from 5 V to 23 V.

pacitor is recommended by the manufacturer. The terminal connections for two types of package are illustrated in Fig. 12-18(b).

The LM 340 data sheet lists the device performance for an output current of 1 A. The tolerance on the output voltage is ±2%, the line regulation is 0.01% per output volt, and the load regulation is 0.3% per amp of load current. The regulator circuit includes current limiting, and a thermal shutdown circuit which protects the device against excessive internal power dissipation.

REVIEW QUESTIONS

12-1. Briefly explain the action of a dc voltage regulator. Write the equations for line regulation, load regulation, and ripple rejection.

12-2. Sketch the circuit of a voltage follower regulator. Explain its operation.

12-3. Briefly discuss the design procedure for a voltage follower regulator.

12-4. Write the equations for line regulation, load regulation, and ripple rejection for a voltage follower regulator.

12-5. Show how a voltage follower regulator should be modified to produce an output voltage greater than the reference voltage. Explain. Also, show how the output voltage can be made adjustable.

12-6. Write the equations for line regulation, load regulation, and ripple rejection for a voltage follower regulator that has an output voltage greater than the reference voltage.

12-7. Briefly discuss the design procedure for a voltage follower regulator with an output voltage greater than the reference voltage.

12-8. Sketch the circuit of a precision voltage regulator. Explain its operation, and discuss how it differs from voltage follower regulators.

12-9. Discuss the performance of the precision voltage regulator.

12-10. Briefly discuss the design procedure for a precision voltage regulator.

12-11. Sketch Darlington-connected transistors, as used in a voltage regulator. Explain.

12-12. Discuss the need for current limiting in a dc voltage regulator. Sketch the circuit and the voltage/current characteristics for a (non-foldback) current limiter. Explain the circuit operation.

12-13. Sketch the circuit and voltage/current characteristics for a foldback current limiter. Explain its operation.

12-14. Briefly discuss the design procedure for both types of current-limiting circuit.

12-15. Show how two voltage regulators may be connected to form a plus-minus supply. Explain.

12-16. Sketch the circuit of a tracking plus-minus supply. Explain its operation.

12-17. Sketch the basic circuit of a 723 integrated circuit dc voltage regulator. Explain.

12-18. Show how a 723 regulator can be used as a positive voltage regulator. Explain the circuit operation, write the equation for output voltage, and state the ranges of output voltage and current that can be obtained.

12-19. Show how a 723 regulator should be connected to function as a negative voltage regulator. Explain.

12-20. Sketch a regulator circuit using an LM217 IC voltage regulator. Explain the circuit operation, write the equation for output voltage, and discuss the required supply voltage.

12-21. Sketch a regulator that uses a LM340 IC voltage regulator. Briefly discuss the LM340 and its performance.

PROBLEMS

12-1. A voltage regulator circuit as in Fig. 12-2 has: $V_s = 25$ V, $R_s = 20\ \Omega$, $R_1 = 470\ \Omega$, $V_z = 15$ V, $Z_z = 10\ \Omega$, and $I_{L(max)} = 75$ mA. Analyze the circuit to determine the line regulation, load regulation, and ripple rejection.

12-2. Design a voltage follower regulator circuit as in Fig. 12-2 to have, $V_o = 10$ V, and $I_{L(max)} = 75$ mA.

12-3. Analyze the circuit designed for Problem 12-2 to determine the line regulation, load regulation, and ripple rejection. The supply source resistance is 15 Ω.

12-4. Design a voltage regulator as in Fig. 12-3(a) to produce a 10 V output with a maximum load current of 100 mA. Use a 741 op-amp.

12-5. Assuming a 12 Ω supply voltage source resistance, analyze the circuit designed for Problem 12-4 to determine the line regulation, load regulation, and ripple rejection.

12-6. Modify the circuit designed for Example 12-4 to make the output voltage adjustable from 8 V to 10 V.

12-7. Design a precision voltage regulator as in Fig. 12-5 to produce an output voltage adjustable from 15 V to 18 V. The maximum load current is to be 150 mA.

12-8. Modify the circuit designed for Problem 12-7 to supply a 300 mA maximum load current.

12-9. Analyze the circuit designed for Problem 12-7 to determine the line regulation, load regulation, and ripple rejection. Assume the supply voltage source resistance is 8 Ω.

12-10. Design a non-foldback current limiting circuit for the regulator designed for Problem 12-2. The short-circuit output current is to be 80 mA.

12-11. The regulator designed for Example 12-4 is to have short-circuit protection to limit the output current to 105 mA. Design a suitable circuit.

12-12. Design a foldback current limiting circuit for the regulator designed for Problem 12-7. The circuit is to become effective when the load current reaches 160 mA, and is to fold back to 80 mA.

12-13. Modify the current limiter in Problem 12-12 to fold back to approximately 55 mA.

12-14. The tracking plus-minus voltage regulator in Fig. 12-12(b) is to produce outputs adjustable from ±10 V to ±14 V. The maximum load current is to be 200 mA. Design the complete circuit.

12-15. Using a 723 IC regulator, design a positive voltage regulator to produce an output voltage adjustable from 8 V to 16 V. Determine the maximum output current that can be supplied.

12-16. Modify the circuit designed for Problem 12-15 to increase the load current to 200 mA.

12-17. Modify the circuit designed for Problem 12-15 to limit the short-circuit output current to 90% of the calculated maximum load current.

12-18. An LM217 IC voltage regulator is to be used to produce an output voltage adjustable from 5 V to 7 V. Sketch the circuit and determine suitable component values.

COMPUTER PROBLEMS

12-19. Write a computer program to design a dc voltage regulator circuit as in Fig. 12-7 with foldback current limiting as in Fig. 12-10. Given: V_o range, I_L, and I_{SC}.

12-20. Write a computer program to analyze the type of dc regulator circuit in Fig. 12-4 to determine line and load regulation, and ripple rejection. Given: all component values, V_s, V_o, and I_L.

LABORATORY EXERCISES

12-1 Series Regulator

1. Construct the series regulator circuit shown in Fig. 12-4, as designed in Example 12-2.
2. Apply a 16 V dc input with a 10 Ω source resistance.
3. Measure the output voltage on a digital voltmeter and compare to the designed quantity.
4. Adjust the input by ±1.6 V. Note the output voltage changes, calculate the line regulation, and compare it to the line regulation calculated in Example 12-3.
5. Connect a 240 Ω, 1 W resistor at the output terminals. Note the output voltage change, calculate the load regulation, and compare it to the load regulation calculated in Example 12-3.
6. Modify the circuit to include the current limiter shown in Fig. 12-9, as designed in Example 12-5.
7. Connect a 1 kΩ, 1 W variable resistor in series with a 100 mA ammeter at the regulator output terminals.

8. Slowly reduce the resistance to zero. Measure the output short-circuit current, and compare to the designed quantity.

12-2 723 Regulator

1. Construct the 723 series regulator circuit shown in Fig. 12-15, as designed in Example 12-7.
2. Apply a 23 V dc input with a 10 Ω source resistance.
3. Measure output voltage on a digital voltmeter and compare to the designed quantity.
4. Adjust the input by ±10%. Note the output voltage changes and calculate the line regulation.
5. Connect a 225 Ω, 2 W resistor at the output terminals. Note the output voltage change and calculate the load regulation.

Glossary

A_v Symbol used for closed-loop gain.

Absolute value circuit Precision full-wave rectifier.

Active filter Filter circuit using an operational amplifier.

All-pass filter Filter circuit that passes all frequencies in a given range without attenuation, but with varying phase shifts.

Averaging circuit Circuit that produces an output which is the average of two or more inputs.

Bandpass filter Filter circuit that passes a specific band of frequencies.

Bandstop filter Filter that rejects a specific band of frequencies.

Beta β Symbol used for amount of feedback from an amplifier output to input.

BIFET op-amp Operational amplifier with field effect transistor input stage and bipolar transistor amplification and output stages.

Bipolar op-amp Operational amplifier with bipolar transistors throughout.

Bode plot Straight-line approximation of the gain/frequency and/or phase/frequency response graph of a circuit.

Clamping circuit Circuit which clamps an output to a particular upper or lower voltage level.

Clipping circuit Circuit which clips off a portion of a waveform.

Closed-loop gain Voltage gain of a circuit with negative feedback, measured from input to output.

Common mode gain Op-amp circuit voltage gain when a common mode input is applied.

Common mode input Voltage applied to both input terminals of an op-amp.

Common mode rejection ratio Ratio of open-loop gain to common mode gain.

Comparator Circuit that compares the level of two inputs.

Crossing detector Circuit which detects the input voltage crossing a specific level.

Current sink Circuit into which a specific current flows in the conventional direction from positive to negative.

Current source Circuit which produces a specific output current in the conventional direction from positive to negative.

Current-to-voltage converter Circuit which produces an output voltage proportional to an input current.

Cutoff frequency Frequency at which the voltage gain of an op-amp circuit has fallen by 3 dB below its normal value.

Decade Ten times increase in frequency.

Difference amplifier Amplifier which produces an output proportional to the difference between two input voltages.

Differential amplifier Same as difference amplifier.

Differential input Input voltage which is measured as the difference between the voltages at the two input terminals of an op-amp.

Differential-input/differential-output amplifier Amplifier with a differential input, and a differential output at two output terminals.

Differentiator Circuit which differentiates an input.

Frequency compensation Use of capacitors and resistors to prevent an op-amp circuit from oscillating.

Full power bandwidth Upper cutoff frequency for an op-amp which has an output that swings between the extremes of the supply voltage.

Gain-bandwidth product Amplifier closed-loop voltage gain multiplied by its upper cutoff frequency.

Gain/frequency response Graph of circuit voltage gain versus frequency plotted to a logarithmic base.

High-pass filter Filter circuit that passes all signal frequencies above a specific frequency.

Hysteresis For an op-amp switching circuit, the difference between the input level to switch the output positively and that to switch it negatively.

Input bias current Current that flows into the input terminals of an op-amp.

Input offset current Difference between the bias currents at the two inputs of an op-amp.

Input offset voltage Op-amp differential input voltage required to set the output to zero.

Input voltage range Maximum positive or negative input voltage that may be applied for linear operation of an op-amp.

Instrumentation amplifier Op-amp differential amplifier with adjustable voltage gain, common mode nulling, and output level shifting.

Integrator Circuit that integrates an applied input.

Inverting amplifier Op-amp circuit that amplifies and inverts the input voltage.

Inverting input Op-amp input terminal that produces an inverted output.

Limiter Same as clipping circuit.

Loop gain Voltage gain of an op-amp circuit from the inverting input to the output and back via the feedback network to the input.

Loop phase shift Phase shift of an op-amp circuit from the inverting input to the output and back via the feedback network to the input.

Low-pass filter Filter circuit that passes all signal frequencies below a specific frequency.

M Symbol for op-amp open-loop voltage gain.

Miller effect Capacitance amplification in an inverting amplifier.

Noninverting amplifier Op-amp circuit that amplifies but does not invert an input voltage.

Noninverting input Op-amp input terminal that produces a noninverted output.

Octave Doubling of frequency.

Offset nulling Adjustment to zero the dc output voltage of an op-amp, to counter the effects of input offset voltage and/or current.

Open-loop gain Actual op-amp voltage gain from input to output (not circuit gain).

Open-loop phase shift Actual op-amp phase shift from input to output (not circuit phase shift).

Output level shifting Altering the dc output voltage level.

Output offset voltage Op-amp dc output voltage due to input offset voltage and/or current.

Output voltage range Maximum positive or negative op-amp output voltage for a given supply voltage.

Output voltage swing Same as output voltage range.

Over compensation Use of more compensation than necessary to stabilize an op-amp circuit.

Peak detector Circuit that produces an output voltage equal to the peak level of an input waveform.

Phase lag circuit Circuit that introduces a phase lag.

Phase lead circuit Circuit that introduces a phase lead.

Phase margin Difference between 360° and the actual loop phase shift around the feedback loop of an op-amp circuit.

Phase response Graph of circuit phase shift versus frequency, plotted to a logarithmic base.

Pole frequency Frequency at which the rate of change of op-amp voltage gain alters to a new rate of change.

Precision rectifier Op-amp circuit that rectifies an applied input with virtually no loss of signal voltage.

Roll-off Fall of circuit voltage gain with frequency change.

Sample-and-hold circuit Circuit that samples an input and holds its output at the sampled level until a new sample is taken.

Saturation voltage Op-amp maximum and minimum output voltage, usually about 1 V below the supply voltage levels.

Schmitt trigger circuit Circuit that produces an output that rapidly changes from one level to another when the input equals the trigger voltage.

Slew rate Maximum rate of change of op-amp output voltage.

Summing amplifier Circuit that produces an output which is the sum of two or more inputs.

Supply voltage rejection Measure of the effect of supply voltage change on op-amp output voltage.

Unity-gain bandwidth High frequency at which the op-amp voltage gain is reduced to unity.

Voltage follower Op-amp circuit which has a high input impedance, low output impedance, and a noninverting unity voltage gain.

Voltage-to-current converter Circuit which produces an output current proportional to an applied voltage.

Zero-crossing detector Circuit that produces an output voltage change when the input voltage arrives at ground level.

Answers to Odd-Numbered Problems

1-1. 749.985 mV, 749.996 mV
1-3. 20 000
1-5. 13.8 V
1-7. 152 Ω
1-9. 9.2 V
1-11. 57.4, 0.017

2-1. ±8 V, ±14 V
2-3. ±0.52 V
2-5. 5 mV
2-7. 0.3 μV
2-9. 3 mV, 6.6 μV
2-11. 1.5×10^{12} Ω, 5×10^{10} Ω
2-13. 2.5×10^{-3} Ω, 0.12 Ω
2-15. 60 μs
2-17. 0.43 μs, 1.7 μs

3-1. 22 kΩ, 4.4 mV
3-3. 94 mV, 300 mV
3-5. (150 kΩ + 12 kΩ), 270 kΩ, 100 kΩ
3-7. (560 kΩ + 39 kΩ), 1 MΩ, 560 kΩ
3-9. 1.8 kΩ, (47 kΩ + 5.6 kΩ), 1.8 kΩ
3-11. 13.3×10^{9} Ω, 11.25×10^{-3} Ω
3-13. 12 kΩ, 1 MΩ, 12 kΩ
3-15. 12 kΩ, 820 Ω
3-17. 1.8 kΩ, 1.8 kΩ, 1.8 kΩ, 560 Ω
3-19. 1 MΩ, 1 MΩ, 1 MΩ, 330 kΩ, 150 kΩ
3-21. 1 MΩ, 1 MΩ, 1 MΩ, (100 kΩ + 10 kΩ)
3-23. 1.8 kΩ, 27 kΩ, 1.8 kΩ, (22 kΩ + 2.7 kΩ) + 5 kΩ variable

4-1. 100 kΩ, 0.15 μF, 0.18 μF
4-3. 188 Hz
4-5. 68 kΩ, 179 Hz, 5.9 kΩ
4-7. 470 kΩ, 470 kΩ, 0.03 μF, 1000 pF, 0.18 μF
4-9. 1 MΩ, 1 MΩ, 10 kΩ + 1.2 kΩ, 0.027 μF, 0.25 μF

4-11. 120 kΩ, 56 kΩ, 560 Ω, 0.2 μF, 0.25 μF
4-13. 101, 33 kΩ, 125 Hz, 3.26 kΩ
4-15. 120 kΩ, 100 kΩ, 820 Ω, 1000 pF, 2 μF, 6 μF
4-17. 6.8 kΩ, 1 MΩ, 3 μF, 1 μF
4-19. −37, 2.7 kΩ, 156 Hz, 262 Ω
4-21. 180 kΩ, 1 MΩ, 180 kΩ, 1 MΩ, 0.03 μF, 4700 pF, 0.18 μF
4-23. 106 Hz, 0.5 V
4-25. 27 kΩ, 27 kΩ, 1.2 μF, 0.5 μF
4-27. 330 kΩ, 330 kΩ, 120 kΩ, 1 MΩ
4-29. 1 kΩ, 68 kΩ, 180 kΩ, 180 kΩ, 8 μF, 1.8 μF

5-1. 500 pF, 1.5 kΩ, 20 pF, 150 kHz, 20 kHz
5-3. 680 Ω, 68 kΩ, 680 Ω, 50 pF, 250 pF, 3.7 V
5-5. 4 MHz, 200 kHz, 800 kHz
5-7. 6 MHz, 2.2 V
5-9. (22 kΩ + 3.3 kΩ), 1 MΩ, 22 kΩ, 3 pF, 200 kHz
5-11. 100 pF, 500 pF, 1000 pF, 2 MHz
5-13. 1.5 kΩ, 180 kΩ, 6 kHz
5-15. 330 Ω, 27 kΩ, 330 Ω, 100 pF, 500 pF, 1000 pF, 1 MHz
5-17. 1 kΩ, 33 kΩ, 1 kΩ, 100 pF, 500 pF, 1000 pF, 2 MHz
5-19. 7.5 MHz
5-21. 3 μs, 39 ns, 0.1 μs
5-23. 4.8 pF
5-25. 0.48 pF
5-27. 354 pF, 0.8 pF
5-29. 330 Ω, 330 pF, 15 MHz
5-31. 4.7 kΩ, 120 pF, 500 pF, 1.5 kΩ, 20 pF

6-1. (120 kΩ + 12 kΩ), 150 kΩ
6-3. 82 Ω, 270 kΩ, 1 MΩ, 500 Ω
6-5. 120 Ω, 9.6 V
6-7. 100 Ω, 4.5 V
6-9. 220 Ω, 7.7 V
6-11. 6.2 V, 270 Ω, (820 kΩ + 100 kΩ), 1 MΩ, 200 kΩ
6-13. (15 kΩ + 560 Ω), 820 Ω
6-15. 900 kΩ, 90 kΩ, 9 kΩ, 1 kΩ, (270 Ω + 50 Ω)
6-17. 1 MΩ, (8.2 kΩ + 1 kΩ), LF 353, ±9 V, 1.8 μF
6-19. 27 kΩ, 2.7 kΩ, 270 Ω, ±18 V
6-21. 10 kΩ, (330 Ω + 100 Ω variable), 10 kΩ, 18 kΩ, 180 kΩ, 18 kΩ, (150 kΩ + 100 kΩ variable)

7-1. 1.8 kΩ, 6.8 kΩ, 1.5 kΩ, 2 μs
7-3. 1.8 kΩ, (1.8 kΩ + 1.8 kΩ), 1.2 kΩ, 1.8 kΩ, 1.8 kΩ, 1.8 kΩ, 560 Ω, 2 μs
7-5. 3.9 kΩ, 2.7 kΩ, (5.6 kΩ + 5.6 kΩ), 5.6 kΩ, 5.6 kΩ, 5.6 kΩ
7-7. IN753, 330 Ω, 3.3 kΩ, 330 Ω

7-9. 4.7 kΩ, 4.7 kΩ, 4.7 kΩ, 1.5 kΩ, 3 μs

7-11. 21 kΩ to 44 kΩ

7-13. R_1 through R_4, and R_1' through R_4' = [(9.4 kΩ to 23.4 kΩ), 3.9 kΩ, 3.9 kΩ, 1.5 kΩ]; R_5 through R_8 = 3.9 kΩ, R_9 = 1 kΩ

7-15. 8.2 kΩ, 1 kΩ, 8.2 kΩ

7-17. 1 MΩ, 1 MΩ, 6000 pF, 200 μA

7-19. 257 μs, 12.6 μs

8-1. 1 kΩ, 18 kΩ, 18 kΩ, 1800 pF

8-3. 6.8 kΩ, 120 kΩ, 120 kΩ, 0.02 μF

8-5. 23 kHz, 0.33 V/μs, 70 kHz, 0.26 V/μs, 3.5 kHz, 0.07 V/μs, 3.5 kHz, 0.07 V/μs

8-7. 39 kΩ, 820 kΩ, 39 kΩ, 5100 pF

8-9. (22 kΩ + 1.2 kΩ), 470 kΩ, 22 kΩ, 0.1 μF

8-11. 0.1 V/μs, 0.1 V/μs, 3 V/μs, 3 V/μs

8-13. ±2.8 V, 300 Hz

8-15. 133 Hz, 0.06 V/μs

8-17. 354 Hz, 0.03 V/μs

9-1. ±14 V, 220 kΩ, 27 kΩ

9-3. 120 kΩ, 330 kΩ, 1.8 kΩ, 3000 pF

9-5. 3.3 kΩ, 1 MΩ, 5.6 kΩ, 0.1 μF

9-7. 1 MΩ, 47 kΩ

9-9. 27 kΩ, 2.7 kΩ, 12 kΩ

9-11. 10 kΩ, (270 kΩ + 10 kΩ)

9-13. 4.7 kΩ, (27 kΩ + 3.3 kΩ)

9-15. 1.8 kΩ, (18 kΩ + 20 kΩ variable), (12 kΩ + 10 kΩ variable)

9-17. 1.3 V/μs

9-19. 330 kΩ, 330 kΩ, 10 kΩ, 0.06 μF

9-21. 190 kΩ, 1 MΩ, 33 kΩ, 0.1 μF

9-23. 15 kΩ, 820 Ω, 120 kΩ, (8.2 kΩ + 10 kΩ variable), 43 pF

10-1. 180 kΩ, (180 kΩ + 22 kΩ), 120 kΩ, 2400 pF

10-3. 1 MΩ, 270 kΩ, 100 kΩ variable, 10 kΩ variable, 8.2 kΩ, 0.01 μF

10-5. 33 kΩ, 1 MΩ, 200 pF

10-7. (a) 620 pF (b) 15 kΩ, 470 kΩ

10-9. 680 Ω, 2.7 kΩ, 10 kΩ variable, (12 kΩ + 1.5 kΩ), 6200 pF

10-11. 220 pF, (100 kΩ + 10 kΩ)

10-13. 82 kΩ, 82 kΩ, 5.6 kΩ, 5.6 kΩ, (68 kΩ + 5.6 kΩ), 15 kΩ, 1 MΩ, 160 pF, 160 pF, 0.27 μF, 910 pF

10-15. 120 kΩ, (1.5 kΩ + 50 kΩ variable), 82 kΩ, 50 kΩ, 82 kΩ, 1500 pF

11-1. 100 kΩ, 100 kΩ, (33 kΩ + 25 kΩ variable), 1000 pF

11-3. (120 kΩ, 120 kΩ, 430 pF), (56 kΩ, 56 kΩ, 1000 pF)

11-5. 68 kΩ, 68 kΩ, 150 kΩ, 330 pF, 680 pF

11-7. 10.5 kΩ ± 1%, 10.5 kΩ ± 1%, 1000 pF, 1.2 MHz

11-9. 16.2 kΩ ± 1%, 32.4 kΩ ± 1%, 32.4 kΩ ± 1%, 1000 pF, 1000 pF, 6 MHz

11-11. 5.76 kΩ ± 1%, 5.6 kΩ, 3.92 kΩ ± 1%, 3.92 kΩ ± 1%, 8.2 kΩ, 1000 pF, 1000 pF

11-13. 650 kHz

11-15. 3.32 kΩ ± 1%, 3.32 kΩ ± 1%, 3.3 kΩ, 0.047 μF, 1000 pF

11-17. 23.7 kΩ ± 1%, 47 kΩ, 47 kΩ, 10.7 kΩ ± 1%, 0.068 μF, 0.068 μF

11-19. 202.5 Hz

11-21. 47 kΩ, 3.09 MΩ ± 1%, 47 kΩ, 47 kΩ, 48.7 kΩ ± 1%, 48.7 kΩ ± 1%, 47 kΩ, 47 kΩ, 1000 pF, 1000 pF

11-23. 127 Hz, 69.5 kHz

12-1. 0.35%, 0.2%, 33 dB

12-3. 0.72%, 0.4%, 29 dB

12-5. 0.31%, 0.27%, 33 dB

12-7. IN757, 390 Ω, 100 kΩ, 120 kΩ, (50 kΩ variable) ‖ (150 kΩ), 100 μF

12-9. 0.0005%, 0.0003%, 87 dB

12-11. 4.7 Ω, 2.7 kΩ

12-13. 9.09 Ω ± 1%, 4.7 kΩ, 1 kΩ, (15 kΩ + 470 Ω)

12-15. 820 Ω, 3.3 kΩ, (5 kΩ variable) ‖ (12 kΩ), 95 mA

12-17. 5.9 Ω ± 1%

LM741 Frequency-Compensated Operational Amplifier

GENERAL DESCRIPTION – The μA741 is a high performance monolithic Operational Amplifier constructed using the Fairchild Planar* epitaxial process. It is intended for a wide range of analog applications. High common mode voltage range and absence of "latch-up" tendencies make the μA741 ideal for use as a voltage follower. The high gain and wide range of operating voltage provides superior performance in integrator, summing amplifier, and general feedback applications.

- NO FREQUENCY COMPENSATION REQUIRED
- SHORT CIRCUIT PROTECTION
- OFFSET VOLTAGE NULL CAPABILITY
- LARGE COMMON-MODE AND DIFFERENTIAL VOLTAGE RANGES
- LOW POWER CONSUMPTION
- NO LATCH UP

ABSOLUTE MAXIMUM RATINGS

Supply Voltage	
Military (741)	±22 V
Commercial (741C)	±18 V
Internal Power Dissipation (Note 1)	
Metal Can	500 mW
DIP	670 mW
Mini DIP	310 mW
Flatpak	570 mW
Differential Input Voltage	±30 V
Input Voltage (Note 2)	±15 V
Storage Temperature Range	
Metal Can, DIP, and Flatpak	−65°C to +150°C
Mini DIP	−55°C to +125°C
Operating Temperature Range	
Military (741)	−55°C to +125°C
Commercial (741C)	0°C to +70°C
Lead Temperature (Soldering)	
Metal Can, DIP, and Flatpak (60 seconds)	300°C
Mini DIP (10 seconds)	260°C
Output Short Circuit Duration (Note 3)	Indefinite

EQUIVALENT CIRCUIT

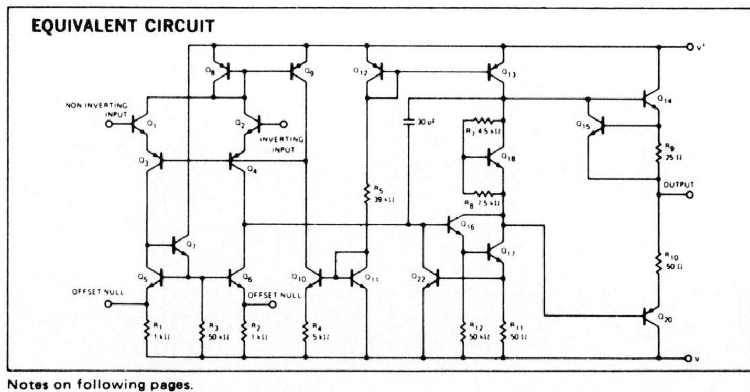

Notes on following pages.

CONNECTION DIAGRAMS

8-LEAD METAL CAN
(TOP VIEW)
PACKAGE OUTLINE 5B

Note: Pin 4 connected to case

ORDER INFORMATION

TYPE	PART NO.
741	741HM
741C	741HC

14-LEAD DIP
(TOP VIEW)
PACKAGE OUTLINE 6A

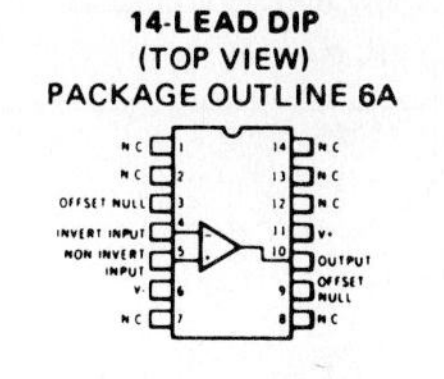

ORDER INFORMATION

TYPE	PART NO.
741	741DM
741C	741DC

10-LEAD FLATPAK
(TOP VIEW)
PACKAGE OUTLINE 3F

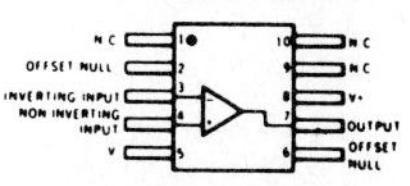

ORDER INFORMATION

TYPE	PART NO.
741	741FM

8-LEAD MINIDIP
(TOP VIEW)
PACKAGE OUTLINE 9T

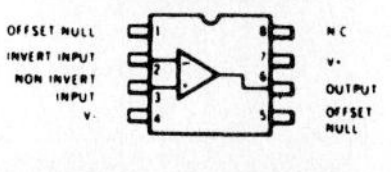

ORDER INFORMATION

TYPE	PART NO.
741C	741TC

*Planar is a patented Fairchild process.

ELECTRICAL CHARACTERISTICS (V_S = ±15 V, T_A = 25°C unless otherwise specified)

PARAMETERS (see definitions)		CONDITIONS	MIN.	TYP.	MAX.	UNITS
Input Offset Voltage		R_S < 10 kΩ		1.0	5.0	mV
Input Offset Current				20	200	nA
Input Bias Current				80	500	nA
Input Resistance			0.3	2.0		MΩ
Input Capacitance				1.4		pF
Offset Voltage Adjustment Range				±15		mV
Large Signal Voltage Gain		R_L > 2 kΩ, V_{OUT} = ±10 V	50,000	200,000		
Output Resistance				75		Ω
Output Short Circuit Current				25		mA
Supply Current				1.7	2.8	mA
Power Consumption				50	85	mW
Transient Response (Unity Gain)	Risetime	V_{IN} = 20 mV, R_L = 2 kΩ, C_L < 100 pF		0.3		μs
	Overshoot			5.0		%
Slew Rate		R_L > 2 kΩ		0.5		V/μs

The following specifications apply for −55°C < T_A < +125°C:

PARAMETERS	CONDITIONS	MIN.	TYP.	MAX.	UNITS
Input Offset Voltage	R_S < 10 kΩ		1.0	6.0	mV
Input Offset Current	T_A = +125°C		7.0	200	nA
	T_A = −55°C		85	500	nA
Input Bias Current	T_A = +125°C		0.03	0.5	μA
	T_A = −55°C		0.3	1.5	μA
Input Voltage Range		±12	±13		V
Common Mode Rejection Ratio	R_S < 10 kΩ	70	90		dB
Supply Voltage Rejection Ratio	R_S < 10 kΩ		30	150	μV/V
Large Signal Voltage Gain	R_L > 2 kΩ, V_{OUT} = ±10 V	25,000			
Output Voltage Swing	R_L > 10 kΩ	±12	±14		V
	R_L > 2 kΩ	±10	±13		V
Supply Current	T_A = +125°C		1.5	2.5	mA
	T_A = −55°C		2.0	3.3	mA
Power Consumption	T_A = +125°C		45	75	mW
	T_A = −55°C		60	100	mW

TYPICAL PERFORMANCE CURVES FOR 741

OPEN LOOP VOLTAGE GAIN AS A FUNCTION OF SUPPLY VOLTAGE

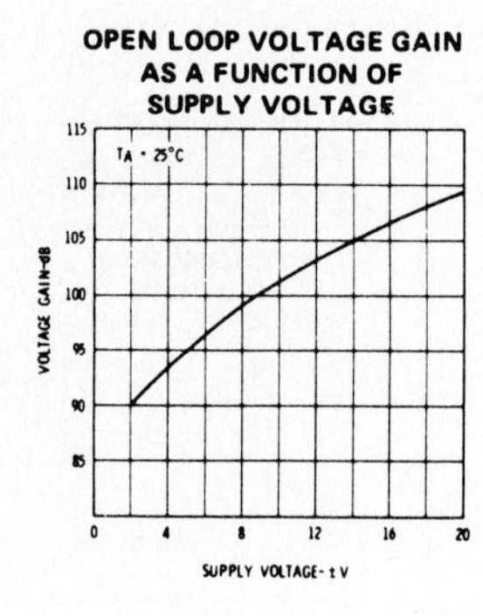

OUTPUT VOLTAGE SWING AS A FUNCTION OF SUPPLY VOLTAGE

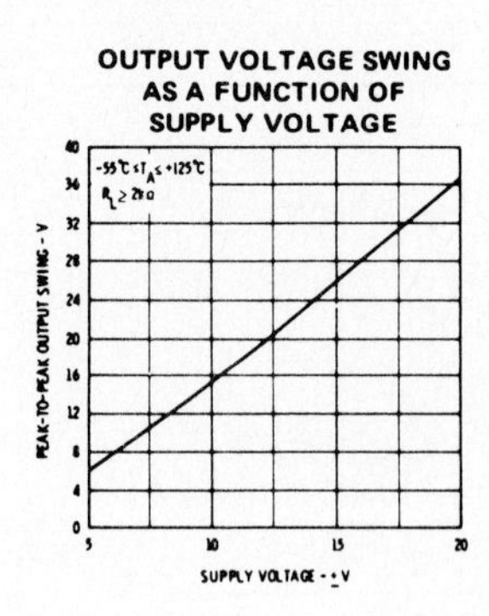

INPUT COMMON MODE VOLTAGE RANGE AS A FUNCTION OF SUPPLY VOLTAGE

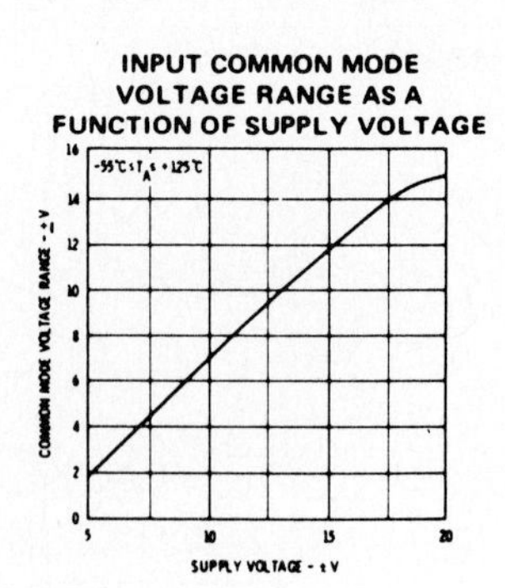

TYPICAL PERFORMANCE CURVES FOR 741A AND 741E

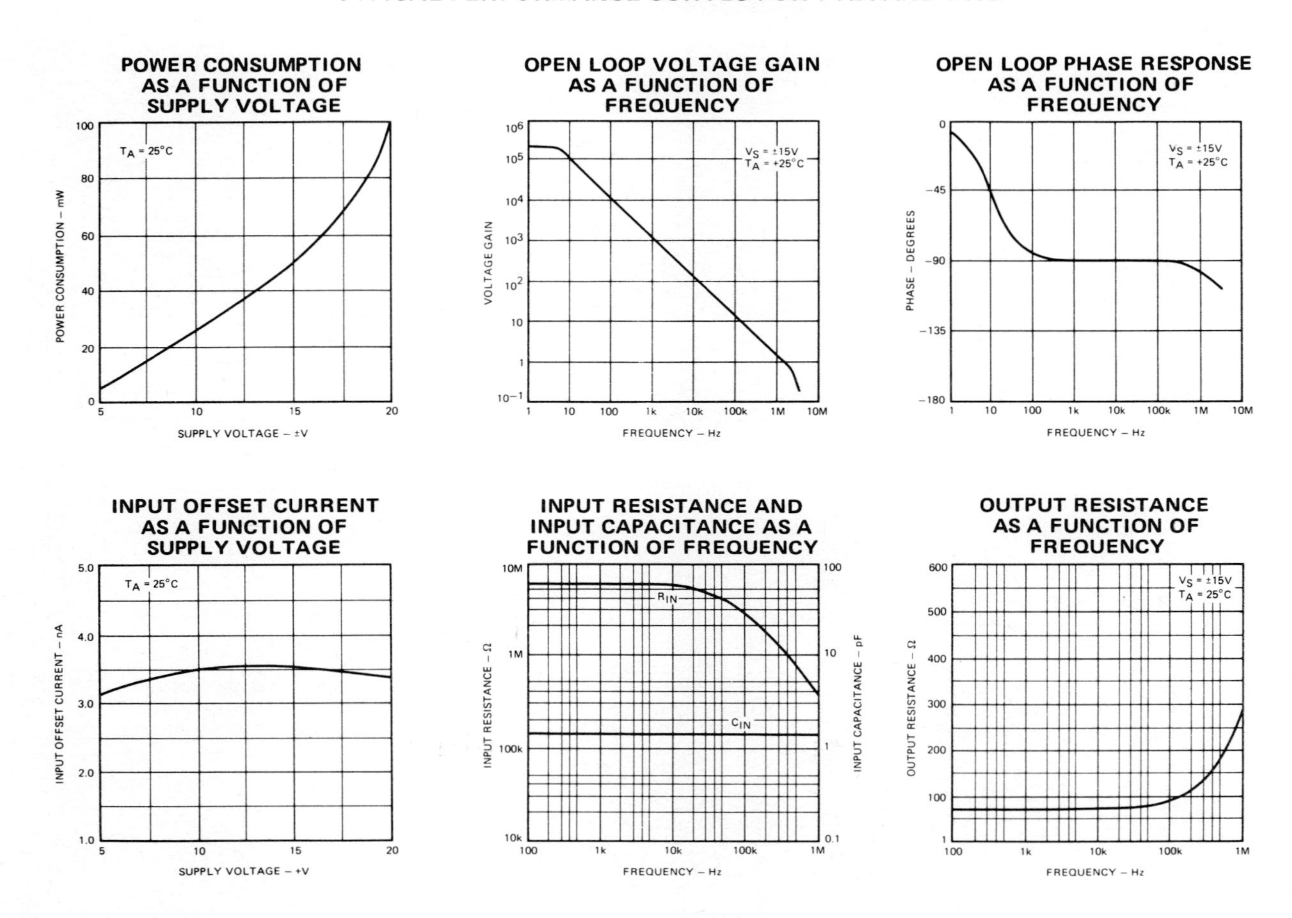

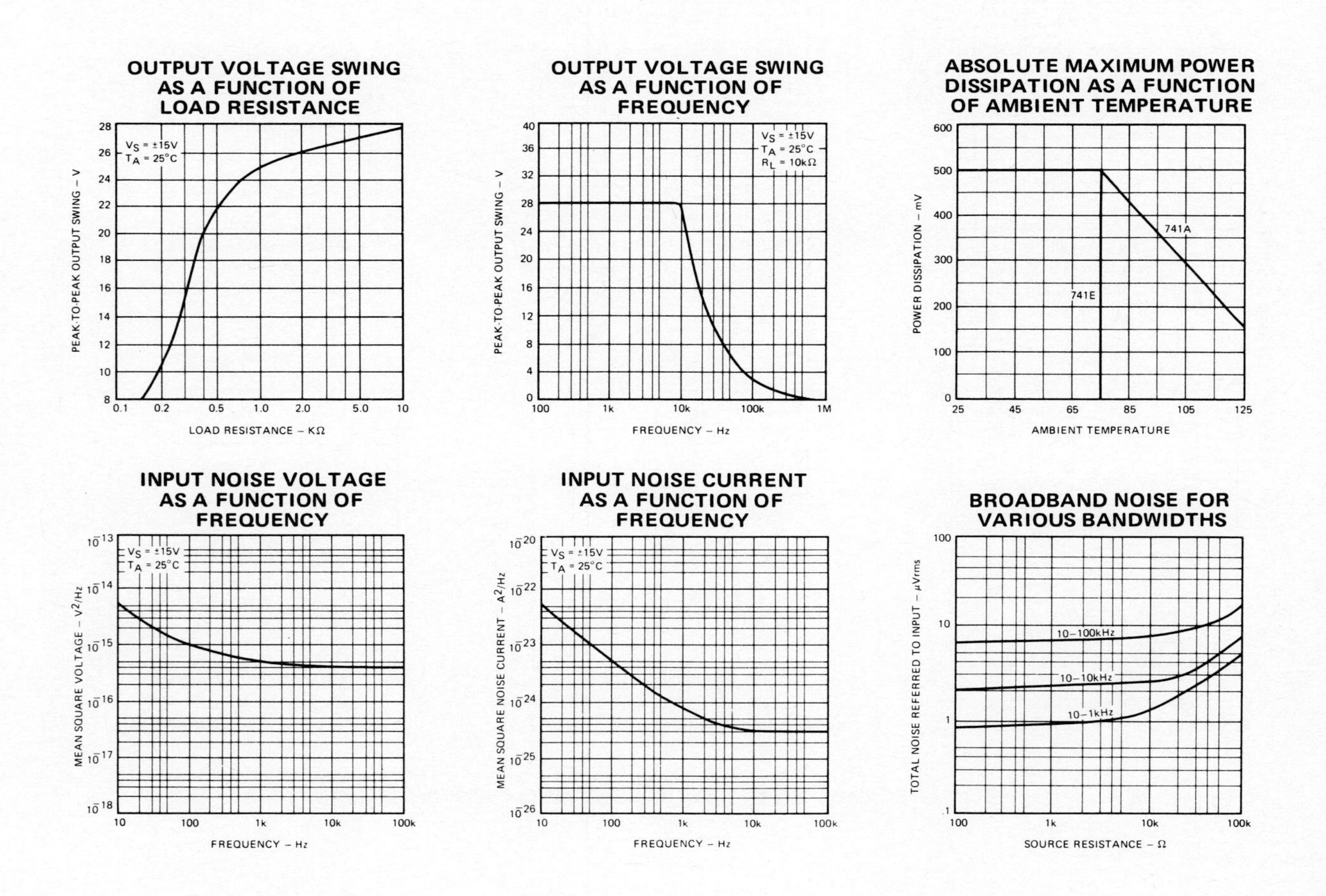
OUTPUT VOLTAGE SWING AS A FUNCTION OF LOAD RESISTANCE
$V_S = \pm 15V$
$T_A = 25°C$
PEAK-TO-PEAK OUTPUT SWING – V
LOAD RESISTANCE – KΩ
28 26 24 22 20 18 16 14 12 10 8
0.1 0.2 0.5 1.0 2.0 5.0 10
OUTPUT VOLTAGE SWING AS A FUNCTION OF FREQUENCY
$V_S = \pm 15V$
$T_A = 25°C$
$R_L = 10k\Omega$
PEAK-TO-PEAK OUTPUT SWING – V
FREQUENCY – Hz
40 36 32 28 24 20 16 12 8 4 0
100 1k 10k 100k 1M
ABSOLUTE MAXIMUM POWER DISSIPATION AS A FUNCTION OF AMBIENT TEMPERATURE
741A
741E
POWER DISSIPATION – mV
AMBIENT TEMPERATURE
600 500 400 300 200 100 0
25 45 65 85 105 125
INPUT NOISE VOLTAGE AS A FUNCTION OF FREQUENCY
$V_S = \pm 15V$
$T_A = 25°C$
MEAN SQUARE VOLTAGE – V^2/Hz
FREQUENCY – Hz
10^{-13} 10^{-14} 10^{-15} 10^{-16} 10^{-17} 10^{-18}
10 100 1k 10k 100k
INPUT NOISE CURRENT AS A FUNCTION OF FREQUENCY
$V_S = \pm 15V$
$T_A = 25°C$
MEAN SQUARE NOISE CURRENT – A^2/Hz
FREQUENCY – Hz
10^{-20} 10^{-22} 10^{-23} 10^{-24} 10^{-25} 10^{-26}
10 100 1k 10k 100k
BROADBAND NOISE FOR VARIOUS BANDWIDTHS
10–100kHz
10–10kHz
10–1kHz
TOTAL NOISE REFERRED TO INPUT – μVrms
SOURCE RESISTANCE – Ω
100 10 1 .1
100 1k 10k 100k

LM108 and LM308 Operational Amplifiers

Absolute Maximum Ratings

	LM108/LM208	LM308
Supply Voltage	±20V	±18V
Power Dissipation (Note 1)	500 mW	500 mW
Differential Input Current (Note 2)	±10 mA	±10 mA
Input Voltage (Note 3)	±15V	±15V
Output Short-Circuit Duration	Indefinite	Indefinite
Operating Temperature Range (LM108)	−55°C to +125°C	0°C to +70°C
(LM208)	−25°C to +85°C	
Storage Temperature Range	−65°C to +150°C	−65°C to +150°C
Lead Temperature (Soldering, 10 seconds)	300°C	300°C

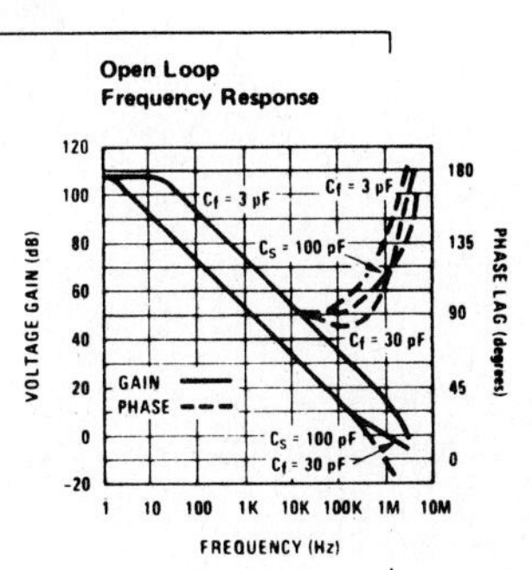

Electrical Characteristics (Note 4)

PARAMETER	CONDITIONS	LM108/LM208 MIN	LM108/LM208 TYP	LM108/LM208 MAX	LM308 MIN	LM308 TYP	LM308 MAX	UNITS
Input Offset Voltage	$T_A = 25°C$		0.7	2.0		2.0	7.5	mV
Input Offset Current	$T_A = 25°C$		0.05	0.2		0.2	1	nA
Input Bias Current	$T_A = 25°C$		0.8	2.0		1.5	7	nA
Input Resistance	$T_A = 25°C$	30	70		10	40		MΩ
Supply Current	$T_A = 25°C$		0.3	0.6		0.3	0.8	mA
Large Signal Voltage Gain	$T_A = 25°C$, $V_S = \pm15V$, $V_{OUT} = \pm10V$, $R_L \geq 10\ k\Omega$	50	300		25	300		V/mV
Input Offset Voltage				3.0			10	mV
Average Temperature Coefficient of Input Offset Voltage			3.0	15		6.0	30	μV/°C
Input Offset Current				0.4			1.5	nA
Average Temperature Coefficient of Input Offset Current			0.5	2.5		2.0	10	pA/°C
Input Bias Current				3.0			10	nA
Supply Current	$T_A = 125°C$		0.15	0.4				mA
Large Signal Voltage Gain	$V_S = \pm15V$, $V_{OUT} = \pm10V$, $R_L \geq 10\ k\Omega$	25			15			V/mV
Output Voltage Swing	$V_S = \pm15V$, $R_L = 10\ k\Omega$	±13	±14		±13	±14		V
Input Voltage Range	$V_S = \pm15V$	±13.5			±14			V
Common-Mode Rejection Ratio		85	100		80	100		dB
Supply Voltage Rejection Ratio		80	96		80	96		dB

Note 1: The maximum junction temperature of the LM108 is 150°C, for the LM208, 100°C and for the LM308, 85°C. For operating at elevated temperatures, devices in the TO-5 package must be derated based on a thermal resistance of 150°C/W, junction to ambient, or 45°C/W, junction to case. The thermal resistance of the dual-in-line package is 100°C/W, junction to ambient.

Note 2: The inputs are shunted with back-to-back diodes for overvoltage protection. Therefore, excessive current will flow if a differential input voltage in excess of 1V is applied between the inputs unless some limiting resistance is used.

Note 3: For supply voltages less than ±15V, the absolute maximum input voltage is equal to the supply voltage.

Note 4: These specifications apply for $\pm5V \leq V_S \leq \pm20V$ and $-55°C \leq T_A \leq 125°C$, unless otherwise specified. With the LM208, however, all temperature specifications are limited to $-25°C \leq T_A \leq 85°C$, and for the LM308 they are limited to $0°C \leq T_A \leq 70°C$.

APPENDIX 1-3

LF353 BIFET Operational Amplifier

Absolute Maximum Ratings

Supply Voltage	±18V
Power Dissipation (Note 1)	500 mW
Operating Temperature Range	0°C to +70°C
$T_{j(MAX)}$	115°C
Differential Input Voltage	±30V
Input Voltage Range (Note 2)	±15V
Output Short Circuit Duration (Note 3)	Continuous
Storage Temperature Range	−65°C to +150°C
Lead Temperature (Soldering, 10 seconds)	300°C

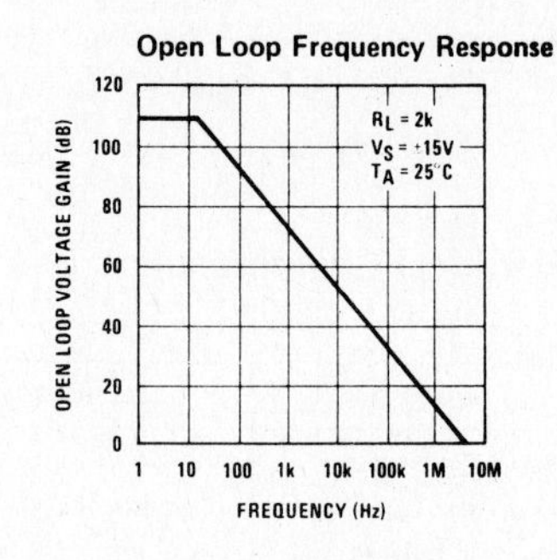

DC Electrical Characteristics (Note 4)

SYMBOL	PARAMETER	CONDITIONS	LF353A			LF353B			LF353			UNITS
			MIN	TYP	MAX	MIN	TYP	MAX	MIN	TYP	MAX	
V_{OS}	Input Offset Voltage	R_S = 10 kΩ, T_A = 25°C		1	2		3	5		5	10	mV
		Over Temperature			4			7			13	mV
$\Delta V_{OS}/\Delta T$	Average TC of Input Offset Voltage	R_S = 10 kΩ		10	20		10	30		10		μV/°C
I_{OS}	Input Offset Current	T_j = 25°C, (Notes 4, 5)		25	100		25	100		25	100	pA
		$T_j \leq$ 70°C			2			4			4	nA
I_B	Input Bias Current	T_j = 25°C, (Notes 4, 5)		50			50	200		50	200	pA
		$T_j \leq$ 70°C			4			8			8	nA
R_{IN}	Input Resistance	T_j = 25°C		10^{12}			10^{12}			10^{12}		Ω
A_{VOL}	Large Signal Voltage Gain	V_S = ±15V, T_A = 25°C V_O = ±10V, R_L = 2 kΩ	50	100		50	100		25	100		V/mV
		Over Temperature	25			25			15			V/mV
V_O	Output Voltage Swing	V_S = ±15V, R_L = 10 kΩ	±12	±13.5		±12	±13.5		±12	±13.5		V
V_{CM}	Input Common-Mode Voltage Range	V_S = ±15V	±11	+15 −12		±11	+15 −12		±11	+15 −12		V V
CMRR	Common-Mode Rejection Ratio	$R_S \leq$ 10 kΩ	80	100		80	100		70	100		dB
PSRR	Supply Voltage Rejection Ratio	(Note 6)	80	100		80	100		70	100		dB
I_S	Supply Current			3.6	5.6		3.6	5.6		3.6	6.5	mA

AC Electrical Characteristics (Note 4)

SYMBOL	PARAMETER	CONDITIONS	LF353A			LF353B			LF353			UNITS
			MIN	TYP	MAX	MIN	TYP	MAX	MIN	TYP	MAX	
	Amplifier to Amplifier Coupling	$T_A = 25°C$, f = 1 Hz–20 kHz (Input Referred)		−120			−120			−120		dB
SR	Slew Rate	$V_S = \pm15V$, $T_A = 25°C$	10	13			13			13		V/μs
GBW	Gain-Bandwidth Product	$V_S = \pm15V$, $T_A = 25°C$	3	4			4			4		MHz
e_n	Equivalent Input Noise Voltage	$T_A = 25°C$, $R_S = 100\Omega$, f = 1000 Hz		16			16			16		$nV/\sqrt{Hz}$
i_n	Equivalent Input Noise Current	$T_j = 25°C$, f = 1000 Hz		0.01			0.01			0.01		$pA/\sqrt{Hz}$

Note 1: For operating at elevated temperature, the device must be derated based on a thermal resistance of 160°C/W junction to ambient for the N package, and 150°C/W junction to ambient for the H package.

Note 2: Unless otherwise specified the absolute maximum negative input voltage is equal to the negative power supply voltage.

Note 3: The power dissipation limit, however, cannot be exceeded.

Note 4: These specifications apply for $V_S = \pm15V$ and $0°C \leq T_A \leq +70°C$. V_{OS}, I_B and I_{OS} are measured at $V_{CM} = 0$.

Note 5: The input bias currents are junction leakage currents which approximately double for every 10°C increase in the junction temperature, T_j. Due to limited production test time, the input bias currents measured are correlated to junction temperature. In normal operation the junction temperature rises above the ambient temperature as a result of internal power dissipation, P_D. $T_j = T_A + \Theta_{jA} P_D$ where Θ_{jA} is the thermal resistance from junction to ambient. Use of a heat sink is recommended if input bias current is to be kept to a minimum.

Note 6: Supply voltage rejection ratio is measured for both supply magnitudes increasing or decreasing simultaneously in accordance with common practice.

APPENDIX 1-4

LM715 Operational Amplifier

ELECTRICAL CHARACTERISTICS FOR 715C (V_S = ±15 V, T_A = 25°C unless otherwise specified)

PARAMETER	CONDITIONS	MIN.	TYP.	MAX.	UNITS
Input Offset Voltage	$R_S \leq 10\ k\Omega$		2.0	7.5	mV
Input Offset Current			70	250	nA
Input Bias Current			0.4	1.5	μA
Input Resistance			1.0		MΩ
Input Voltage Range		±10	±12		V
Common Mode Rejection Ratio	$R_S \leq 10\ k\Omega$	74	92		dB
Supply Voltage Rejection Ratio	$R_S \leq 10\ k\Omega$		45	400	μV/V
Large Signal Voltage Gain	$R_L \geq 2\ k\Omega$, V_{OUT} = ±10 V	10,000	30,000		
Output Resistance			75		Ω
Supply Current			5.5	10	mA
Power Consumption			165	300	mW
Acquisition Time (Unity Gain)	V_{OUT} = +5 V		800		ns
Settling Time (Unity Gain			300		ns
Transient Response (Unity Gain) — Risetime	V_{IN} = 400 mV		30	75	ns
Transient Response (Unity Gain) — Overshoot			25	50	%
Slew Rate	Av = 100		70		V/μs
	Av = 10		38		V/μs
	Av = 1 (non-inverting)	10	18		V/μs
	Av = 1 (inverting)		100		V/μs
The following apply for 0°C ≤ T_A ≤ +70°C:					
Input Offset Voltage	$R_S \leq 10\ k\Omega$			10	mV
Input Offset Current	T_A = +70°C			250	nA
	T_A = 0°C			750	nA
Input Bias Current	T_A = +70°C			1.5	μA
	T_A = 0°C			7.5	μA
Large Signal Voltage Gain	$R_L \geq 2\ k\Omega$, V_{OUT} = ±10 V	8,000			
Output Voltage Swing	$R_L \geq 2\ k\Omega$	±10	±13		V

TYPICAL PERFORMANCE CURVES FOR 715 AND 715C

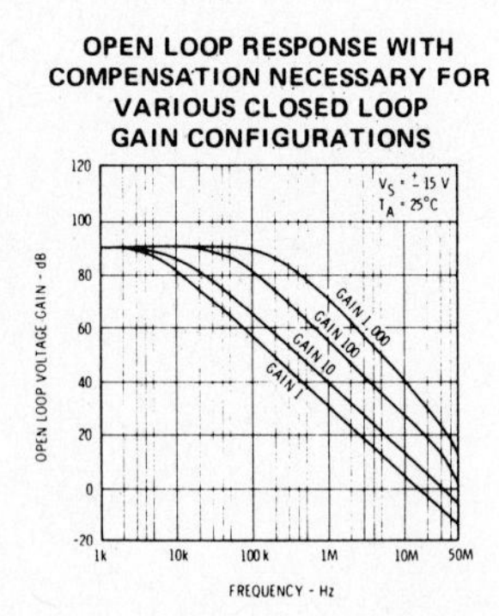

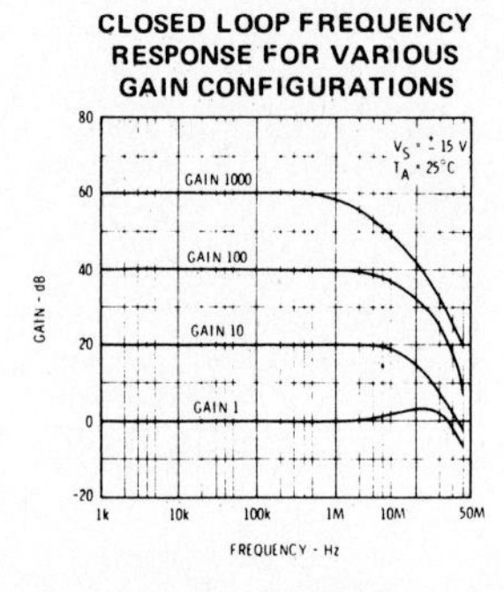

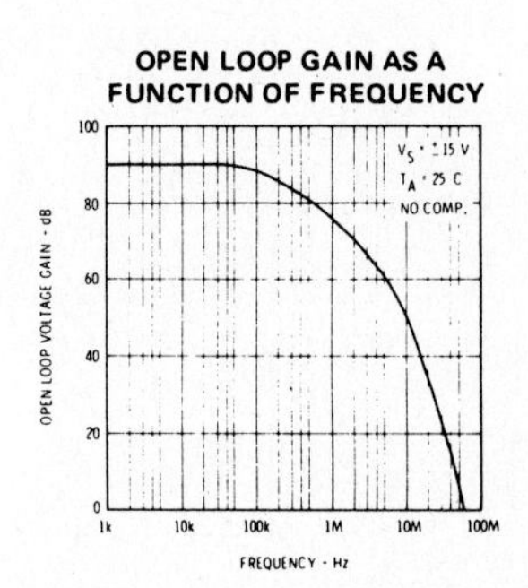

LM709 Operational Amplifier

ELECTRICAL CHARACTERISTICS (T_A = +25°C, ±9 V ≤ V_S ≤ ±15 V unless otherwise specified)

PARAMETER (see definitions)		CONDITIONS	MIN.	TYP.	MAX.	UNITS
Input Offset Voltage		R_S ≤ 10 kΩ		0.6	2.0	mV
Input Offset Current				10	50	nA
Input Bias Current				100	200	nA
Input Resistance			350	700		kΩ
Output Resistance				150		Ω
Supply Current		V_S = ±15 V		2.5	3.6	mA
Power Consumption		V_S = ±15 V		75	108	mW
Transient Response	Risetime	V_S = ±15 V, V_{IN} = 20 mV, R_L = 2 kΩ, C_1 = 5 nF, R_1 = 1.5 kΩ, C_2 = 200 pF, R_2 = 50Ω C_L ≤ 100 pF			1.5	μs
	Overshoot				30	%
The following specifications apply for −55°C ≤ T_A ≤ +125°C:						
Input Offset Voltage		R_S ≤ 10 kΩ			3.0	mV
Average Temperature Coefficient of Input Offset Voltage		R_S = 50Ω, T_A = +25°C to +125°C		1.8	10	μV/°C
		R_S = 50Ω, T_A = +25°C to −55°C		1.8	10	μV/°C
		R_S = 10 kΩ, T_A = +25°C to +125°C		2.0	15	μV/°C
		R_S = 10 kΩ, T_A = +25°C to −55°C		4.8	25	μV/°C
Input Offset Current		T_A = +125°C		3.5	50	nA
		T_A = −55°C		40	250	nA
Average Temperature Coefficient of Input Offset Current		T_A = +25°C to +125°C		0.08	0.5	nA/°C
		T_A = +25°C to −55°C		0.45	2.8	nA/°C
Input Bias Current		T_A = −55°C		300	600	nA
Input Resistance		T_A = −55°C	85	170		kΩ
Input Voltage Range		V_S = ±15 V	±8.0			V
Common Mode Rejection Ratio		R_S ≤ 10 kΩ	80	110		dB
Supply Voltage Rejection Ratio		R_S ≤ 10 kΩ		40	100	μV/V
Large Signal Voltage Gain		V_S = ±15 V, R_L ≥ 2 kΩ, V_{OUT} = ±10 V	25,000		70,000	V/V
Output Voltage Swing		V_S = ±15 V, R_L ≥ 10 kΩ	±12	±14		V
		V_S = ±15 V, R_L ≥ 2 kΩ	±10	±13		V
Supply Current		T_A = +125°C, V_S = ±15 V		2.1	3.0	mA
		T_A = −55°C, V_S = ±15 V		2.7	4.5	mA
Power Consumption		T_A = +125°C, V_S = ±15 V		63	90	mW
		T_A = −55°C, V_S = ±15 V		81	135	mW

OPEN-LOOP FREQUENCY RESPONSE FOR VARIOUS VALUES OF COMPENSATION

FREQUENCY RESPONSE FOR VARIOUS CLOSED-LOOP GAINS

OUTPUT VOLTAGE SWING AS A FUNCTION OF FREQUENCY FOR VARIOUS COMPENSATION NETWORKS

Reprinted with permission of National Semiconductor Corporation. (Not authorized for use as critical components in a life support system.)

APPENDIX 1-6

723 Precision Voltage Regulator

	Min	Typ	Max
Continuous voltage V_{CC^+} to V_{CC^-}			40 V
Input-output voltage differential			40 V
Input voltage range	9.5 V		37 V
Output voltage range	2 V		37 V
Output current			150 mA
Standby current		2.3 mA	3.5 mA
Reference voltage	6.95 V	7.15 V	7.35 V
Current from V_{ref}			15 mA
Maximum power dissipation at or below 25°C free air temperature:			
Metal can			800 mW
DIP			1000 mW

APPENDIX 2-1

Typical Standard Resistor Values

10% Tolerance

Ω	Ω	Ω	kΩ	kΩ	kΩ	MΩ	MΩ
—	10	100	1	10	100	1	10
—	12	120	1.2	12	120	1.2	12
—	15	150	1.5	15	150	1.5	15
—	18	180	1.8	18	180	1.8	18
—	22	220	2.2	22	220	2.2	22
2.7	27	270	2.7	27	270	2.7	—
3.3	33	330	3.3	33	330	3.3	—
3.9	39	390	3.9	39	390	3.9	—
4.7	47	470	4.7	47	470	4.7	—
5.6	56	560	5.6	56	560	5.6	—
6.8	68	680	6.8	68	680	6.8	—
8.2	82	820	8.2	82	820	8.2	—

1% Tolerance

Basic values above 10 ohms may be multiplied by any power of 10 e.g. x1, x10, x100, etc.

1.00	6.04	11.0	20.0	36.5	66.5	121	221
1.10	6.19	11.3	20.5	37.4	68.1	124	226
1.21	6.34	11.5	21.0	38.3	69.8	127	232
1.30	6.49	11.8	21.5	39.2	71.5	130	237
1.50	6.65	12.1	22.1	40.2	73.2	133	243
1.62	6.81	12.4	22.6	41.2	75.0	137	249
1.82	6.98	12.7	23.2	42.2	76.8	140	255
2.00	7.15	13.0	23.7	43.2	78.7	143	261
2.21	7.32	13.3	24.3	44.2	80.6	147	267
2.43	7.50	13.7	24.9	45.3	82.5	150	274
2.67	7.68	14.0	25.5	46.4	84.5	154	280
3.01	7.87	14.3	26.1	47.5	86.6	158	287
3.32	8.06	14.7	26.7	48.7	88.7	162	294
3.57	8.25	15.0	27.4	49.9	90.9	165	301
3.92	8.45	15.4	28.0	51.1	93.1	169	309
4.32	8.66	15.8	28.7	52.3	95.3	174	316
4.75	8.87	16.2	29.4	53.6	97.6	178	324
4.99	9.09	16.5	30.1	54.9	100	182	332
5.11	9.31	16.9	30.9	56.2	102	187	340
5.23	9.53	17.4	31.6	57.6	105	191	348
5.36	9.76	17.8	32.4	59.0	107	196	357
5.49	10.0	18.2	33.2	60.4	110	200	365
5.62	10.2	18.7	34.0	61.9	113	205	374
5.76	10.5	19.1	34.8	63.4	115	210	383
5.90	10.7	19.6	35.7	64.9	118	215	392

5% Tolerance

Ω	Ω	Ω	kΩ	kΩ	kΩ	MΩ	MΩ
	10	100	1	10	10	1	10
	11	110	1.1	11	110	1.1	11
	12	120	1.2	12	120	1.2	12
	13	130	1.3	13	130	1.3	13
	15	150	1.5	15	150	1.5	15
	16	160	1.6	16	160	1.6	16
	18	180	1.8	18	180	1.8	18
	20	200	2	20	200	2	20
	22	220	2.2	22	220	2.2	22
	24	240	2.4	24	240	2.4	
2.7	27	270	2.7	27	270	2.7	
3	30	300	3	30	300	3	
3.3	33	330	3.3	33	330	3.3	
3.6	36	360	3.6	36	360	3.6	
3.9	39	390	3.9	39	390	3.9	
4.3	43	430	4.3	43	430	4.3	
4.7	47	470	4.7	47	470	4.7	
5.1	51	510	5.1	51	510	5.1	
5.6	56	560	5.6	56	560	5.6	
6.2	62	620	6.2	62	620	6.2	
6.8	68	680	6.8	68	680	6.8	
7.5	75	750	7.5	75	750	7.5	
8.2	82	820	8.2	82	820	8.2	
9.1	91	910	9.1	91	910	9.1	

Potentiometers

Ω	Ω	kΩ	kΩ	kΩ	MΩ
10	100	1	10	100	1
20	200	2	20	200	2
			25	250	
50	500	5	50	500	
				750	

APPENDIX 2-2

Typical Standard Capacitor Values

pF	pF	pF	pF	μF	μF	μF	μF	μF	μF	μF
5	50	500	5000		0.05	0.5	5	50	500	5000
—	51	510	5100		—	—	—	—	—	—
—	56	560	5600		0.056	0.56	5.6	56	—	5600
—	—	—	6000		0.06	—	6	—	—	6000
—	62	620	6200		—	—	—	—	—	—
—	68	680	6800		0.068	0.68	6.8	—	—	—
—	75	750	7500		—	—	—	75	—	—
—	—	—	8000		—	—	8	80	—	—
—	82	820	8200		0.082	0.82	8.2	82	—	—
—	91	910	9100		—	—	—	—	—	—
10	100	1000		0.01	0.1	1	10	100	1000	10000
—	110	1100		—	—	—	—	—	—	
12	120	1200		0.012	0.12	1.2	—	—	—	
—	130	1300		—	—	—	—	—	—	
15	150	1500		0.015	0.15	1.5	15	150	1500	
—	160	1600		—	—	—	—	—	—	
18	180	1800		0.018	0.18	1.8	18	180	—	
20	200	2000		0.02	0.2	2	20	200	2000	
22	220	2200		—	0.22	2.2	22	—	—	
24	240	2400		—	—	—	—	240	—	
—	250	2500		—	0.25	—	25	250	2500	
27	270	2700		0.027	0.27	2.7	27	270	—	
30	300	3000		0.03	0.3	3	30	300	3000	
33	330	3300		0.033	0.33	3.3	33	330	3300	
36	360	3600		—	—	—	—	—	—	
39	390	3900		0.039	0.39	3.9	39	—	—	
—	—	4000		0.04	—	4	—	400	—	
43	430	4300		—	—	—	—	—	—	
47	470	4700		0.047	0.47	4.7	47	—	—	

Index

108 op-amp, 104, 335
213 and 217 regulators, 317
308 op-amp, 335
340 regulators, 318
353 op-amp, 336
709 op-amp, 106, 110, 339
715 op-amp, 102, 338
723 regulator, 313, 317
741 op-amp, 7, 101, 110

A

Absolute value circuit, 169
AC amplifiers, 67
A_{cm}, 23
AC voltmeter, 145
Acquisition time, 187
Active filters, 259
All-pass circuit, 260
Amplifier
  AC, 67
  buffer, 42
  capacitor-coupled, 67
  difference, 5, 57, 80
  differential, 5, 57, 80, 151
  direct-coupled, 37
  instrumentation, 151, 154, 156
  inverting, 15, 50, 78, 85
  noninverting, 13, 46, 74, 75, 81
  summing, 54, 56
Amplitude stabilization, 247
Astable multivibrator, 227
A_v, 13, 15, 108
Averaging circuit, 55

B

Band reject filter, 286
Bandpass filter, 276, 283
Bandstop filter, 286
Bandwidth, 107, 112, 113
Barkhausen criteria, 91, 244
Beta (β), 29
Bias currents, 38
Biasing, 38
BIFET op-amp, 41
Bipolar op-amp, 3
Bode plot, 95
Buffer amplifier, 42
Butterworth filter, 267, 271

C

Capacitor-coupled
  amplifiers, 67
  crossing detector, 215
  difference amplifier, 80
  inverting amplifier, 78
  noninverting amplifier, 74, 75
  voltage follower, 68
Characteristic frequency, 199, 205
Chebyshev filter, 267, 271
Clamping circuits, 178, 180
Clipping circuit, 172, 176
Closed-loop bandwidth, 107
Closed-loop gain (A), 13, 15, 108
CMRR, 24
Common mode
  gain (A), 23
  input, 23, 60
  rejection ratio (CMRR), 24
  voltages, 60
Compensation, 90
Complementary emitter follower, 8
Constant current circuit, 7
Crossing detectors, 213
Current
  amplifier, 143
  limiting, 306, 309
  mirror, 9
  sinks, 140
  sources, 137, 140
  -to-voltage converter, 142
Cutoff frequency, 107, 115

D

Darlington circuit, 303
DC amplifiers, 37
DC voltage regulators, 292
DC voltmeter, 144
Dead zone circuit, 174
Decade, 33
Derivative, 195
Difference amplifier, 5, 57, 80
Differential input resistance, 59
Differential input voltage, 3, 12
Differential-input/output amplifier, 151
Differentiating circuit, 194
Diode reverse recovery time, 165, 221
DIP package, 4
Dual-in-line package, 4
Duty cycle, 241

E

Emitter follower, 45
Error amplifier, 298

F

Feedback factor (β), 29
Feedforward compensation, 101
Filters
  band reject, 286
  bandpass, 267, 283
  bandstop, 286
  biquad, 283
  Butterworth, 267, 271
  Chebyshev, 267, 271
  high-pass, 269, 270, 274
  low-pass, 264, 266, 272
  notch, 286
  state-variable, 283
Fold-back current limiting, 309
Frequency compensation
  feedforward, 101
  load capacitance, 122
  manufacturer's, 101
  Miller effect, 100
  overcompensation, 105
  phase-lag, 98
  phase-lead, 99
  stray capacitance, 119
  switching circuits, 213
  Z_{in} mod, 123
Frequency limitations, 32
Frequency response, 93
Full-power bandwidth, 113
Full-wave rectifiers, 167

G

Gain-bandwidth product, 111
Gain/frequency response, 93

H

Half-wave rectifier, 163
Heat sink, 296
High Z_{in} noninverting amplifier, 75
High Z_{in} voltage follower, 71
High-pass filter, 269, 270, 274
Hysteresis, 220

I

IC circuit techniques, 8
Input
  bias current, 3, 27, 38
  impedance, 3, 29
  offset current, 27
  offset voltage, 26
  voltage range, 21, 211
Instrumentation amplifier, 151, 154, 156
Integrating circuit, 200
Inverter, 214
Inverting
  amplifier, 15, 50, 78, 85
  input, 2
  Schmitt trigger, 218

L

Level detector, 215
Level shifting, 60
Limiting circuits, 172
Line regulation, 294
Load capacitance, 120
Load regulation, 294
Loop gain, 92
Loop phase shift, 92
Low-pass filter, 264, 266, 272
Lower trigger point, 218

M

M, 12
Matched transistors, 5
Metal can, 4
Miller effect, 100
Monostable multivibrator, 229

N

Narrow-band filter, 280, 282
Negative feedback, 29
Noninverting
  amplifier, 13, 46, 74, 75, 81
  input, 2
  Schmitt trigger, 222
Nonlinear circuits, 210
Nonsaturating rectifier, 164
Notch filter, 286

O

Octave, 93
Offset
  current, 27
  nulling, 27
  voltage, 26
Ohmmeter, 148
Op-amp symbol, 2
Open-loop gain, *M*, 4, 12, 36
Open-loop phase shift, 260
Oscillator
  amplitude stabilization, 247, 251
  output controls, 253
  phase shift, 244
  Wein bridge, 249
Output
  impedance, 30
  level shifting, 60
  offset voltage, 26
  saturation voltage, 211
  voltage range, 22, 211
  voltage swing, 22, 211
Over compensation, 105

P

Packaging, 4
Peak clipper, 172
Peak detectors, 182
Phase
  lag circuit, 98, 260
  lead circuit, 99, 263
  margin, 95, 117
  response, 93
  shift circuits, 260
  shift oscillator, 244
Plus-and-minus supplies, 5, 312
Pole frequency, 95
Potential divider, 39
Power supply rejection ratio (PSRR), 25
Precision
  clamping circuit, 180
  clipping circuit, 176
  current sink, 140
  current source, 140
  peak detector, 180
  rectifiers, 163
  voltage regulator, 302
  voltage source, 135
PSRR, 25

Q

Q factor, 280

R

Rectifier circuits, 163
Resistor tolerance, 28
Reverse recovery time, 165, 221
Ripple rejection, 295
Rise time, 115

S

Sample-and-hold circuits, 185, 188
Sampling time, 187
Saturating rectifier, 163
Saturation voltage, 3, 211
Schmitt trigger circuit, 218, 222
Series-pass transistor, 295
Short-circuit current, 31
Short-circuit protection, 306
Signal generators, 237
Single-polarity supply, 81
Slew rate, 31, 112, 114, 115, 212
Stability, 81, 95, 97, 126
State-variable filter, 283

Stray capacitance, 116, 119
Summing circuit, 54, 56

T

Terminals, 2
Tracking regulator, 313
Triangular/rectangular wave generator, 238
t_{rr}, 165, 221
Two-output rectifier, 166

U

Unity-gain bandwidth, 111
Upper cutoff frequency, 78, 108
Upper trigger point, 218

V

Virtual earth, 15
Virtual ground, 15
Voltage
  dc regulators, 292
  follower, 9, 41, 45, 68, 71, 81
  level detector, 215
  output range, 22, 211
  sources, 133
  -to-current converter, 137, 144
Voltage regulator
  adjustable, 298
  current limiting, 306, 309
  IC, 313
  precision, 302
  voltage follower, 295
Voltmeters, 144

W

Wein bridge oscillator, 249
Wide-band filter, 280

Z

Zener diode, 172, 293
Zero crossing detector, 213
Z_{in}, 29
Z_{in} mod, 123